Petrology of Polygenic Mafic-Ultramafic Massifs of the East Sakhalin Ophiolite Association

*Sobolev Institute of Geology and Mineralogy,
Russian Academy of Sciences (Siberian Branch),
Novosibirsk, Russia*

Petrology of Polygenic Mafic-Ultramafic Massifs of the East Sakhalin Ophiolite Association

F.P. Lesnov

Scientific editor
Professor V.N. Sharapov

CRC Press
Taylor & Francis Group
Boca Raton London New York

CRC Press is an imprint of the
Taylor & Francis Group, an **informa** business

A BALKEMA BOOK

Originally published in Russian under the title: "Ф.П. Леснов. Петрология полигенных мафит-ультрамафитовых массивов Восточно-Сахалинской офиолитовой ассоциации".
© Academic Publishing House "Geo", Novosibirsk, 2015.

CRC Press
Taylor & Francis Group
6000 Broken Sound Parkway NW, Suite 300
Boca Raton, FL 33487-2742

First issued in paperback 2020

© 2017 Taylor & Francis Group, London, UK
CRC Press is an imprint of Taylor & Francis Group, an Informa business

Typeset by MPS Limited, Chennai, India

No claim to original U.S. Government works

ISBN 13: 978-0-367-57401-7 (pbk)
ISBN 13: 978-1-138-02974-3 (hbk)

Library of Congress Cataloging-in-Publication Data

Names: Lesnov, F. P. (Feliks Petrovich), author. | Sharapov, V. N. (Viktor Nikolaevich), editor.
Title: Petrology of polygenic mafic-ultramafic massifs of the East Sakhalin ophiolite association / F.P. Lesnov; scientific editor, V.N. Sharapov.
Other titles: Petrologīīa poligennykh mafit-ul'tramafitovykh massivov Vostochno-Sakhalinskoǐ ofiolitovoǐ assotsiatsii. English
Description: Leiden, The Netherlands : CRC Press/Balkema, 2017. | Includes bibliographical references and index.
Identifiers: LCCN 2016044165 (print) | LCCN 2016053084 (ebook) | ISBN 9781138029743 (hardcover : alk. paper) | ISBN 9781315227139 (Ebook) | ISBN 9781315227139 (ebook)
Subjects: LCSH: Convergent margins—North Pacific Ocean. | Island arcs—North Pacific Ocean. | Plate tectonics—North Pacific Ocean.
Classification: LCC GC87.2.N66 .L4713 2017 (print) | LCC GC87.2.N66 (ebook) | DDC 552.09164/5—dc23
LC record available at https://lccn.loc.gov/2016044165

**Visit the Taylor & Francis Web site at
http://www.taylorandfrancis.com**

**and the CRC Press Web site at
http://www.crcpress.com**

This book is dedicated to my beloved sons,
George and Oleg, and granddaughter Emma

Table of contents

Acknowledgements

The author received help and advice from geologists and analysts in the collection and treatment of rock samples and other materials considered in the monograph, as well as in the discussion concerning the manuscript. The author thanks V.V. Slodkevich, V.T. Sheiko, V.G. Gal'versen, A.N. Rechkin, V.F. Evseev, V.N. Korolyuk, A.T. Titov, V.S. Palesskii, I.V. Nikolaeva, O.A. Koz'menko, V.Yu. Kiseleva, S.A. Sergeev, A.A. Stepashko, V.G. Tsimbalist, L.S. Denisova, and many others. Special thanks go to LN-Cosmetics Scientific and Production Laboratory and its CEO N.V. Lesnova for financial support of the publication.

Reviewed by:

Prof. *A.I. Chernyshov*, Doctor of Geology and Mineralogy
Prof. *G.B. Fershtater*, Doctor of Geology and Mineralogy

Translators: *N. Kopteva and A. Tsikoza*

Editor's foreword

This monograph gives an important insight into the geodynamics of evolution of systems of epicontinental volcanic arcs on the northwestern margin of the Pacific. It is shown that the regional faults dissecting the eastern margin of the Asian continent into tectonic segments were characterized by complicated cyclical evolution. The collision component of that process (formation of ophiolite nappes and protrusions) was followed by the periodic opening of sutures and the emplacement of basic mantle melts. The lack of systematic geological, petrochemical, and other data on igneous cycles of the Sea of Okhotsk Plate, close to those for mafic-ultramafic massifs of Sakhalin, did not permit demonstrating their space and time variations. Undoubtedly, as new data accumulate, these results will supplement the knowledge of the dynamics of formation of the Kurile volcanic arc and backarc basin. Along with available data on volcanic arcs of the Kamchatka Peninsula, the statistically processed results of geological, petrochemical, geochemical, mineralogical, and isotope–geochronological studies given in the monograph will help to present the geodynamic evolution of the Sea of Okhotsk sector of the considered part of the ocean–continent transition as a more complete picture.

Prof. *V.N. Sharapov*, Doctor of Geology and Mineralogy

Introduction

Petrologic studies of mafic-ultramafic massifs of different types in continental folded areas, midocean ridges, and platform structures have remained topical for decades. The knowledge of conditions of their formation is continually supplemented by new information on the structure of such massifs, including data on the composition of their rocks and minerals, obtained by modern analytical methods. Some approaches to the study of the structural position and geologic structure of mafic-ultramafic massifs have changed, along with views on the spatiotemporal relations among the constituent rock associations. The quantitative characteristics of major- and trace-element distribution in rocks and rock-forming and accessory minerals, as well as the isotope parameters of rocks, have been refined. Also, methods for processing of numerical data and graphical interpretation in analytical studies have been improved. All the foregoing helped to establish new cause–effect relationships among rock, mineral, and ore formation processes; that created conditions for improvement of petrogenetic and geodynamic reconstructions of evolution of deep-seated material and its sources, as well as prerequisites for developing more substantiated models for mantle magmatism.

Through the 20th century and till now, the idea of genetic unity and contemporaneous formation of ultramafic and mafic rocks making up such massifs has prevailed in interpretation of data on the structural position, geologic structure, petrographic composition, and ore content of mafic-ultramafic massifs and in determination of conditions of their formation. This view is based on the theory of intrachamber differentiation and gravitative crystallization fractionation of deep-seated mafic melts [Wager, Brown, 1970; Yoder, 1983; Sharkov, Bogatikov, 1985; Latypov, 2009; and others]. It is one of the fundamentals of plate tectonics which explains the genesis of the so-called banded, or cumulate, series of rocks contained in the majority of complicated mafic-ultramafic massifs which belong to ophiolite associations [Coleman, 1979].

In the last few decades, geological surveying and targeted studies have been carried out on mafic-ultramafic massifs localized in folded structures of different ages, which are eroded to different depths and differ in internal structure, petrographic composition, and many other features. It has been found that such massifs are usually associations of heterogeneous rocks. Reliable evidence has been obtained that ultramafic and gabbroid bodies belonging to the massifs differ not only in petrographic composition but also in the sources of material, as well as in the time and mechanism of its penetration into the crust. In the course of time, all these data created prerequisites for development of a new petrogenetic concept of mafic-ultramafic magmatism [Pinus et al., 1973]. It was based mainly on the results of observations showing that ultramafic bodies in mafic-ultramafic massifs are genetically autonomous tectonic blocks or protrusions of different extents, marking deep-seated fault zones. These protrusions moved to the upper crust much earlier than mafic mantle melts were emplaced along the same fault zones and formed gabbroid intrusions cutting the protrusions. It was proposed to recognize complicated mafic-ultramafic massifs of this type as *polygenic mafic-ultramafic massifs* [Lesnov, 1981b, 1984, 1986].

Mafic-ultramafic massifs occur widely in all ophiolite associations in folded areas of Russia. As yet, they are studied mainly by small- and medium-scale mapping, and for most of them researchers have made no detailed petrological, geochemical, and isotope–geochronological studies by modern analytical methods and theoretical approaches. Products of mafic-ultramafic magmatism are particularly widespread in folded structures of different ages in eastern Russia. Regional structure–tectonic studies show that numerous mafic-ultramafic massifs belonging to ophiolite associations are concentrated as discontinuous belts and their branches among fold–block structures along the western coast of the Pacific, from Chukchi Peninsula and Koryakia to Kamchatka Peninsula, Sakhalin Island, Primorye, isles of Japan, and Southeast Asia. The study of these massifs is important both for better understanding of geodynamic evolution of the crust and crustal ophiolite associations in the transition zone from the Pacific to the Asian continent and for a more substantiated assessment of their ore potential.

Therefore, it is also necessary to carry out detailed petrological studies of mafic-ultramafic massifs in the ophiolite association of Sakhalin Island. The N–S East Sakhalin branch of an ophiolite belt ~220 km in length has the highest content of mafic-ultramafic bodies. It includes several large mafic-ultramafic massifs and many small ultramafic and mafic bodies. The ultramafic bodies in this region were combined into the Ivashka complex, and the gabbroid bodies, into the Berezovka complex [Kovtunovich *et al.*, 1971]. Later V.V. Slodkevich [1975a] assigned complicated mafic-ultramafic massifs to the peridotite–pyroxenite–norite association, whereas ultramafic bodies were assigned to the gabbro–peridotite association. Regional geological survey data show that the most suitable massifs for detailed petrological studies are the Berezovka, Shel'ting, and Komsomol'sk massifs along the eastern coast of Sakhalin Island and the South Schmidt massif in the northern part of the island; they all belong to the Mesozoic East Sakhalin ophiolite association.

The publication is based on data from the author's field and laboratory studies, made intermittently from 1972 to 2011. Along with his own rock collections, the author used the rock samples, graphics, and analytical data provided courtesy of Slodkevich and geologists of the Sakhalin Geological Exploration Expedition: V.T. Sheiko, V.G. Gal'versen, V.F. Evseev, and A.N. Rechkin. Analytical studies of these collections were largely supported by the Sobolev Institute of Geology and Mineralogy (Novosibirsk) and the Russian Foundation for Basic Research (grant no. 05-09-0091).

The studies whose results served as the basis for this monograph were carried out by the following methods: (1) traverse observations accompanied by the refinement of the boundaries of geologic bodies and by sampling; (2) petrographical examination of the rocks with a choice of samples suitable for petrochemical, mineralogical, geochemical, and isotope analyses; (3) determination of the chemical composition of the rocks by a wet chemical technique and the X-ray fluorescence (XRF) method with an ARL 9900XP spectrometer (Thermo-Electron, USA–Switzerland) at the Analytical Center of the Sobolev Institute of Geology and Mineralogy; (4) determination of the chemical composition of minerals by electron probe microanalysis (EPMA) using a JXA-8100 Camebax-Micro microanalyzer at the same center; (5) determination of rare earth and trace elements in the rocks and minerals by instrumental (INAA) and radio-chemical (RNAA) neutron activation analyses at the same center; (6) determination of rare earth and trace elements in the rocks by inductively coupled plasma mass spectrometry (ICP-MS) with an Element mass spectrometer (Finnigan MAT, Germany) at the Analytical Center of the Sobolev Institute of Geology and Mineralogy and with an ELAN DRC II mass spectrometer (Perkin Elmer, USA) at the Analytical Center of the Kosygin Institute of Tectonics and Geophysics (Khabarovsk); (7) determination of trace elements in the rocks by synchrotron-radiation XRF analysis at the Analyt-ical Center of the Sobolev Institute of Geology and Mineralogy (here and in points 8–15); (8) determination of rare earth and trace elements in coexisting rock-forming and accessory minerals by laser ablation inductively coupled plasma mass spectrom-etry (LA ICP-MS) using an Element mass spectrometer (Finnigan MAT, Germany) with a UV Laser Probe attachment (Finnigan MAT, Germany); (9) determination of the chemical composition and trace elements of chromitites by inductively coupled plasma atomic-emission spectroscopy (ICP-AES) with an IRIS Advantage analyzer (Intertex, USA); (10) determination of PGE and Re in the rocks and chromitites by ICP-MS with an Element mass spectrometer (Finnigan MAT, Germany); (11) deter-mination of PGE, Au, and Ag in the chromitites by the atomic-absorption method; (12) study of morphology and determination of the chemical composition of PGE minerals in the chromitites using a LEO 1430VP scanning electron microscope (SEM); (13) determination of the ratios ^{87}Rb/^{86}Sr and ^{87}Sr/^{86}Sr in the rocks by the solid-phase thermal-ionization method with an MI1201T thermal-ionization mass spectrometer (TIMS); (14) determination of the chemical composition of zircons by X-ray spec-troscopy with a JXA-8100 Camebax-Micro microanalyzer; (15) determination of rare earth and trace elements in zircons by LA ICP-MS with an Element mass spectrometer (Finnigan MAT, Germany); and (16) study of the optical properties and morphology and determination of the isotopic age of zircons by the U–Pb method with a SHRIMP II secondary-ion mass-spectrometer (SIMS) at the Center of Isotopic Research of VSEGEI (St. Petersburg).

The author's main goals were (1) to present the first and, if possible, complete geological and petrological description of mafic-ultramafic massifs of the East Sakhalin ophiolite association, based on results of comprehensive study, including study by modern analytical methods, and (2) to obtain a more complete substantiation of the earlier proposed model of polygenic formation of such massifs as the most important components of ophiolite associations.

Chapter 1

Topical problems of petrology of mafic-ultramafic massifs

Further efforts toward the structural and petrological interpretation of each ophiolite occurrence will permit a stricter and more reliable interpretation of ophiolites in general.

R.G. Coleman

Over the last few decades, key problems of petrology of mafic-ultramafic complexes (massifs) of different types in different crustal structures have been discussed, with emphasis on massifs belonging to ophiolite associations. The aspects concerned were their structural position; bedding conditions; internal structure; spatial and age relations between ultramafic and gabbroid bodies; rock, mineral, major- and trace-element, and isotope compositions; time and possible conditions of generation of melts and their emplacement into the crust; and composition of deep-seated sources of material. It has been established that mafic-ultramafic massifs can be considerably different in structural position, bedding conditions and age, size, morphology, internal structure, relationship between the volumes of ultramafic rocks and gabbroids at the present-day erosion level, petrographic composition, petrochemical and isotope–geochemical characteristics, and metallogeny. It has been found that ophiolite associations and constituent mafic-ultramafic massifs have a belt-like structure and mark zones of long-lived deep-seated faults and feathering faults; at the present-day erosion level, most of these massifs are elongated lens-shaped objects with steeply inclined contacts. On the other hand, massifs at the intersection of faults which experienced fold–block deformations are often of irregular or subisometric shape and show gentle bedding. The length of massifs is from few hundreds of meters to tens and hundreds of kilometers, and their width is from few tens of meters to few tens of kilometers. Massifs can outcrop in areas of fractions of a square kilometer to several thousands of square kilometers. By relationships between the areas of ultramafic- and mafic-rock outcrops, massifs vary from almost "monogene" ultramafic bodies and complex mafic-ultramafic bodies to almost "monogene" gabbroid bodies.

In the 1960s–1970s, plutonic- and volcanic-rock complexes were classified in Russian igneous geology; that was quite important during large-scale geological mapping of the country's territory. The results of these studies were summarized by Yu.A. Kuznetsov [1964], who combined mafic-ultramafic massifs and complexes into two igneous associations: gabbro-pyroxenite–dunite and ultrabasic. According to Kuznetsov, the massifs of the first association formed owing to the multistage emplacement of melts and their differentiation; the massifs of the second association, in which gabbroids occur in extremely small quantities, crystallized from magma which was a mixture of olivine and enstatite crystals. Using facies analysis, S.S. Zimin [1973] summarized available geological and petrological data on mafic-ultramafic complexes located in the Far East and on the folded framing of the Pacific. He assigned these massifs to three igneous associations (dunite–harzburgite, dunite-wehrlite–pyroxenite, and olivinite–wehrlite), explaining the formation of different types of ophiolite series mainly by the heterogeneous composition of the original "paleomantle" in the horizontal direction.

Petrology of the so-called layered mafic-ultramafic massifs, usually localized within cratons, has long been considered by many researchers based on the model of differentiation of mafic melts [Wager and Brown, 1970; Cox et al., 1982; Yoder, 1983; Sharkov and Bogatikov, 1985; Yaroshevskii et al., 2006; Latypov, 2009; Lavrenchuk and Latypov, 2009; and others]. The same approach was applied in the solution of problems related to the genesis of complex mafic-ultramafic massifs in folded areas [Balykin et al., 1986; Ledneva et al., 2009; Mekhonoshin and Kolotilina, 2009; Nikolaev, 2009; Khain and Remizov, 2009, and others]. The "cumulate" model of formation of layered mafic-ultramafic massifs was substantiated with geological-mapping data as well as the results of physical experiments and numerical simulation with calculations of the compositions of parental melts for individual massifs, based on estimations of their average weighted chemical compositions [Lavrenchuk and Latypov, 2009, and others].

The fundamental review by K.G. Cox et al. [1982] covered an extremely wide range of issues of igneous petrology, with a focus on the prevailing role of magma differentiation. However, it almost lacked data on the genesis of complex mafic-ultramafic massifs in ophiolite associations; note that ultramafic bodies were defined as "ultramafic intrusions". Concentrically zoned Pt-bearing pyroxenite–dunite massifs, widespread in the Urals, were studied by O.K. Ivanov [1997]; he made the conclusion that they are multistage hypabyssal intrusive–postmagmatic complexes formed in settings of ancient island arcs with the differentiation of basaltoid magma.

Mafic-ultramafic massifs are usually localized on the framing of differently aged stratified structures: from almost unmetamorphosed sedimentary–volcanic strata to their high-grade metamorphic analogs. The primary sections of the strata were often disturbed by later plication and disjunctions. Contacts between massifs and host rocks are very often faulted and overlain by loose sediments. On the other hand, the study of contacts between massifs and host rocks, as well as those between ultramafic and gabbroid bodies within massifs, is important for understanding their genesis. If the contacts between gabbroid and ultramafic rocks are well-exposed, they vary from eruptive, with gabbroid bodies intruding into ultramafic ones, to extremely gentle, with a gradual transition from gabbroid to ultramafic bodies; in this case it is always unclear in what sequence the ultramafic and gabbroid bodies formed.

During field studies of many mafic-ultramafic massifs, we found that ultramafic bodies usually border on the host strata along faults whose planes have inclinations varying from near-vertical and steep to gentle. These faults are accompanied by foliated and brecciated (mélange) zones of highly serpentinized ultramafic rocks and host rocks, which are of different thicknesses. Sometimes coarse or fine platy jointing is retained in serpentinites at such contacts. Contacts between gabbroid bodies within mafic-ultramafic massifs and host rocks vary from abrupt and crosscutting to gradual. In some cases such contacts are characterized by irregular alternation of sill-like gabbroid bodies with sheet-like or lens-shaped host rock xenoliths.

In the second half of the 20th century, geophysical and petrological studies of ophiolite associations and constituent mafic-ultramafic massifs, which are localized within midocean ridges and continental folded areas, led to a change from the geosynclinal paradigm of formation of crustal structures to the new, geodynamic, one. This paradigm was presented and approved at the Penrose Conference and then systematically described by R.G. Coleman [1979] as the concept of plate tectonics. It was based on the hypothesis that ophiolite associations, which are widespread in both midocean ridges and continental folded structures, generally belong to oceanic-type crust and have a "stratified" structure. According to Coleman, the complete section of an ophiolite association consists of the following rock complexes overlying one another: (1) metamorphic peridotites (tectonites); (2) cumulate peridotites passing up the section into layered gabbro, which often include plagiogranite differentiates in the upper part; (3) complex of parallel dikes varying in composition from basalts to keratophyres; and (4) complex of pillow lavas interbedded with pelagic sediments. It was stressed that the main components of the ophiolite section are often separated by faults and these sections can be incomplete and isolated; also, this rough stratigraphic sequence is clearly proven at many places, so that there are, apparently, no objections to the idea of its existence [Coleman, 1979]. Afterward, for several decades, many researchers tried to interpret results of their studies in accordance with this scheme for the structure of ophiolite associations, which has become almost classical.

The geologic structure, petrology, and stages of evolution of ophiolite associations in the Polar Urals and West Sayan, in which mafic-ultramafic massifs are particularly widespread, were described in detail by N.L. Dobretsov *et al.* [1977]. Regarding the West Sayan ophiolites, the authors made the conclusion that ultrabasic rocks and gabbroids, including the gabbro-diabase dike complex and the rocks of the Chinge Formation, form a sublayered ophiolite series; the ultrabasic and gabbroid rocks are in sublayered and intrusive relations simultaneously (gabbro formed later); and the intrusive contacts of the ultrabasic rocks and gabbro with the rocks of the Chinge Formation have not been reliably detected anywhere.

However, our field and analytical data on geology of ophiolite associations and mafic-ultramafic massifs belonging to them, as well as data by some other petrologists, clearly show that the structure of many massifs far from always agrees with the models based on the concepts of gravitative crystallization differentiation of mafic melts or plate tectonics. First and foremost, these data show that mafic-ultramafic massifs, like ophiolite associations in general, are considerably more diverse in their structure–tectonic position, internal structure, petrographic composition, and other features than it was presumed before; also, ultramafic and gabbroid bodies localized in a morphologically united massif are genetically autonomous and their formation is nonsimultaneous. Besides that, our data mostly do not confirm the presence of "stratification" in mafic-ultramafic massifs or ophiolite associations in general. Discrepancy between observations in many of these massifs, on the one hand, and plate tectonics, on the other, was discussed in the literature repeatedly, including the works on the massifs of the Anadyr'–Koryak folded system [Pinus *et al.*, 1973], Mongolia and entire Central Asia [Pinus *et al.*, 1976, 1979, 1984], and other regions [Dilek and Robinson, 2003; Herzberg, 2004]. Some aspects of geology of ophiolite associations and constituent mafic-ultramafic massifs concentrated in the zone of transition from Asian continent to Pacific were discussed in the works by many petrologists and tectonists [Velinskii, 1979; Vysotskii *et al.*, 1998; Chekhov, 2000; Ishiwatari *et al.*, 2003], including the works concerned with massifs on Sakhalin Island [Raznitsyn, 1975, 1978, 1982; Rechkin *et al.*, 1975; Rozhdestvenskii, 1975; Slodkevich, 1975a,b, 1977; Slodkevich and Lesnov, 1976; Bekhtol'd and Semenov, 1978; Rozhdestvenskii and Rechkin, 1982; Rechkin, 1984; Lesnov, 1986; Ostapenko, 2003; Grannik, 2008; Stepashko and Lesnov, 2013]. The recently discussed subduction model for ophiolite formation suggests a relationship between ophiolite genesis and crust destruction in backarc basins [Shervais, 2001; Herzberg, 2004; Dilek and Polat, 2008].

The conditions of formation of the so-called banded complexes, or transition zones, usually bordering on ultramafic and gabbroid bodies, remain a key problem related to the genesis of mafic-ultramafic massifs in ophiolite associations. These zones, extremely inhomogeneous in their internal structure, petrographic composition, and other features, are defined in plate tectonics as complexes of cumulate peridotites passing into layered gabbro [Coleman, 1979]. Also, Coleman pointed out that a considerable part of the ophiolite section is composed of rocks formed by the fractional crystallization of basaltic magma, which produced an ultramafic series and ended with the crystallization of leucocratic derivates. This view on the genesis of "banded complexes" is held by many geologists, who draw analogies with the so-called layered mafic-ultramafic massifs (e.g., the Skaergaard intrusion). Note that the mechanism of fractional crystallization of basaltoid melts is regarded as reliable evidence which does not require any additional confirmation. Based on the study of Mongolian ophiolite associations, supporters of the concept of plate tectonics presumed that these associations are underlain everywhere by ultramafic rocks and overlain by pyroxenites and gabbroids, which are, in turn, overlain by pillow lavas; all these structures make up a stratified ophiolite section [Zonenshain and Kuz'min, 1978].

However, there exist many contradictions and inconsistencies when banded complexes within mafic-ultramafic massifs are interpreted based on the "cumulate" model for their formation, as shown by geological and structural studies of a great number of mafic-ultramafic massifs belonging to ophiolite associations in different regions of the former Soviet Union and in Mongolia. There is abundant evidence for the nonsimultaneous formation of ultramafic and gabbroid bodies in such massifs, strong effect of mafic magmas on older ultramafic rocks, and contact–reaction origin of the rocks of the so-called banded complexes [Lesnov, 1976, 1979, 1980, 1981a–d, 1982, 1984, 1985, 1986a,b, 2007e, 2009c, 2011a; Pinus *et al.*, 1976, 1979; Slodkevich and Lesnov, 1976; Lesnov *et al.*, 1982; Balykin *et al.*, 1991].

The study of Mongolian mafic-ultramafic massifs showed that they are all confined to large faults and feathering faults, which often separate folded systems of different ages and their blocks, or they are traced within these structures [Pinus *et al.*, 1979]. Also, evidence was provided that gabbroid bodies in such massifs intruded into spatially related earlier ultramafic protrusions, whereas ultramafic rocks experienced different degrees of alteration under the effect of mafic melts and their fluids. It was proposed to assign complex mafic-ultramafic bodies of this type to *polygenic* mafic-ultramafic massifs.

The earlier intrusion of ultramafic bodies compared to spatially related gabbroid bodies and the contact–reaction origin of gabbro-pyroxenite–wehrlite zones along their contacts were hypothesized by researchers of mafic-ultramafic massifs of the Urals [Shteinberg *et al.*, 1986]; Kuyul massif in Koryakia [Ledneva and Matukov, 2009]; Kroton massif on the Kamchatka Peninsula [Sidorov, 2009]; Dyukali massif in Primorye [Izokh, 1965; Romanovich, 1973]; Horoman, Ogiwara, Oshima, Yakuno, and Oeyama massifs in Japan [Miyashiro, 1966; Ultrabasic..., 1967; Hirano, 1977; Ishiwatari, 1985, 1991; Kurokawa, 1985]; Lanzo massif in the Western Alps [Piccardo *et al.*, 2005]; and the mafic-ultramafic complexes making up the Mid-Atlantic Ridge [Pertsev *et al.*, 2009].

During the study of ophiolite complexes of the Polar Urals, V.R. Shmelev [2009] made the following conclusions: (a) the ultramafic massifs he studied showed no evidence for stratification; (b) the ultramafic rocks are often abnormally enriched in LREE and, partly, MREE, as well as in some other incompatible elements (Cs, Rb, Ba, Sr, etc.), which are concentrated as epigenetic film phases within microfractures; (c) the Polar Urals ultramafic rocks are divided into two geochemical types: with ordinary and abnormal (negative Eu anomalies) REE patterns; and (d) the mantle ultramafic rocks of the Polar Urals are polygenic, and they formed owing to the long-lasting evolution of mantle material, with stages of partial melting (depletion), fluid–magmatic remobilization, and metamorphism in different geodynamic settings.

Petrochemical methods are still widely used in studying mafic-ultramafic massifs. A summary of analytical data on some better studied massifs showed the bi- or polymodal structure of histograms for the occurrence of the main components in the rocks of these massifs. Note that some occurrence maxima correspond to the composition of ultramafic rocks; others, to the composition of gabbroids; and the minima between them, to the composition of rocks of the so-called banded complexes. Also, the chemical compositions of rocks of mafic-ultramafic massifs are discrete on the diagrams CaO/Al_2O_3–$FeO_{tot}/(FeO_{tot} + MgO)$. These petrochemical features of ultramafic rocks and gabbroids within mafic-ultramafic massifs are considered to be an additional criterion for their genetic peculiarity [Lesnov, 1985, 1986]. Besides that, some petrochemical parameters of restite ultramafites, such as the Mg/Si ratio, are used to assess the degree of partial melting of upper-mantle sources during the formation of ultramafic restites [Glebovitskii *et al.*, 2009]. The indicator properties of some other petrochemical parameters (CaO/Al_2O_3, Fe/Si, and Cr/Si) used to determine the conditions of formation of restite ultramafites were pointed out by B.A. Bazylev [2003].

To study the conditions of formation of mafic-ultramafic massifs in ophiolite associations, researchers analyze the regularities of distribution of groups of microimpurities with contrasting geochemical properties, such as REE and PGE. The data obtained suggest an inverse relationship between these elements owing to their differently directed fractionations during the partial melting of upper-mantle sources, with the generation of basaltic melts and the formation of ultramafic restites, and during the contamination and crystallization of these melts with the formation of orthomagmatic and hybrid gabbroids [Lesnov, 2007, 2009a,b, 2010, 2012b].

Also, note the work by O.M. Glazunov [1979], concerned with interaction between basaltoid melts and host rocks as a cause of the diverse composition of gabbroids. It was pointed out that the $^{87}Sr/^{86}Sr$ ratio is an important indicator of the degree of contamination of such melts with the material of the enclosing rocks and that gabbroids associated with ultrabasic rocks are characterized by extremely low values of this parameter (close to 0.7010).

Uranium–lead isotope dating of zircon is increasingly used in modern petrology for study of the conditions of formation of mafic-ultramafic massifs and for spatiotemporal reconstructions. A study of zircon from different rocks of ultramafic-mafic massifs in continental and midocean-ridge ophiolite associations showed that this mineral usually occurs in several generations. Four main genetic types of zircon were recognized: relict, xenogenic, syngenetic, and epigenetic. They differ not only in isotopic age but also in the morphology of grains, their optical properties, and trace-element composition [Savelieva *et al.*, 2007; Bortnikov *et al.*, 2008; Batanova *et al.*, 2009; Krasnobaev *et al.*, 2009; Malitch *et al.*, 2009; Lesnov *et al.*, 2010; Skolotnev *et al.*, 2010].

According to the polygenic model of formation of mafic-ultramafic massifs, they include structure–compositional elements which occupy a certain position and differ in structure and composition as well as in the formation time and conditions [Lesnov, 1981a–d, 1982, 1984, 1985, 1986a,b, 2007e, 2009c, 2011a]: (1) protrusion of ultramafic restites; (2) gabbroid intrusion cutting the protrusion; (3) contact–reaction zone along the boundaries between gabbroid intrusion and ultramafic protrusion; and (4) contact–reaction zone along the boundaries between gabbroid intrusion and host rocks. The criteria for the multistage formation of mafic-ultramafic massifs (first and foremost, later formation of gabbroid intrusions compared to ultramafic protrusions) are as follows: (1) ultramafic xenoliths in rocks of gabbroid intrusions; (2) crosscutting gabbroid bodies (stocks, dikes, sills, veins, and veinlets) in rocks of ultramafic protrusions; (3) contact–reaction zones at the boundaries between gabbroid and ultramafic bodies; (4) discrete petrochemical composition of ultramafic rocks and gabbroids; and (5) time gap between the formation of ultramafic restites and the emplacement of gabbroid intrusions, detected by isotope dating of zircon from different rocks of these massifs.

As it will be shown below, the geological model of multistage formation of mafic-ultramafic massifs, which received additional substantiation, is based on the evidence that the leading mechanism of formation of all the various petrographic types of rocks in these massifs was not *gravitative crystallization differentiation* of upper-mantle basaltoid melts; it was *magma–metasomatic integration* of these melts, on the one hand, and the substance of older protrusions of restite ultramafites, as well as the host terrigenous volcanics of these massifs and their metamorphic derivates, on the other.

***

The review of topical problems of petrology of mafic-ultramafic massifs in ophiolite associations has shown that the genesis of these igneous complexes is studied using a wide range of methods and approaches, which permit developing genetic models of formation of such massifs [Lesnov, 2013]. Nevertheless, there are many disputable questions related to the structure and conditions of formation of mafic-ultramafic massifs and ophiolite associations in general. The concept of gravitative crystallization fractionation of mafic melts is still often used as a basis for studies on this subject and resulting petrogenetic reconstructions of formation of the entire diversity of petrographic types of rocks making up these complex massifs. However, the alternative model for the formation of mafic-ultramafic massifs has been proposed and substantiated. It is based on evidence for the nonsimultaneous emplacement of restitic ultramafic protrusions and gabbroid intrusions as well as for the magma–metasomatic action of mafic melts and their fluids on older ultramafic restites with the formation of hybrid rocks. For example, such evidence was obtained during study of mafic-ultramafic massifs of the East Sakhalin ophiolite association, and it will be considered in detail in the following chapters.

Chapter 2

Structural position of mafic-ultramafic massifs of the East Sakhalin ophiolite association

According to regional structural and tectonic reconstructions, numerous mafic-ultramafic massifs belonging to ophiolite associations are concentrated in Paleozoic, Mesozoic, and younger folded structures along the northwestern shore of the Pacific, from Chukchi Peninsula to the islands of Japan; they are of various structures and compositions. Certainly, a comprehensive study of these massifs is interesting for solving general problems related to the geodynamic evolution of the crust and upper mantle and the formation of ophiolite associations in this large segment of the Earth; also, it is interesting for assessment of the ore potential of these igneous complexes. Published data show that mafic-ultramafic massifs of Sakhalin Island mark the wide zone of the long-lived deep-seated fault traced along the Sea of Okhotsk shore of the island, from northern peninsulas to Aniva Bay.

According to the earliest structural and tectonic summaries by P.N. Kropotkin and K.A. Shakhvarstova [1965], the Sakhalin mafic-ultramafic massifs are localized within the East Sakhalin anticlinorium, which is part of the Hokkaido–Sakhalin folded system. Together with many tens of massifs in folded structures of the Chukchi and Kamchatka Peninsulas, Koryakia, Primorye, and Japan, the Sakhalin massifs were included in the ophiolite belts traced along the Asian continent–Pacific transition zone [Semenov, 1967; Bogdanov and Khain, 2000]. In the later geodynamic reconstructions [Khanchuk, 2006], the mafic-ultramafic massifs on the eastern shore of Sakhalin Island were included in the Nabil' terrane, which was assigned to the Sakhalin–Kamchatka orogen. As shown by geophysical observations, the fault zone along the Sea of Okhotsk shore of Sakhalin Island is a N–S-trending transpressional-fault structure with a steep westward inclination; the band of fault-line dislocations here sometimes reaches 7 km in width [Vasilenko and Prytkov, 2012].

The East Sakhalin branch of the Sakhalin–Kamchatka orogen (total length of ~220 km) includes quite many small and several large bodies varying in composition from essentially ultramafic to mafic-ultramafic and essentially gabbroid. The essentially ultramafic massifs were combined into the Ivashka igneous complex; the complex mafic-ultramafic massifs, in the Berezovka igneous complex [Kovtunovich et al., 1971]. Afterward V.V. Slodkevich [1975a, 1977] combined the mafic-ultramafic massifs into the Shel'ting igneous complex, assigned to the peridotite–pyroxenite–norite facies, whereas the massifs of the Ivashka complex were assigned to the gabbro-peridotite facies.

From data on the disjunction tectonics of northeastern Sakhalin Island, V.S. Rozhdestvenskii [1975] made the conclusion that an important role in the regional

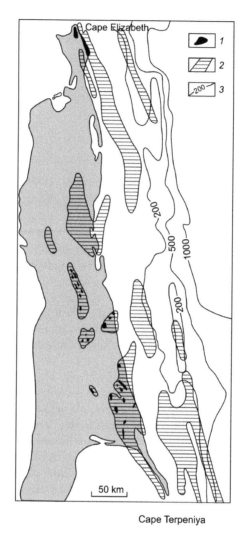

Figure 2.1   Position of mafic-ultramafic massifs in the northern part of Sakhalin Island (modified after [Rozhdestvenskii, 1987]).
1 – serpentinite, peridotite, dunite, and gabbroid bodies; *2* – positive magnetic anomalies; *3* – isobaths and their marks.

structure is played by a system of strike-slip faults, normal faults, and reverse fault–thrusts, to which the formation of Cenozoic folded structures is related. According to his observations, the present-day structure of Sakhalin Island was formed by horizontal compression from east-northeast to west-southwest (i.e., near-perpendicularly to the strike of present-day folded and fault structures). Later Rozhdestvenskii [Rozhdestvenskii and Rechkin, 1982; Rozhdestvenskii, 1987] distinguished two ophiolite belts within Sakhalin Island: one in the central part and the other in the eastern one (Figs. 2.1, 2.2). The ophiolites in the central part of the island are significantly

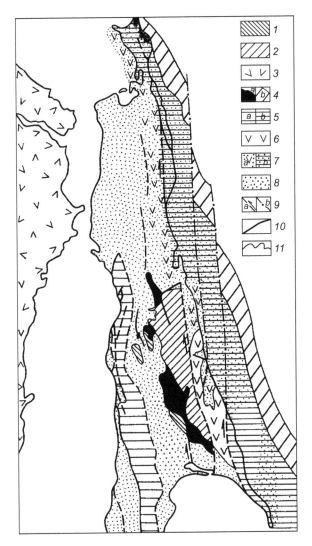

Figure 2.2 Position of structure-facies zones within northern Sakhalin Island (after [Rozhdestvenskii, 1987]).

1 – Triassic–Lower Cretaceous deposits of the Western zone (Daldagan Group); 2 – Triassic–Upper Cretaceous deposits of the Eastern zone; 3 – rocks of the Sikhote-Alin' volcanic belt; 4 – ophiolite–metamorphic belts: *a* – on land (Central Sakhalin suture zone), *b* – estimated from geophysical data under water and overlying sediments; 5 – Upper Cretaceous terrigenous sediments: *a* – West Sakhalin forearc basin, *b* – East Sakhalin forearc basin; 6 – Upper Cretaceous volcanics (paleovolcanic arc of East Sakhalin Island); 7 – Upper Cretaceous deposits of East Sakhalin, overlain by Cenozoic deposits: *a* – volcanic, *b* – terrigenous; 8 – Cenozoic deposits; 9 – deep faults (strike-slip faults): *a* – on land, *b* – in the Sea of Okhotsk water area; 10 – transverse faults; 11 – stratigraphic boundaries.

overlapped by Paleozoic–Mesozoic rocks, while those in the eastern part are only partly overlapped by Cretaceous siliceous–volcanic–terrigenous eugeosynclinal strata. It was pointed out that the ultramafic rocks making up the massifs of both belts are similar in chemical composition to Alpine-type peridotites. Also, according to Rozhdestvenskii, the mafic-ultramafic massifs in both ophiolite belts are localized within the outlines of aeromagnetic anomalies elongated in the north-northwestward direction, and similarly sized or larger anomalies of this kind are detected east of Sakhalin Island, beneath the Sea of Okhotsk water area.

Having interpreted the structural position of the mafic-ultramafic bodies on northern Sakhalin Island, based on models of plate tectonics, Yu.N. Raznitsyn [1975] drew the conclusion that the ultramafic, gabbroid, and basaltoid bodies occur in the area as oceanic-mantle and crustal plates thrust over the edges of continental structures in the Late Cretaceous–Tertiary. That was due to obduction directed away from extension zones in deep-water basins of marginal seas. Raznitsyn pointed out that the ultramafic bodies on the Schmidt Peninsula form a chain ~47 km in length and 4 km in width; as seen from aeromagnetic data, they are traced in the southward direction, where they are overlain by Neogene deposits. This chain of ultramafic bodies includes the South Schmidt massif, the largest one on Sakhalin Island. It is dissected by a system of high- and low-angle faults accompanied by crush and mélange zones. In later works [1978, 1982], Raznitsyn applied an analogous approach in the description of the structural position and geologic structure of the fragment of the East Sakhalin ophiolite association including the Berezovka and Shel'ting mafic-ultramafic massifs (Fig. 2.3).

The structural position and petrographic composition of the Sakhalin ophiolite association by the example of the South Schmidt and Berezovka massifs were considered by A.N. Rechkin et al. [1975]. Two differently aged structure–compositional complexes were distinguished: Late Paleozoic–Early Cretaceous and Late Cretaceous. Based on the data by G.V. Pinus et al. [1973] and F.P. Lesnov et al. [1973], they divided the gabbroids forming these massifs into two genetic types (ortho- and paramagmatic) and made the conclusion that the wehrlites, hornblende lherzolites, and pyroxenites in these massifs formed owing to the impact of mafic magmas on ultramafic rocks. Later Rechkin [1984] returned to the issue of the structural position of ophiolites on Sakhalin Island, which he divided into two types: (1) those localized within pre-Cretaceous eugeosynclinal rocks and Paleozoic–Early Mesozoic metamorphic schists and (2) those occurring as an ultrabasic–gabbroid–radiolarite association within Upper Cretaceous volcanics and eugeosynclinal rocks. According to A.V. Rikhter [1985], the oldest basement rocks on Sakhalin Island are metaophiolites, whereas the oldest paleontologically constrained sediments are limestones (probably, of Silurian age). Also, this researcher presumed that the central and eastern parts of the island have an imbricated-thrust structure, including ophiolite allochthons.

Having studied the geologic structure, petrographic composition, and petrochemical features of the rocks of the Shel'ting massif, A.F. Bekhtol'd and D.F. Semenov [1978] presumed that the contacts between ultrabasic rocks and gabbroids in the massif, like those between the massif and host volcanosedimentary rocks, pass along faults everywhere, but there might have been a time gap between the formation of the ultramafic and gabbroid bodies. Semenov [1982] presented a general description of the igneous associations of the Pacific folded area by the example of Sakhalin Island. Considering

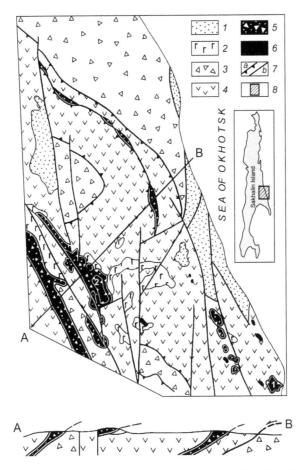

Figure 2.3 Tectonic sketch map of the southeastern part of the East Sakhalin Mountains (after [Raznitsyn, 1982]).
1 – Neogene rocks, undivided; *2* – plagiogranites, granodiorites, gabbro; *3* – Berezovka formation (olistostrome, Late Santonian–Danian?); *4* – Bogataya and Rakitinskaya formations (Coniacian?–Santonian); *5* – serpentinite mélange; *6* – basic–ultrabasic complex; *7* – faults: *a* – thrusts, *b* – steeply dipping; *8* – position of the region described by the tectonic sketch map (inset). I – Shel'ting massif; 2 – Berezovka massif.

the geology of the Sakhalin ophiolite association, Bekhtol'd and Semenov [1990] and V.M. Grannik [2008] pointed out that these rocks occur in five areas of the island: Schmidt Peninsula, East Sakhalin Mts., Nabil' and Susunai Ridges, and Tonin–Aniva Peninsula. Also, they stressed that the vast majority of outcrops of the ophiolite association is dissected by faults of NW strike, which is typical of the structure of the lower stage in the entire region. In Grannik's view [2008], the ophiolites were exposed to the day surface by the Alpine activity of pre-Cenozoic faults; note that at some places, ophiolite plates are thrust over Cenozoic sediments and most of the ultramafic bodies have tectonic contacts with the gabbroid ones.

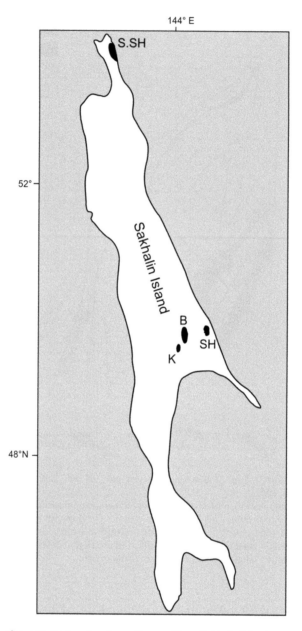

Figure 2.4 Position of investigated mafic-ultramafic massifs on Sakhalin Island.
Massifs (out of scale): B – Berezovka; K – Komsomol'sk; SH – Shel'ting; S.SH – South Schmidt.

Lesnov *et al.* [2009c] summarized data on the structure, geology, and petrographic composition of mafic-ultramafic massifs of the East Sakhalin ophiolite association. The Berezovka, Shel'ting, Komsomol'sk, and South Schmidt massifs (Figs. 2.4, 2.5) were studied in most detail. As it will be shown below, they differ in structural position,

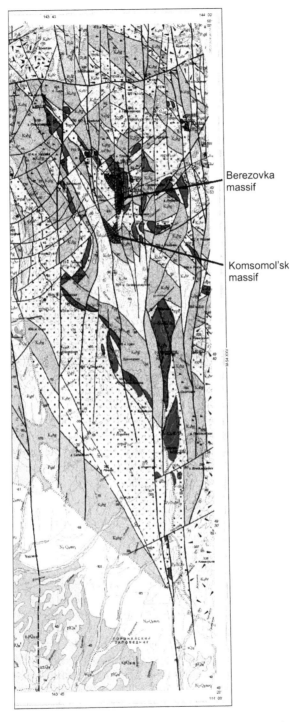

Berezovka massif

Komsomol'sk massif

Figure 2.5 Position of the Berezovka, Komsomol'sk, and other mafic-ultramafic and mafic massifs in structures of the East Sakhalin Mountains (fragment of the State Geological Map of the Russian Federation, 1:200,000. Sheet M-54-XXIV) (after [Explanatory note ..., 2009]).

size, morphology, internal structure, the ratio between the areas occupied by ultra-mafic rocks and gabbroids, and petrographic types of rocks as well as petrochemical, geochemical, and mineralogic features, etc.

Many issues of the tectonics, structure, age, petrographic composition, and gene-sis of the Sakhalin mafic-ultramafic massifs are debatable and remain poorly studied. According to some geologists [Starozhilov, 1990; Zharov, 2004], the present-day struc-ture of the island formed by accretion, with the leading role of transverse compression, and it is divided into three structural provinces: (1) structures of the West Sakhalin Mts., (2) structures of the East Sakhalin Mts., and (3) the system of intermontane basins which separates them. Besides that, it is presumed that the oldest rocks on the island outcrop within the East Sakhalin Mts., where they are the Paleozoic basement rocks of the Susunai metamorphic complex and make up Early Jurassic intrusive complexes. The rock complexes of this basement are overlain by strata of Cretaceous–Paleogene basalts, siliceous limestones, and shales. Recently the idea has been developed that the Sakhalin mafic-ultramafic massifs formed during the evolution of an island arc [Ishiwatari *et al.*, 2003]. Now the genetic unity of all the Sakhalin ultramafic rocks is presumed. It is hypothesized that no complete section of the oceanic crust is preserved within the island, but it can be reconstructed based on data on the structure and com-position of the largest mafic-ultramafic massifs and associated volcanic–terrigenous complexes. Also, it is believed that the lower part of the ophiolite section is preserved only within the South Schmidt ultramafic massif; the Shel'ting massif corresponds to the above-lying fragment of this section, in which the rocks of the so-called cumulate complex are added to dunites and harzburgites; the Berezovka massif might correspond to the uppermost oceanic crust.

However, as emphasized by I.Yu. Lobodenko [2010], there is no generalized structure–tectonic model for the Sakhalin Island region or different structures of the regional ophiolite association which would be accepted by most researchers.

Based on new information, it is attempted in the next chapters to specify and sup-plement existing views on the internal structure, composition, and time and conditions of formation of mafic-ultramafic massifs of the East Sakhalin ophiolite association. This will favor the least contradictory genetic modeling of formation of such massifs on and outside Sakhalin Island.

* * *

The various data obtained by many researchers from geological mapping and regional targeted studies in the last few decades have shown that numerous mafic-ultramafic massifs are concentrated within Paleozoic, Mesozoic, and younger fold–block struc-tures along the western Pacific shore, on the segment from Chukchi Peninsula through Koryakia, Kamchatka Peninsula, Sakhalin Island, and Primorye to the islands of Japan and Southeast Asia. They form belts of different strikes and lengths and their branches, which mark zones of long-lived deep-seated faults along the boundaries between Asian continent and Pacific. One of the fragments of this long belt within Sakhalin Island includes tens of small and several large mafic-ultramafic massifs. The Berezovka, Shel'ting, Komsomol'sk, and South Schmidt massifs, which are best stud-ied, show many differences in structural position, size, morphology, geologic structure, occurrence of ultramafic and mafic rocks at the present-day erosion level, etc.

Until recently these massifs were explored mainly during small- and medium-scale geological mapping and, partly, targeted studies. Therefore, there still exist different views on the structural position, structure, and conditions of formation of mafic-ultramafic massifs and the entire ophiolite association of Sakhalin Island. It remains important to carry out a detailed petrological study of these massifs by modern analytical methods. The solution of this problem will favor the least contradictory modeling of their formation both on and outside Sakhalin Island.

Chapter 3

Geologic structure of mafic-ultramafic massifs of the East Sakhalin ophiolite association

Based on the published data of some researchers and the author's data, we present the generalized geologic structure of the Berezovka, Shel'ting, Komsomol'sk, and South Schmidt mafic-ultramafic massifs, which are considered typical objects within the East Sakhalin ophiolite association.

3.1 BEREZOVKA MASSIF

The Berezovka mafic-ultramafic massif is one of the best studied massifs in the East Sakhalin ophiolite association. It is localized in the eastern spurs of the Central Ridge of Sakhalin Island, on the watersheds of the Zloveshchaya, Geran', and Berezovka Rivers. Its size is 1.5×4.5 km; outcrops ($\sim$6.7 km^2) consist of $\sim$90% ultramafic rocks, and the rest are gabbroids (Fig. 3.1). At a distance from the massif, there are considerably smaller ultramafic and gabbroid bodies, which might be its tectonic outliers.

The first information on the geologic structure and petrography of the massif was provided by Yu.M. Kovtunovich, geologist of the Sakhalin Geological Survey Expedition, in reports on medium-scale geological surveying (1959–1961). Later geological survey within the massif was done by V.T. Sheiko, I.I. Gritsenko, and V.V. Slodkevich. In 1972 the author carried out traversing studies in the area and collected rock samples; at the same time, some samples were obtained from the collections of Sheiko and Slodkevich. Based on the rock collections, a petrographic description of the Berezovka massif was carried out, and part of the analytical studies were made [Slodkevich and Lesnov, 1976; Lesnov *et al.*, 1998a]. In 2009 the collection was supplemented by the samples provided courtesy of V.G. Gal'versen and V.F. Evseev, geologists of the Sakhalin Expedition.

The Berezovka massif is localized within an intricately dislocated, compositionally heterogeneous complex of terrigenous volcanics (presumably, Cretaceous ones). It was divided into the Bogataya and Rakitinskaya Formations, according to the geological survey [Explanatory note ..., 2009]. The Bogataya Formation consists of mafic and intermediate volcanics metamorphosed to the greenschist facies: trachyandesite-basalt, andesite, and trachyandesite lavas, as well as their tuffs, tuff breccias, and tuffaceous rocks, interbedded with sandstones and siltstones. The total thickness of the Bogataya Formation is 800–900 m. The rocks of the Rakitinskaya Formation above in the section are more widespread within the massif; these are variegated pyroclasts interbedded

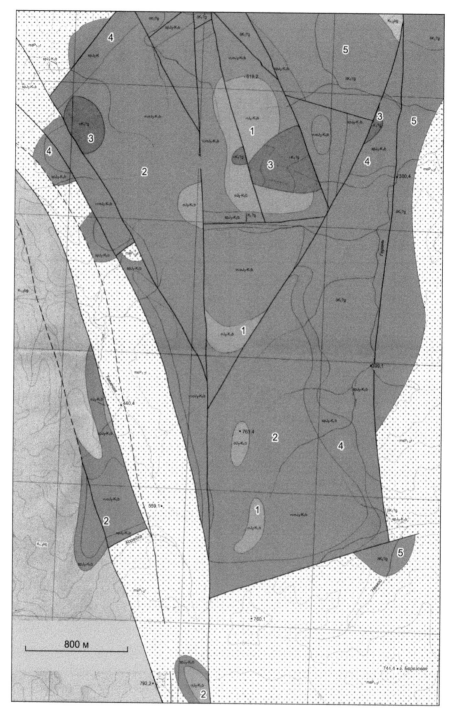

Figure 3.1 Geological sketch map of the Berezovka massif (according to A.N. Rechkin and V.G. Gal'versen, with modifications).

1 – dunites, harzburgites, lherzolites, their serpentinized varieties; *2* – plagioclase-bearing peridotites alternating with banded and lens-shaped bodies of wehrlites, clinopyroxenites, websterites, orthopyroxenites, cortlandites (schriesheimites), olivine gabbro and gabbronorites, troctolites and anorthosites; *3* – gabbronorites, gabbro, norites; *4* – serpentinites, which form the peripheral and near-fault line zones of the massif; *5* – hornblende gabbro and gabbro-diorites, diorites, quartz diorites, very rarely – tonalites.

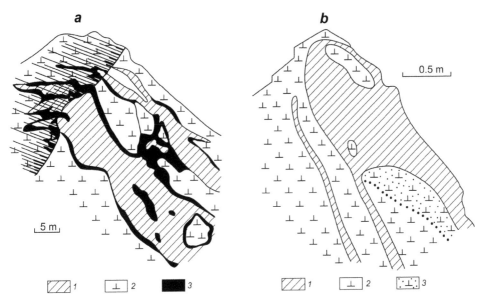

Figure 3.2 Sketches of bedrock outcrops of the Berezovka massif at the contact between gabbroid intrusion and ultramafic protrusion (made by V.V. Slodkevich).
a) gabbro veins (1) are injected into serpentinized harzburgites (2) and contain their xenoliths; the harzburgites are nonuniformly enriched in plagioclase (3) along the contact with gabbro; *b)* veins of clinopyroxene-containing troctolite (1) pierce serpentinite (2) after harzburgite, transformed into clinopyroxenite (3) at some sites; serpentinite xenoliths in the troctolite vein partly or completely transform into clinopyroxenite.

with jaspers, limestones, and radiolarites. The estimated total thickness of this formation is 1400 m. The massif and its framing are dissected by differently directed faults, along which the rock blocks separated by them were displaced in different directions. Almost everywhere the faults dissect the primary contacts of the massif with the host terrigenous volcanics. These faults are mostly accompanied by zones of rocks which experienced intense dynamic metamorphism and hydrothermal alteration, from few meters to tens of meters in thickness. Observations in mapping trenches show that the planes of the faults passing along the eastern and western contacts of the massif are inclined at 80° eastward in the northern part of the massif to 25° in its southern part. Albite–carbonate–zoisite and zoisite–talc–chlorite rocks, as well as cataclastic and carbonated diabases, injected with micropegmatite veinlets at some places, occur in narrow zones along the boundaries of the massif with the host rocks. Gabbroid and pyroxenite veins are observed at some places in the ultramafic rocks near their contacts with the gabbroid bodies (Fig. 3.2).

As stated above, the Berezovka massif consists of predominant ultramafic rocks and minor gabbroids. The ultramafic rocks are divided into two composition–genetic types. Type 1 includes restite harzburgites, lherzolites, or, less often, dunites, and serpentinites formed after them, which make up a protrusion. They have a homogeneous

Figure 3.3 Cross section of a sample of banded olivine gabbro (here and in Figs. 3.4–3.13, the Berezovka massif, gabbroid intrusion–ultramafic protrusion contact, is shown). Small lens-shaped xenoliths of serpentinized peridotite are observed. Scaled down to half size.

Figure 3.4 Cross section of a sample of banded melanocratic olivine gabbro. One can see lens-shaped and irregularly shaped xenoliths of serpentinized peridotites containing plagioclase porphyroblasts. Half size.

texture, structure, and quantitative mineral composition. Type 2 is hybrid (paramagmatic) ultramafic rocks making up a contact–reaction zone along the boundaries of the ultramafic protrusion and the gabbroid bodies cutting them. These rocks vary widely in texture, structure, and quantitative mineral composition: plagioperidotites, websterites, orthopyroxenites, clinopyroxenites, and their olivine- and plagioclase-bearing varieties. The same varieties of ultramafic rocks form xenoliths in gabbro. Note that the hybrid ultramafic rocks often have taxitic, including parallel-banded, textures (Figs. 3.3–3.13).

The gabbroids of the Berezovka massif are divided into three composition–genetic types. Orthomagmatic gabbroids (olivine-free gabbronorites, gabbro, and, less often, norites, characterized by a stable quantitative mineral composition and a massive

Figure 3.5 Cross section of a sample in which pinching-out veins of melanocratic gabbro pierce serpentinite after peridotite.
Gabbro contains small serpentinite xenoliths. Minute clinopyroxene and plagioclase porphyroblasts are present in serpentinite.

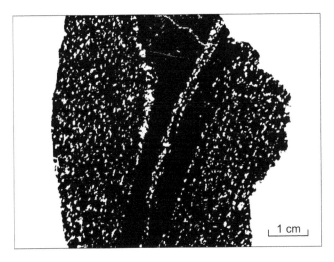

Figure 3.6 Cross section of a sample of banded olivine gabbro.
One can see lens-shaped and banded xenoliths of serpentinites after peridotites, which contain small plagioclase porphyroblasts and veinlets. At one of the xenolith contacts the content and sizes of plagioclase grains are somewhat increased.

structure) are considered a product of crystallization of uncontaminated upper-mantle mafic melts. Hybrid (paramagmatic) gabbroids (olivine and olivine-containing meso-, melano-, and leucocratic gabbro, gabbronorites, and, more rarely, troctolites and anorthosites) with a variable quantitative mineral composition and taxitic, including

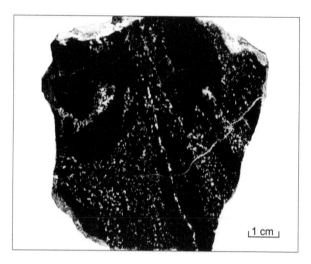

Figure 3.7 Cross section of a sample of melanocratic olivine gabbro.
The sample contains banded and irregularly shaped xenoliths of serpentinites after peridotites. In the center there is an elongated serpentinite xenolith dissected by a thin plagioclase–pyroxene veinlet.

Figure 3.8 Cross section of a plagiowehrlite sample.
Small relict xenoliths of serpentinites after peridotites are observed. In the center one can see a dissecting anorthosite veinlet, with clinopyroxene grains concentrated at the periphery.

parallel-banded, textures are regarded as a product of crystallization of upper-mantle mafic melts nonuniformly contaminated with the material of an ultramafic protrusion. Hybrid (paramagmatic) gabbroids (amphibole- and quartz-containing gabbro, gabbro-diorites, diorites, and quartz diorites) with a variable quantitative mineral composition

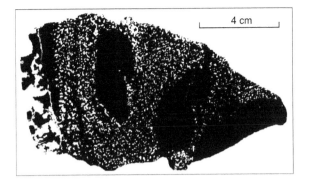

Figure 3.9 Cross section of a sample of taxitic olivine gabbro.
Olivine gabbro contains rounded xenoliths of serpentinites, injected with gabbroid veinlets. In the left part of the sample olivine gabbro and the serpentinite xenolith it contains are dissected by a vein of pegmatoid gabbro.

Figure 3.10 Cross section of a sample of banded plagiowehrlite.
The plagiowehrlite texture is characterized by the subparallel orientation of small relict xenoliths of serpentinites after peridotites (black) containing porphyroblastic accumulations of clinopyroxene and plagioclase grains.

Figure 3.11 Cross section of a sample of banded olivine gabbro.
Gabbro contains imbricate lens-shaped xenoliths of serpentinites (black) with porphyroblastic plagioclase segregations. At some places along the xenolith boundaries the content of plagioclase grains increases.

Figure 3.12 Cross section of a sample of coarse-banded olivine gabbro (right), in which an anorthosite vein (center) alternates with flattened serpentinite xenoliths (black).

and taxitic textures are considered a result of crystallization of upper-mantle mafic melts nonuniformly contaminated with the material of different host rocks.

The ultramafic xenoliths in the gabbroids, gabbroid and pyroxenite apophyses in the ultramafic rocks, and the contact–reaction zone suggest that gabbroid intrusions formed later than ultramafic protrusions. The widespread occurrence of hybrid varieties of ultramafic rocks and gabbroids in the Berezovka massif testifies to highly intense interaction of mafic melts and their fluids with the substance of the preexisting ultramafic protrusion and enclosing rocks. The presented description of the geologic structure and petrographic composition of the Berezovka massif, as well as the contact relations between the ultramafic and gabbroid bodies, permits viewing the massif as a typical polygenic mafic-ultramafic complex in ophiolite associations.

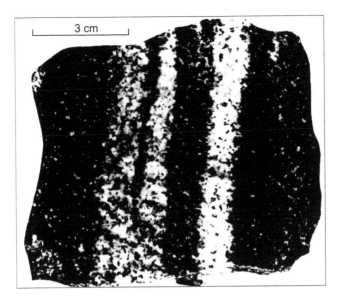

Figure 3.13 Cross section of a sample of parallel-banded ultramafic-mafic rock.
The rocks in the alternating "bands" vary in quantitative mineral composition from plagiowehrlite (dark gray sites) to leucocratic gabbro and anorthosite (light sites).

3.2 SHEL'TING MASSIF

The Shel'ting massif is localized on the shore of the Sea of Okhotsk, near Shel'ting Cape. It was first described by S.S. Darbidyan in 1953; in 1961–1963 it was mapped by Yu.M. Kovtunovich and V.T. Sheiko. Information on the geologic structure and petrographic composition of the massif was later published by G.P. Vergunov [1964], V.V. Slodkevich [1975b], Yu.M. Raznitsyn [1978], A.F. Bekhtol'd and D.F. Semenov [1990], and S.V. Vysotskii *et al.* [1998]. Using the rock collection kindly provided by Slodkevich, F.P. Lesnov [1988] made petrographic, petrochemical, and geochemical studies of the Shel'ting massif.

The massif (1.75 × 2.25 km in size) outcrops in an area of ~1.8 km^2 (Fig. 3.14). About 70% of this area is occupied by ultramafic rocks, and the rest, by gabbro. Vergunov [1964], having observed an alternation of near-vertical "sheet-like" dunite and peridotite bodies near Shel'ting Cape, presumed that the dunites are older than the peridotites. Slodkevich [1975b] described the Shel'ting massif as an Early Miocene layered pluton within Late Cretaceous dislocated strata of the eugeosynclinal type. This researcher distinguished three elements in the massif structure: quenched zone; lower, lateral, and upper marginal groups; and a ternary layered series. According to his data, the massif is composed of dunites, harzburgites, lherzolites, enstatitites, gabbronorites, norites, gabbro, and gabbro-diorites. Based on measurements of optical constants, Slodkevich determined the compositions of rock-forming minerals and made the conclusion that the rocks of the massif are characterized by "cryptic layering". Also, he noted that harzburgite blocks are localized in pyroxenites at some places.

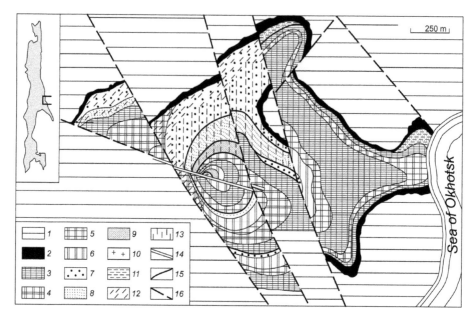

Figure 3.14 Geological sketch map of the Shel'ting massif (after [Slodkevich, 1975]).
1 – pebble gravel–siliceous–clayey deposits of the Uchir Formation (Upper Cretaceous);
2 – rocks of the quenched zone: gabbro, gabbro-diorites, and diorites; *3–9* – rocks
of the layered series and marginal groups: *3* – interbedded peridotite–dunite unit;
4 – peridotites; *5* – interbedded pyroxenite–peridotite unit; *6* – pyroxenites; *7* – norites;
8 – gabbronorites; *9* – gabbro, gabbro-diorites, quartz diorites; *10* – rocks of the intermediate horizon (tonalites, gabbro-plagioclasites, and plagioclasites). *11–13* – marginal groups:
11 – lower; *12* – lateral; *13* – upper; *14* – microdiorite dikes; *15* – boundaries of the
structural units of the pluton; *16* – tectonic faults.

Raznitsyn [1978], having explored the geology of the Shel'ting massif, stressed structure–tectonic aspects. In his view, the massif is a thin (200–250 m) near-horizontal "plate" which overlies the siltstones of the Uchir Formation and consists of almost undeformed massive and banded harzburgites, enstatitites, gabbronorites, and troc-tolites. In some outcrops Raznitsyn found serpentinized wehrlites with segregations of green spinel. On the eastern flank of the massif, within the ultramafic rocks, he described a tectonic block (20×30 m in size) of silicified basalts. The researcher pointed out that the banding in the harzburgites is due to the alternation of 5 to 30 cm thick band-like segregations enriched or, on the contrary, depleted in grains of enstatite and subordinate diopside. Measurements of the dips and strikes of the banding in the ultramafic rocks revealed no regularity. Besides that, Raznitsyn noted that the base of the ultramafic "sheet" is a 50 to 60 m thick serpentinite mélange and a series of faults divides the massif into several blocks which experienced differently directed displacements.

Bekhtol'd and Semenov [1978] compiled a schematic geologic section of the Shel'ting massif, with five horizons differing in petrographic composition (Fig. 3.15).

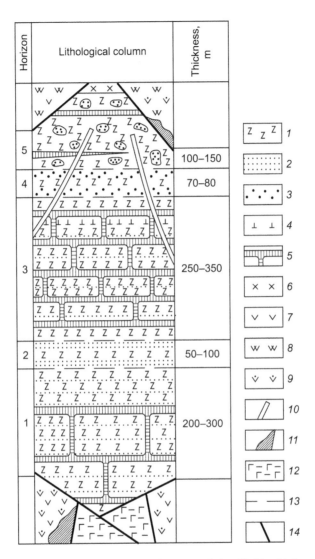

Figure 3.15 Schematic geologic section of the Shel'ting massif, after [Bekhtol'd, Semenov, 1978]. *1* – harzburgites, *2* – dunites, *3* – lherzolites, *4* – orthopyroxenite horizon, *5* – orthopyroxenite veins, *6* – gabbro-plagioclasites, *7* – norites, *8* – gabbronorites, *9* – hornblende gabbro, *10* – microgabbro and dolerite dikes, *11* – endocontact dolerites, *12* – volcanosedimentary rocks of the Uchir Formation (Upper Cretaceous), *13* – boundaries between horizons, *14* – faults.

According to their estimates, the areas of outcrops of the rock varieties in the massif structure are related as follows: 35% serpentinized harzburgites and lherzolites; 10% dunites; 20% orthopyroxenites and olivine orthopyroxenites; 25% norites and gabbronorites; and 10% hornblende gabbro. It was noted that ultramafic and gabbroid bodies usually have tectonic contacts and the Shel'ting ultramafic rocks are similar in chemical composition to the Alpine-type ultramafic rocks making up the South Schmidt massif, but the ultramafic and gabbroid bodies, most likely, formed at different times.

As observed by Vysotskii *et al.* [1998], ultramafic rocks make up the lower part of the section of the Shel'ting massif (~120 m in thickness); they are overlain by pyroxenites and then by gabbroids—mainly olivine-free gabbro and gabbronorites. These researchers presumed that the Shel'ting and Berezovka massifs are coeval intrusive derivates of boninitic lavas and all their rock varieties resulted from "layering" of boninitic magma emplaced in the setting of an ensimatic island arc.

Having summarized the available data on the structure and composition of the Shel'ting massif and results of his own petrographic studies of the collection provided by Slodkevich, Lesnov [1988] distinguished three composition–structural elements which are closely related in space but genetically independent: (1) protrusion of restite ultramafites (harzburgites, lherzolites, dunites, their serpentinized varieties, and serpentinites); (2) orthomagmatic-gabbroid intrusion cutting the ultramafic protrusion (olivine-free gabbronorites, gabbro, and, less often, norites); and (3) contact–reaction zone composed of hybrid ultramafic rocks (orthopyroxenites, olivine orthopyroxenites, and websterites) and hybrid gabbroids (olivine gabbronorites and anorthosites). At some places the hybrid ultramafic rocks and gabbroids occur as vein bodies within the restite ultramafites. All the above-mentioned features suggest that the Shel'ting massif, like the Berezovka massif, is a polygenic mafic-ultramafic complex (Fig. 3.16).

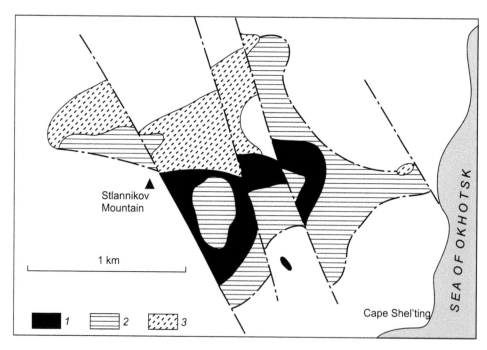

Figure 3.16 Structural scheme of the Shel'ting massif (after [Lesnov, 1988]).
1 – dunites, pyroxene-containing dunites, harzburgites, lherzolites, serpentinites; *2* – wehrlites, plagiowehrlites, clinopyroxenites, websterites, and their olivine- and plagioclase-containing varieties; *3* – gabbronorites, norites, gabbro, olivine-containing and olivine gabbro, and, less often, troctolites, anorthosites, and hornblende gabbro.

3.3   KOMSOMOL'SK MASSIF

The Komsomol'sk massif (0.5–1.0×3.5 km in size) is localized ~2 km south-southwest of the Berezovka massif. It was first mapped by V.T. Sheiko, V.V. Vasil'ev, A.I. Donets, and N.V. Galushko during medium-scale geological surveying. As presumed by Raznitsyn [1982], the massif is a near-horizontal "sheet" elongated for 2 km north-westward, ≤ 500 m in width, and 200–250 m in thickness (Fig. 3.17). According to this author, the massif is "thrust" over the flyschoid sediments of the Berezovka Formation and the siliceous volcanics of the Bogataya Formation. In the upper reaches of the Uzkii Creek, Raznitsyn observed a near-horizontal contact of ultrabasic rocks with interbedded flyschoid strata of the Berezovka Formation sandstones and siltstones, whose beds are inclined at 70° eastward. He noted the presence of zones of brecciated serpentinites, which he defined as mélanges, along the contacts between ultramafic and host rocks. Near Mt. Komsomol'skaya, Raznitsyn observed an outcrop of amphi-bolized gabbronorites 70×50 m in size (probably, a tectonic block). Somewhat away from the Komsomol'sk massif, in the host sediments, several small tectonic blocks of serpentinites and stock-like gabbroid bodies are localized.

Based on published data and results of study of the rock collection provided by Sheiko, as well as on the experience of studying the Berezovka massif, Lesnov [1988] came to the conclusion that the Komsomol'sk massif is a fault-bounded ultramafic protrusion of lherzolites, dunites, and serpentinites intruded by a stock-like gabbroid body. As observed by the predecessors, a thin discontinuous contact–reaction zone of olivine-containing hybrid gabbroids and hybrid ultramafic rocks (wehrlites and

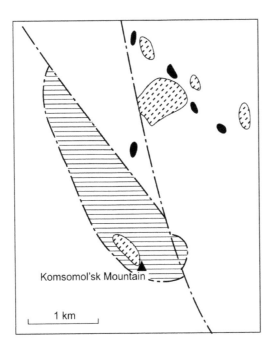

Komsomol'sk Mountain

1 km

Figure 3.17   Structural scheme of the Komsomol'sk mafic-ultramafic massif (after [Lesnov, 1988]). See legend in Fig. 3.16.

pyroxenites) is localized along the boundaries of this gabbroid stock with the ultramafic rocks. It might have resulted from transformation of restite ultramafites under the effect of mafic melts and their fluids. This suggests that the Komsomol'sk massif is a polygenic complex [Lesnov, 1988].

3.4 SOUTH SCHMIDT MASSIF

The South Schmidt massif is the largest essentially ultramafic body on Sakhalin Island. It is localized in the northern part of the island, on the Schmidt Peninsula. The massif is elongated from north to south, 1–3 × 19 km in size, the total area of outcrops being ~42 km^2 (Fig. 3.18). The earliest information on the massif was provided by V.G. Krasnov, A.Ya. Chalykh, and B.A. Naumenko in reports on results of

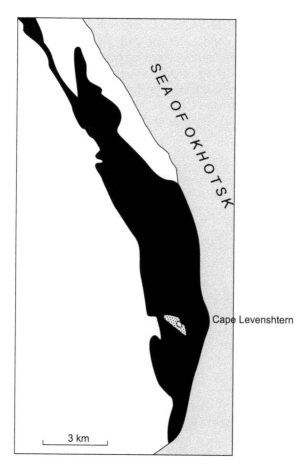

SEA OF OKHOTSK

Cape Levenshtern

3 km

Figure 3.18 Structural scheme of the South Schmidt mafic-ultramafic massif (after [Lesnov, 1988]). See legend in Fig. 3.16.

medium-scale geological mapping. Later it was described in [Raznitsyn, 1975; Rechkin *et al.*, 1975; Rechkin, 1984; Rozhdestvenskii, 1988; Lesnov and Agafonov, 1976; Lesnov *et al.*, 1976; Lesnov, 1988] from results of targeted studies.

According to Raznitsyn [1975], the ultramafic, gabbroid, and basaltoid bodies in the ophiolite association near the Schmidt Peninsula occur as "sheets" thrust over continental structures during Late Cretaceous–Tertiary obduction. This author pointed out that, according to aeromagnetic data, ultramafic bodies are traced southward from the Schmidt Peninsula beneath Neogene sediments. The massif is composed of serpentinized harzburgites, lherzolites, dunites, and minor wehrlites and pyroxenites. It is dissected by a system of high- and low-angle faults, along which crush (mélange) zones are localized.

Rechkin [1984] presumed that the structural position of the South Schmidt massif is determined by its confinement to the Longri fault, which belongs to the zone of a deep disjunction, traced along the Sea of Okhotsk shelf of Sakhalin Island. According to his data, the massif has tectonic contacts with the Jurassic–Early Cretaceous host volcanics. Based on results of airborne geophysical studies, Rechkin showed that ultramafic outcrops on land are traced east of the South Schmidt massif, beneath the water area of the Sea of Okhotsk. The ultramafic rocks in the massif are serpentinized lherzolites and minor harzburgites and dunites intruded by the small hypabyssal gabbroid stocks of the Tomi intrusive complex. One of such stocks (0.4×1.2 km in size) near Levenshtern Mountain is overlapped by Jurassic–Cretaceous volcanics. The ultramafic rocks are injected with orthopyroxenite and websterite vein bodies of different thicknesses. At the gabbroid–ultramafite contacts, rocks which experienced intense dynamic metamorphism are localized; they are pierced with a network of natrolite and albitite veins. Schlieren-like bodies of massive low-alumina chromitites were detected at some places within the lherzolites and dunites.

Rozhdestvenskii [1988], sharing the views of Raznitsyn [1975, 1982] on the allochthonous bedding of ophiolites on the Schmidt Peninsula, disagreed with the statement that the ophiolite "flakes" in the southeastern part of the peninsula show gentle bedding. Having considered results of measurements of the dips and strikes of banding in peridotites, Rozhdestvenskii made the conclusion that the South Schmidt massif is a protrusion emplaced along the Longri fault zone. Based on aeromagnetic and airborne gravity surveys [Sychev, 1966], the researcher stated that the ultramafic protrusion is a near-vertical block expanding with depth.

By V.M. Grannik [2008], the South Schmidt massif was interpreted as a near-horizontal allochthonous "sheet" >500 m in thickness, overlying Upper Cretaceous sediments. Its lower part consists of a serpentinite mélange several tens of meters in thickness, composed of brecciated and foliated apoharzburgite and, less often, apodunite talcose serpentinites. In the mélange Grannik observed blocks 1–10 m in size, made up of unaltered harzburgites and lherzolites as well as hornblende gabbro, diorites, and rodingites. The upper part of the ultramafic "sheet", ~ 400 m in thickness, is composed predominantly of considerably serpentinized massive harzburgites, including banded varieties. Wehrlite and enstatite bodies occur in subordinate quantity. Small gabbroid bodies and chromitite segregations are found at some places in the ultramafic rocks of the massif, dissected by numerous high-angle faults. Gold-containing prehnite–zeolite–chlorite mineralization is locally observed in the gabbroids.

Using results of aeromagnetic and gravity surveys, R.M. Yurkova [2012] makes the conclusion that the South Schmidt massif is a near-vertical dunite–harzburgite diapir, which is the oldest complex of the Sakhalin ophiolite association.

In our view, based on the published data of the earlier researchers and our own field observations and analytical data, the South Schmidt mafic-ultramafic massif is an essentially ultramafic polygenic mafic-ultramafic body within the East Sakhalin ophiolite association [Lesnov and Agafonov, 1976; Lesnov et al., 1976; Lesnov, 1988]. It is localized in Jurassic–Cretaceous terrigenous volcanics and includes a large, steeply dipping protrusion of restite ultramafites and the small hypabyssal gabbroid stocks which cut it. The ultramafic restites are predominant serpentinized lherzolites and minor harzburgites and dunites. Few vein bodies of hybrid ultramafic rocks— orthopyroxenites (enstatitites) and websterites—are localized within the ultramafic restites. At some places along the ultramafite–gabbroid contacts, there are thin cata-clastic zones pierced with a network of natrolite and albitite veinlets. The latter are locally bluish gray because of the presence of fine-fibrous segregations of alkaline amphibole from the crossite–rhodusite group [Lesnov et al., 1976].

∗ ∗ ∗

All the available data suggest that the Berezovka, Shel'ting, Komsomol'sk, and South Schmidt mafic-ultramafic massifs in the East Sakhalin ophiolite association are polygenic plutonic complexes. The Berezovka massif includes an ultramafic-restite protrusion, the small intrusive bodies of orthomagmatic gabbroids cutting the pro-trusion, and a compositionally heterogeneous thick contact–reaction zone of hybrid ultramafic rocks and hybrid gabbroids. The latter contain hybrid ultramafic xenoliths of different sizes and shapes. Near the contacts with the host terrigenous volcanics, the gabbroid massif consists of hybrid gabbroids: hornblende gabbro, gabbro-diorites, diorites, and quartz diorites. In the ultramafic rocks, there are thin chromitite veins and schlieren. The Shel'ting massif is composed of several fragments of an ultramafic protrusion, the gabbroid intrusion cutting the protrusion, and a contact–reaction zone of hybrid ultramafic rocks—orthopyroxenites and websterites. The Komsomol'sk mas-sif is a protrusion of ultramafic restites and their serpentinized varieties, bounded by high- and low-angle faults. In its southern part, the protrusion is cut by a small gab-broid stock, with a narrow discontinuous contact–reaction zone of hybrid wehrlites and olivine gabbro at the periphery. The South Schmidt massif is the largest protru-sion of ultramafic restites (lherzolites, subordinate harzburgites and dunites, and their serpentinized varieties) on Sakhalin Island. It is intruded by small shallow gabbroid stocks. In the ultramafic restites, there are few vein bodies of hybrid ultramafic rocks (websterites and wehrlites) and schlieren-like chromitite segregations of different sizes.

The presented data on the geologic structure of four mafic-ultramafic massifs from the East Sakhalin ophiolite association suggest that the ultramafic bodies belonging to the massifs had intruded along fault zones as tectonic blocks (protrusions) before the mafic melts which formed the gabbroid intrusions spatially related to the protrusions were emplaced through these zones. Therefore, it is justified to consider all these massifs polygenic plutonic complexes.

Chapter 4

Petrography of mafic-ultramafic massifs of the East Sakhalin ophiolite association

... But for the diversity of igneous rocks there would be no science dealing with igneous petrogenesis ...

G.M. Brown

It was shown that the Berezovka, Shel'ting, Komsomol'sk, and South Schmidt mafic-ultramafic massifs (East Sakhalin ophiolite association) consist of highly diverse ultramafic rocks and gabbroids. The petrographic description of rocks from these massifs was made based on their examination under an optical microscope with regard to the rock nomenclature accepted in the Petrographic Code of Russia [2009]. Both the author's descriptions of rocks and those by petrographers of the Sakhalin Geological Exploration Expedition [Explanatory note . . ., 2009] were used. In total, >20 petrographic varieties of ultramafic and mafic rocks were recognized and described.

The Berezovka massif consists of harzburgites, lherzolites, dunites, their serpentinized varieties and serpentinites, wehrlites, websterites, ortho- and clinopyroxenites, olivinites, schriesheimites (cortlandites), plagioclase-containing ultramafic rocks, gabbronorites, gabbro and olivine-containing gabbro, troctolites, hornblende gabbro, gabbro-diorites, diorites, and quartz diorites. The Shel'ting massif is composed of harzburgites, lherzolites, dunites, their serpentinized varieties and serpentinites, wehrlites, websterites, orthopyroxenites, olivine- and plagioclase-containing pyroxenites, gabbronorites, gabbro, norites and olivine-containing norites, troctolites, anorthosites, and hornblende gabbro. In the Komsomol'sk massif, the following rocks are observed: dunites, lherzolites, their serpentinized varieties and serpentinites, wehrlites, pyroxenites, olivine gabbronorites, gabbronorites, gabbro, and gabbro-pegmatites. The rocks of the South Schmidt massif are harzburgites, lherzolites, dunites, wehrlites, websterites, orthopyroxenites, gabbro, and gabbro-diabases. A brief petrographic description of the above ultramafic rocks and gabbroids is presented below. The rocks are divided into the following petrogenetic groups: (1) restitic (orthomagmatic) ultramafic rocks, (2) hybrid (paramagmatic) ultramafites, (3) orthomagmatic gabbroids, (4) hybrid (paramagmatic) gabbroids of endocontact zones with ultramafites, and (5) hybrid (paramagmatic) gabbroids of endocontact zones with enclosing rocks.

4.1   RESTITIC (ORTHOMAGMATIC) ULTRAMAFIC ROCKS

The restitic (orthomagmatic) ultramafic rocks are harzburgites, lherzolites, and dunites, which were earlier defined as Alpine-type ultramafites; in plate tectonics, they are called "metamorphic peridotites".

Harzburgites are the most widespread restitic ultramafic rocks in the Berezovka, Shel'ting, and South Schmidt massifs. These are differently serpentinized medium-grained massive dark green to black rocks with 30–50% olivine, 50–65% orthopyroxene, 0–5% clinopyroxene, and 2–5% Cr-spinel. Secondary minerals are serpentine (5–50%), magnetite (≤2%), and, sometimes, talc. The harzburgites are often cut by pyroxenite veins and veinlets (Fig. 4.1).

Lherzolites are detected in the Berezovka, Komsomol'sk, and South Schmidt massifs. These are medium-grained dirty green to dark green rocks with a massive (less often, indistinctly banded) structure. They have a panidiomorphic-granular texture passing into cellular texture in serpentinized varieties. The quantitative mineral composition of these rocks is as follows: 15–20% olivine, 10–15% orthopyroxene, 10–60% clinopyroxene, 2–5% Cr-spinel, 5–40% serpentine, and 1–2% magnetite (Fig. 4.2).

Dunites are observed in the Berezovka, Shel'ting, Komsomol'sk, and South Schmidt massifs in minor quantities. These are differently serpentinized medium-grained massive dark green to black rocks. They have a cellular texture with relics

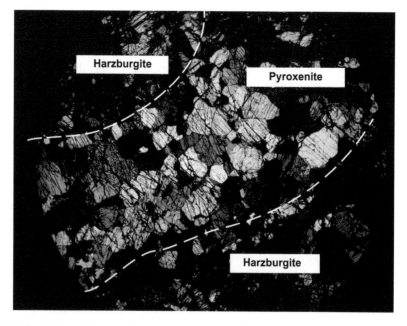

Figure 4.1   Microphotograph of weakly serpentinized harzburgite cut by a clinopyroxenite veinlet (sample 1596, Berezovka massif).
Hereafter, in Figs. 4.2–4.6, 4.8, and 4.10–4.12, samples are from the Berezovka massif. Micrographs in Figs. 4.1–4.4 and 4.6–4.12 were taken using an optical microscope at a 20x magnification (parallel Nicol prism).

of panidiomorphic-granular texture. These rocks consist of dramatically predominant olivine grains (sometimes, with ≤5% clinopyroxene admixture), and Cr-spinel is an accessory phase (1–5%). Schlieren- and vein-like segregations of massive and thick-disseminated chromitites are sometimes detected in the dunites. Orthomagmatic ultramafites near faults are usually intensely serpentinized and pass into serpentinites which have experienced different degrees of cleavage, brecciation, and foliation.

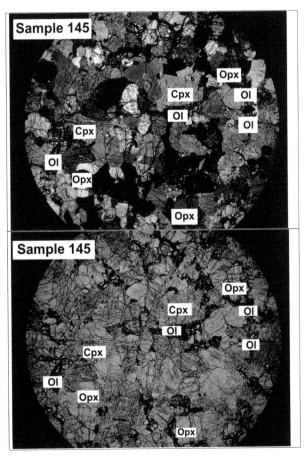

Figure 4.2 Microphotograph of lherzolite (sample 145, Berezovka massif).
Olivine (Ol) forms xenomorphic, often chain-like segregations, cut by magnetite-serpentine veinlets, with polygonal fragments of preserved unaltered mineral grains in the intervals between them. Orthopyroxene (Opx) occurs as subisometric prismatic grains with straight or sinuous edges, often with subparallel thin platy clinopyroxene segregations (exsolution structures). Prismatic clinopyroxene (Cpx) grains, which almost do not undergo secondary alterations, show a network of joints. The contents of olivine, orthopyroxene, and clinopyroxene are ∼20%, ∼20%, and ∼60%, respectively.

4.2 HYBRID (PARAMAGMATIC) ULTRAMAFIC ROCKS

The hybrid (paramagmatic) ultramafites are wehrlites, websterites, ortho- and clinopyroxenites, and olivine ortho- and clinopyroxenites (less often, olivinites and amphibole peridotites (schriesheimites)). All these rocks sometimes contain a plagioclase admixture. Many of the above rocks have a highly variable quantitative mineral composition, taxitic texture, and parallel-banded structure. Rocks of this group usually form contact–reaction zones at the boundaries of ultramafic protrusions and gabbroid intrusions cutting them, vein bodies cutting restitic ultramafites, and xenoliths in hybrid gabbroids.

Minor quantities of *wehrlites* and *plagiowehrlites* are observed in all the massifs under consideration. They are medium-grained or inequigranular rocks consisting of

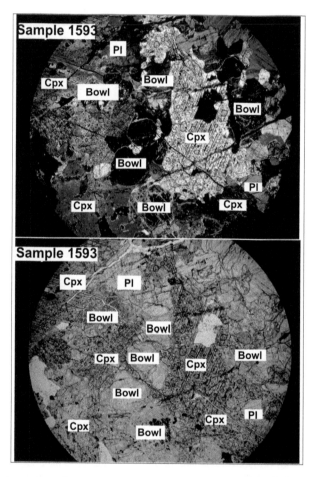

Figure 4.3 Microphotograph of plagioclase-containing wehrlite (sample 1593, Berezovka massif). Subisometric and anhedral grains of olivine are almost completely replaced by brownish yellow bowlingite (Bowl). Clinopyroxene (Cpx) is present as prismatic and xenomorphic grains of different sizes, often with sinuous or jagged edges, filled with a network of adjacent joints. Plagioclase (Pl) is observed as rare small, weakly altered angular grains. The contents of olivine, clinopyroxene, and plagioclase are ~25%, ~65%, and ~10%, respectively.

40–70% olivine grains (predominantly, xenomorphic ones), which are usually partly or completely replaced by serpentine or bowlingite, and 25–60% larger prismatic or xenomorphic clinopyroxene segregations. At some places the wehrlites contain an admixture of amphibole, which partly or completely replaces the clinopyroxene grains; also, xenomorphic plagioclase segregations are detected. Accessory phases are Cr-spinel or, in some cases, sulfide minerals (Fig. 4.3).

Websterites, olivine websterites, and plagiowebsterites are medium-grained (more rarely, coarse- or fine-grained) grayish brown rocks composed of prismatic ortho- and clinopyroxene grains in variable quantities (Fig. 4.4). Xenomorphic olivine and

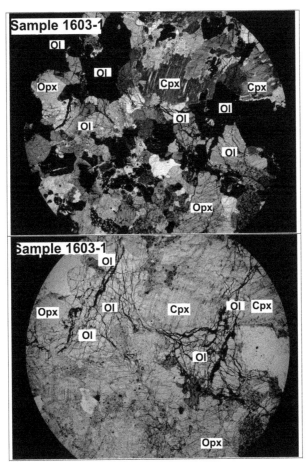

Figure 4.4 Microphotograph of olivine websterite (sample 1603-1, Berezovka massif). Olivine (Ol), almost epigenetically unaltered, occurs as highly xenomorphic, often amoeboid and thread-like segregations dissected in different directions by thin fractures into many small acute-angled "blocks". Large, rarely medium-sized crystals of orthopyroxene (Opx) with sinuous and angular contours are cut by joints. Clinopyroxene (Cpx) is present as large and many smaller xenomorphic grains. As seen from the slight bending of joints in the large clinopyroxene grains (above center), the rock has experienced weak brittle–ductile deformation. The contents of olivine, orthopyroxene, and clinopyroxene are ~35%, ~25%, and ~40%, respectively.

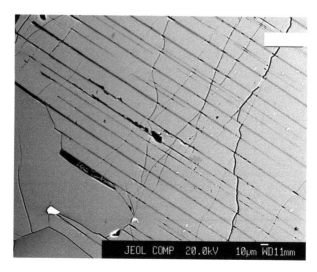

Figure 4.5 Microphotograph of an exsolution structure (lamellae of clinopyroxene partly replaced by amphibole) in orthopyroxene from olivine websterite (sample 140, Berezovka massif). The picture was taken using a scanning electron microscope. The chemical composition of these minerals is given in Table 8.8.

plagioclase segregations are sometimes observed as an admixture. The orthopyroxene grains contain exsolution structures consisting of lamellae of clinopyroxene, sometimes replaced by amphibole (Fig. 4.5).

Orthopyroxenites are medium- or coarse-grained light gray or dark gray rocks with a panidiomorphic-granular texture. Some varieties show a parallel-banded structure formed by an alternation of medium- and coarse-grained "bands". These rocks are composed of prismatic orthopyroxene grains, with segregations of clinopyroxene and accessory Cr-spinel as an admixture. In some cases orthopyroxene is partly replaced by serpentine or talc.

Olivine orthopyroxenites (sometimes, with a clinopyroxene admixture) are medium-grained greenish dark gray rocks with a panidiomorphic-granular texture. They consist of predominant orthopyroxene and variable quantities of olivine (5–10%), clinopyroxene (≤3%), amphibole (≤5%), and accessory magnetite (≤2%).

Clinopyroxenites, olivine clinopyroxenites, and plagioclase clinopyroxenites are medium-grained or inequigranular (often, taxitic) dark gray rocks. Clinopyroxene usually makes up >60%; olivine, whose xenomorphic segregations are partly or completely replaced by bowlingite, is often detected in minor amounts (up to 7–10%); and plagioclase is observed (Fig. 4.6). Some varieties contain xenomorphic segregations of plagioclase partly or completely replaced by secondary products and elongated prismatic segregations of amphibole forming pseudomorphs after clinopyroxene grains (Fig. 4.7). The plagioclase-enriched clinopyroxenites are

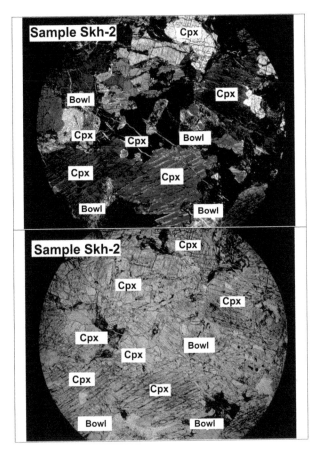

Figure 4.6 Microphotograph of coarse olivine clinopyroxenites (sample Skh-2, Berezovka massif). Olivine, present as small grains with jagged edges (<10% of the rock volume), is completely replaced by brownish yellowish bowlingite (Bowl). Clinopyroxene (Cpx) occurs as unaltered large (~6 mm) and smaller xenomorphic grains with sinuous edges, which are often cut by adjacent joints. Clinopyroxene content is ≤90%.

assigned to *gabbro-pyroxenites*. Accessory minerals in the clinopyroxenites are magnetite, iron sulfides, titanomagnetite, zircon, apatite, and titanite, which is replaced by leucoxene pseudomorphs. In the restitic ultramafites and their serpentinized varieties clinopyroxenite veins and veinlets are observed, at whose exo-contacts the ultramafic rocks transform into fine-fibrous chrysotile serpentinite (Fig. 4.8).

Olivinites are rare in the studied massifs. They have a medium-grained, panidiomorphic-granular, or cellular–platy–netted texture and consist of grains of

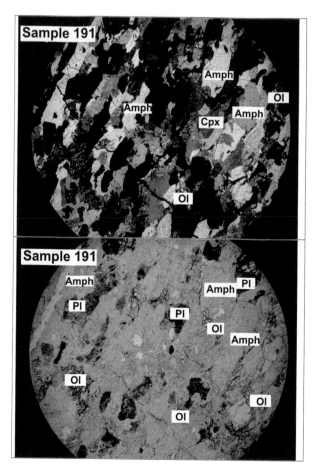

Figure 4.7 Microphotograph of clinopyroxenite containing plagioclase, amphibole, and olivine (sample 191, Komsomol'sk massif).
Olivine (Ol) is present as xenomorphic segregations with a multidirectional network of cracks filled with serpentine and magnetite. Between cracks there are angular fragments in which olivine has not undergone secondary alterations. The angular grains of clinopyroxene (Cpx) are in a matrix of predominant elongated prismatic grains of amphibole (Amph). Plagioclase (Pl) is observed as elongated and xenomorphic segregations completely replaced by secondary minerals. The long axes of the mineral grains have a subparallel orientation. Olivine, clinopyroxene, amphibole, and plagioclase make up 15–20%, <10%, 60%, and ~10%, respectively.

serpentinized olivine (≤90%) and occasional ortho- and clinopyroxene segregations (≤5%). Chrome-magnetite and magnetite (≤5%) are accesory minerals.

Amphibole peridotites (schriesheimites, cortlandites) are dark green, almost black, with a massive structure and a panidiomorphic-granular texture. They are composed of predominant olivine and minor clinopyroxene, which is completely or significantly replaced by amphibole.

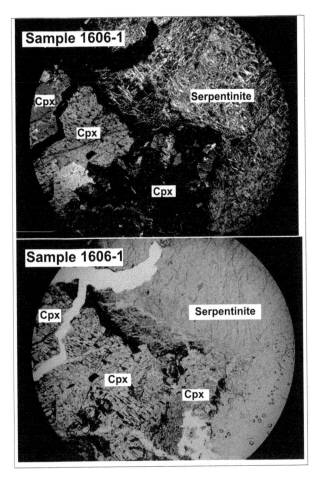

Figure 4.8 Microphotograph of a fragment of the contact of a clinopyroxenite vein (left) with chrysotile serpentinite (right) (sample 1606-1, Berezovka massif).
Clinopyroxenite consists of differently sized prismatic and xenomorphic segregations of clinopyroxene (Cpx), which have not undergone secondary alterations. Serpentinite is an aggregate of fibrous chrysotile segregations 0.1–1.5 mm in length. The direct clinopyroxenite–serpentinite contact has a sinuous line. The gaping crack in the upper left part of the image shows defects caused by the production of the thin section.

4.3 ORTHOMAGMATIC GABBROIDS

Orthomagmatic gabbroids include gabbronorites, gabbro, and, less often, norites and gabbro-pegmatites.

Gabbronorites are medium- to fine-grained rocks with a massive structure and trachytoid texture (Fig. 4.9). They consist of prismatic idiomorphic grains of predominant plagioclase ($\leq 60\%$) and subordinate grains of orthopyroxene and clinopyroxene, which is partly replaced by amphibole. Some varieties contain serpentinized olivine

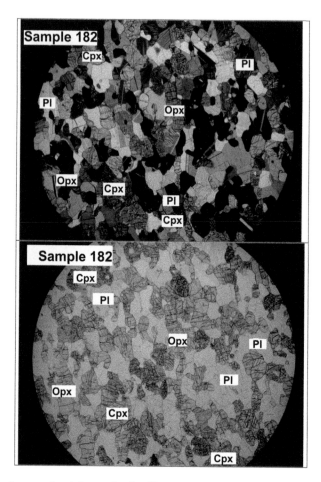

Figure 4.9 Microphotograph of fine-grained gabbronorite with a trachytoid texture (sample 182, Shel'ting massif).

Unaltered elongated prismatic grains of orthopyroxene (Opx) with rounded edges are idiomorphic with respect to grains of clinopyroxene (Cpx) and plagioclase (Pl). Orthopyroxene grains are cut by adjacent joints oriented along their long axis and by rare sinuous cracks perpendicular to their long axis. The orthopyroxene grains contain thin clinopyroxene lamellae (exsolution structures). Clinopyroxene is present as prismatic and xenomorphic grains which have not undergone secondary alterations. Elongated clinopyroxene grains, like the orthopyroxene grains, have a subparallel long axis. Plagioclase is observed as long- and short-prismatic grains and their intergrowths, not subjected to secondary alterations. Their long axes are subparallel to the long axes of the orthopyroxene and clinopyroxene grains. Orthopyroxene, clinopyroxene, and plagioclase make up ~30%, ~30%, and ~40%, respectively.

and secondary hydrogarnet as an admixture. Accessory minerals include spinel, iron sulfides, magnetite, and zircon.

Gabbro are medium-grained meso- and leucocratic rocks composed of 70–75% plagioclase, ≤25% clinopyroxene, and ≤3% hornblende (sometimes, with occasional

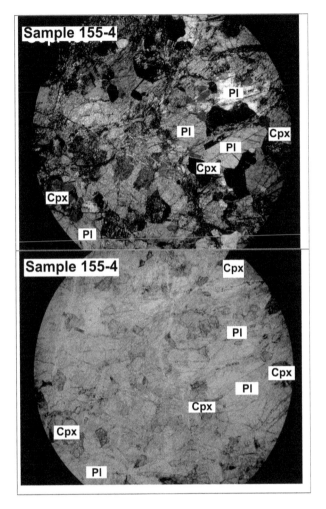

Figure 4.10 Microphotograph of medium-grained gabbro (sample 155-4, Berezovka massif).
The rock consists of ~20% clinopyroxene (Cpx) and ~80% plagioclase (Pl). Small
and, rarely, larger prismatic grains of clinopyroxene are partly replaced by amphibole
along microcracks. Plagioclase occurs as larger prismatic grains with polysynthetic twins,
replaced by a finely dispersed saussurite agregate by 15–30%.

orthopyroxene segregations). Plagioclase is partly or completely replaced by a saussurite or micaceous aggregate, whereas orthopyroxene is differently replaced by bastite.
Zircon is present in these rocks as an accessory mineral (Fig. 4.10).

Norites are a rare gabbroid variety. These are medium- and coarse-grained
greenish dirty gray rocks with a panidiomorphic-granular texture and, sometimes,
a parallel-banded structure. The rocks consist of ≤85% orthopyroxene and ~15%
or more plagioclase. Orthopyroxene is replaced by amphibole in the microcracks
and narrow rims of grains. Hornblende occurs as tabular, elongated, and irregularly

shaped grains. Sometimes small xenomorphic quartz segregations are observed in the grain interstices of other minerals. Plagioclase is often replaced by a saussurite aggregate.

Gabbro-pegmatites are coarse- to giant-grained gray to light gray rocks with a poikilitic texture. They consist of large crystals of predominant plagioclase containing ≤30% poikilitic inclusions of hornblende. These rocks most often occur as veinlets, veins, or lenticular segregations in ultramafic rocks and gabbroids. Some gabbro-pegmatite bodies have a narrow rim of white or pink hydrogarnet grains. A considerable amount of plagioclase in these bodies is usually replaced by a sericite–saussurite aggregate.

4.4 HYBRID (PARAMAGMATIC) GABBROIDS FROM ENDOCONTACT ZONES WITH ULTRAMAFIC ROCKS

Hybrid (paramagmatic) gabbroids of endocontact zones with ultramafites are gabbronorites, olivine–amphibole gabbronorites, olivine gabbro, troctolites, and anorthosites differently enriched in olivine.

Olivine gabbronorites are medium-grained meso- and melanocratic rocks with a massive or banded structure and a gabbroid texture. Their total olivine and ortho- and clinopyroxene contents reach 60%, and the rest are plagioclases. Olivine is usually serpentinized, chloritized, or talcose, whereas clinopyroxene is partly or completely replaced by actinolite. Magnetite, titanite, apatite, and zircon are accessory minerals.

Olivine–amphibole gabbronorites are mainly medium-grained melanocratic rocks, generally with a distinct parallel-banded structure. They have "interlayers" and lenticular segregations 0.5–5 cm in thickness, with a varying quantitative mineral composition (%): olivine (15–20), clinopyroxene (10–15), orthopyroxene (3–5), amphibole (40–50), plagioclase (10–15), and secondary hydrogarnet (5–10). Accessory minerals are spinel and magnetite (≤2%) as well as sulfides (1–2%).

Olivine gabbro are medium-grained meso- and, less often, melanocratic rocks of massive or banded structure and hypidiomorphic-granular texture. The content of olivine is up to 15–25%, and it is usually partly serpentinized, chloritized, or talcose. Clinopyroxene grains (30–40%) are differently replaced by actinolite or, sometimes, chlorite. Plagioclase grains (up to 45–50%) are partly or completely replaced by a saussurite aggregate. As a rule, these rocks contain secondary magnetite (Fig. 4.11).

Troctolites are rare in the studied massifs. These are medium-grained or inequigranular greenish gray rocks. They consist of variable amounts of predominantly xenomorphic segregations of plagioclase and olivine. The olivine segregations sometimes have kelyphitic rims.

Anorthosites are medium- to coarse-grained greenish light gray rocks. In some cases they show a parallel-banded structure due to the alternation of thin "interlayers" of dramatically predominant plagioclase (95–98%) and still thinner "interlayers", which contain not only plagioclase but also up to 2–5% clinopyroxene. Accessory magnetite is sometimes observed. Anorthosites form predominantly thin veins and veinlets injected into ultramafic rocks and gabbroids (Fig. 4.12).

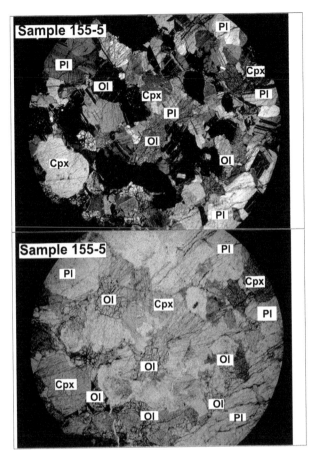

Figure 4.11 Microphotograph of medium-grained gabbro making up the "bands" interspersed with plagiowebsterite "bands" (sample 155-5, Berezovka massif).
Anhedral grains of olivine (Ol) with sinuous and jagged edges are cut by a dense network of microcracks into small angular fragments not subjected to secondary alterations. The grains of clinopyroxene (Cpx) are more idiomorphic than those of olivine and plagioclase (Pl), and they often have joints. Plagioclase occurs as prismatic grains of different sizes, idiomorphic with respect to the olivine grains and xenomorphic with respect to the clinopyroxene grains. Polysynthetic twins of different widths are usually observed in the plagioclase grains. The contents of olivine, clinopyroxene, and plagioclase are ~25%, ~30%, and ~45%, respectively.

4.5 HYBRID (PARAMAGMATIC) GABBROIDS OF ENDOCONTACT ZONES WITH ENCLOSING ROCKS

Hybrid (paramagmatic) gabbroids of endocontact zones with enclosing rocks include hornblende and quartz-containing gabbro, gabbro-diorites, diorites, and biotite-containing and quartz diorites.

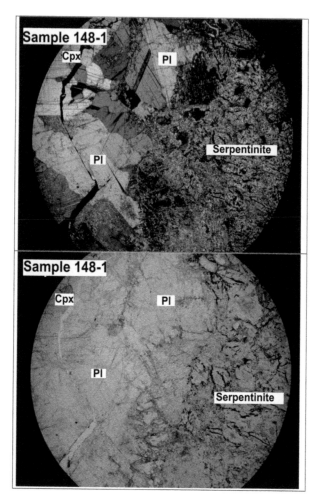

Figure 4.12 Microphotograph of veins of clinopyroxene-containing anorthosite (left) with serpentinite (right) (sample 148-1, Berezovka massif).
Anorthosite is composed of very weakly altered prismatic grains of plagioclase (Pl) with polysynthetic twins.

Hornblende and quartz-containing gabbro and gabbro-diorites are fine- and medium-grained mesocratic massive rocks with a prismatic-granular texture. They are composed of 50–55% hornblende, 40–45% plagioclase, and scarce quartz segregations and sometimes contain a K-feldspar impurity. Magnetite, ilmenite, and zircon are accessory minerals. Secondary minerals include actinolite and chlorite (which replace hornblende) as well as epidote and a saussurite aggregate (which replace plagioclase). The K-feldspar impurity is usually pelitized, whereas ilmenite is partly or completely replaced by leucoxene.

Diorites are medium-grained rocks with a massive or, sometimes, parallel-banded structure. They consist of 50–65% plagioclase, 40–45% hornblende, and rare quartz segregations. Plagioclase forms tabular or elongated grains measuring 0.1 to 2–3 mm.

Quartz occurs as xenomorphic segregations 0.05–1.5 mm in size. Zircon is an accessory mineral. Plagioclase in the diorites is saussuritized or, less often, chloritized, whereas hornblende is replaced by actinolite and chlorite.

Biotite-containing diorites are medium-grained rocks with a massive or, sometimes, maculose structure and a hypidiomorphic-granular texture. These rocks consist of plagioclase and hornblende grains and rare biotite segregations. Secondary minerals, which form after plagioclase, are chlorite (replacing biotite), epidote, saussurite, and sericite. Accessory minerals include occasional titanomagnetite, magnetite, titanite, apatite, and zircon segregations.

Quartz diorites are medium-grained hypidiomorphic-granular rocks, which sometimes show a banded structure. They are composed of 45–50% plagioclase, 25–35% hornblende, ≤15% quartz, and accessory magnetite (2%) and apatite. Plagioclase occurs as wide tabular polysynthetic twins 0.3–3 mm in size. Hornblende forms elongated and irregularly shaped idiomorphic grains 0.2–7 mm in size. Xenomorphic quartz segregations 0.5–5 mm in size fill the grain interstices of amphibole and plagioclase. Zircon is an accessory mineral. Plagioclase is usually strongly saussuritized, whereas hornblende is partly replaced by chlorite and epidote.

* * *

More than 20 petrographic varieties of ultramafic and gabbroid rocks are recognized in the studied massifs. Their main minerals, present in variable ratios, are olivine, ortho- and clinopyroxene, plagioclase, and primary and secondary amphiboles. Biotite, K-feldspar, and quartz are secondary minerals. Finally, Cr-spinel, magnetite, titanomagnetite, zircon, apatite, titanite, and sulfides are accessory minerals. The studied rock varieties are divided into five groups: (1) restitic (orthomagmatic) ultramafic rocks, (2) hybrid (paramagmatic) ultramafic rocks, (3) orthomagmatic gabbroids, (4) hybrid (paramagmatic) gabbroids of endocontact zones with ultramafites, and (5) hybrid (paramagmatic) gabbroids of endocontact zones with enclosing rocks. Restitic ultramafites and orthomagmatic gabbroids are mostly more homogeneous in texture, structure, and quantitative mineral composition than hybrid ultramafic rocks and gabbroids.

Chapter 5

Petrochemistry of rocks of mafic-ultramafic massifs of the East Sakhalin ophiolite association

Petrochemical studies have been a principal approach for compositional classification of igneous rocks, for revealing their conditions of formation based on composition, and verification of petrogenetic models. Special attention to the approach in igneous petrography was paid by numerous non-Russian and Russian geologists, including Zavaritskii [1961], who designed one of the formerly most popular methods of graphic interpretation of petrochemical data. A considerable contribution to the development of petrochemical studies and their genetic interpretation with focus on statistical analysis was made by Belousov [1982], Kutolin [1972], and other geologists.

The recent decades have seen X-ray fluorescence analysis, a more efficient analytical method, come into use, replacing the formerly popular wet chemical method. In the past, interpretation of petrochemical data relied widely on conversion of chemical analyses to normative mineralogic composition (the CIPW method) and Zavaritskii petrochemical coefficients with vector diagrams.

The diversity of petrographic rock types composing the mafic-ultramafic massifs of the East Sakhalin ophiolite association and the inhomogeneity of the quantitative mineral composition account for considerable variations in their chemical compositions. For several decades, petrochemical studies of the mafic-ultramafic massifs of Sakhalin Island were conducted along with structural and petrographic studies, geologic surveys, and subject studies. One of the earliest works [Vergunov, 1964] summarized available petrochemical data for the mafic-ultramafic massifs of Sakhalin Island and the Kuriles. Later Slodkevich [1977] published average chemical component contents for major varieties of mafic-ultramafic rocks belonging to the massifs of the Sokolovsk, Vodopadnensk, Peskovsk, and Schmidt igneous complexes. For graphic interpretation of petrochemical data, V.V. Slodkevich used Zavaritskii–Sobolev vector diagrams and $MgO - FeO_{tot} - (Na_2O + K_2O)$ ternary plots without normalizing the results to dry matter. The diagrams showed distinct compositional differences between ultramafic and gabbroid rocks. It was clear, for example, that ultramafic rocks had considerably higher MgO content and lower FeO_{tot} and alkali contents than gabbroid ones.

Later some petrochemical issues of the Sakhalin Island mafic-ultramafic massifs were discussed in geological survey reports and in publications by A.N. Rechkin, A.F. Bekhtol'd, D.F. Semenov, Yu.N. Raznitsyn, and S.V. Vysotskii. Petrochemical data for the Sakhalin Island mafic-ultramafic massifs were also given in our works [Lesnov, 2009e; Lesnov and Stepashko, 2010; Stepashko and Lesnov, 2011, 2013]. The below summary of the petrochemical characteristics of the regional mafic-ultramafic massifs involved about 90 chemical analyses of major rock varieties, performed on samples from our collections and cited from some works (Table 5.1). Most of our rock samples were analyzed using the X-ray fluorescence method. Table 5.1 characterizes chemical compositions of about 20 petrographic rock varieties from different massifs. The data were processed and interpreted taking into account that the rocks belong to five petrogenetic groups.

Earlier petrochemical comparative analyses, especially those of ultramafic rocks, were not usually normalized to "dry matter", that is, to 100 wt.%. This introduced errors to actual concentrations of chemical components. The errors were especially huge when analyzing highly serpentinized ultramafic rocks, affecting most drastically the contents of the major components (SiO_2, MgO, CaO, Al_2O_3, and FeO_{tot}). Hence, today all primary petrochemical results are normalized to 100%. Besides, as will be shown below, it is preferable to use weight ratios of chemical component contents instead of absolute weights for classification and graphical interpretation of data [Lesnov, 2007c].

5.1 VARIATIONS IN THE CONTENTS OF MAJOR CHEMICAL COMPONENTS IN PETROGENETIC ROCKS GROUPS

A general summary of differences among chemical compositions of rock varieties from the studied massifs and their petrogenetic groups was obtained by plotting component contents on two-dimensional diagrams. As can be seen, the locations of compositional points on the MgO–SiO_2 diagrams indicate the discreteness of chemical compositions of restitic ultramafic (lherzolites, harzburgites, and dunites) (Fig. 5.1, 1) and orthomagmatic gabbroid (gabbronorites and gabbro) (Fig. 5.1, 3) rocks. The last diagram shows the inverse relationship between MgO and SiO_2 contents. The compositional points in Fig. 5.1, 2 place hybrid ultramafic rocks (wehrlites, pyroxenites, etc.) with respect to their chemical compositions between those of restitic ultramafic rocks and orthomagmatic gabbroid rocks. The diagram also shows that some varieties of paramagmatic ultramafic rocks (e.g., orthopyroxenites) have notably higher SiO_2 content than orthomagmatic gabbroid rocks. Similar compositional differences between petrogenetic rock groups can be seen in Fig. 5.2 in the MgO–FeO_{tot} diagram. Considering the locations of the compositional points for the pool of the analyzed samples from the massifs in question, we can assume a direct relationship between the contents of the components (Fig. 5.2, 4). Distinct chemical differences between restitic ultramafic rocks (Fig. 5.3, 1) and orthomagmatic gabbroid rocks are shown in the MgO–CaO diagram (Fig. 5.3, 3). Note that most of the restitic ultramafic rocks samples contain ≤3 wt.% CaO; higher concentrations were found for only some samples from the Berezovka massif. Hybrid ultramafic rocks can be divided into three groups according to the CaO content: (1) high (15–22 wt.%), (2) medium (8–13 wt.%), and (3) low (<5 wt.%) (Fig. 5.3, 2). The discreteness of chemical compositions of restitic and hybrid ultramafic rocks vs. orthomagmatic gabbroid rocks is also apparent from the MgO–Al_2O_3 diagram (Fig. 5.4). While only some ultramafic rocks contain >4 wt.% Al_2O_3, nearly all orthomagmatic gabbroid rocks contain >10 wt.%. The location of the compositional points in the CaO–TiO_2 diagram indicates low Ti content in both ultramafic and gabbroid rocks from the studied massifs; only a few samples

Table 5.1 Chemical compositions of rocks from mafic-ultramafic massifs of the East Sakhalin ophiolite association, wt.%.

Sample	Rock	SiO$_2$	TiO$_2$	Al$_2$O$_3$	Fe$_2$O$_3$	FeO	MnO	MgO	CaO	Na$_2$O	K$_2$O	P$_2$O$_5$	LOI	Total
1	2	3	4	5	6	7	8	9	10	11	12	13	14	15
Berezovka massif														
Sh-4193*	Dunite	40.20	0.10	N.d.	0.60	10.00	0.12	37.00	1.61	0.20	0.11	0.08	8.43	100.78
G-1610-2*	Olivinite serpentinized	37.40	0.02	0.39	7.10	3.39	0.17	38.40	0.30	<0.50	<0.10	<0.50	12.00	99.17
G-1596-4*	Olivinite	34.70	<0.1	0.15	7.57	2.31	0.14	40.20	0.19	N.d.	N.d.	N.d.	N.d.	N.d.
G-1596-6*	Olivinite serpentinized	34.00	0.02	0.38	7.63	2.53	0.37	39.60	0.37	<0.50	<0.01	<0.05	14.20	98.88
133	Harzburgite	40.37	0.02	0.46	17.01	N.d.	0.24	30.61	2.76	0.07	N.d.	0.04	8.23	99.82
1610-2	Apoharzburgite serpentinite	37.52	0.01	0.77	11.37	N.d.	0.18	36.79	0.59	<0.02	N.d.	0.01	12.35	99.60
1610-3	Apoharzburgite serpentinite	36.35	N.d.	0.69	11.67	N.d.	0.18	37.04	0.21	<0.02	0.01	0.01	13.54	99.70
Sh-6320*	Lherzolite	45.40	0.05	1.15	1.80	7.20	0.12	29.60	5.74	1.71	0.15	0.04	7.94	100.81
S-138	Lherzolite	42.81	0.06	2.18	11.2	N.d.	0.16	26.30	9.67	0.11	<0.01	0.04	6.92	99.40
1596	Lherzolite	42.77	0.03	1.93	9.93	N.d.	0.16	30.59	5.01	0.08	0.03	0.02	9.24	99.80
145	Lherzolite	47.25	0.06	1.88	12.02	N.d.	0.20	26.65	8.92	0.12	N.d.	0.04	2.32	99.47
142	Plagiolherzolite	38.73	N.d.	4.04	7.59	N.d.	0.14	36.89	3.67	0.08	0.11	0.04	8.79	99.97
139	Wehrlite	43.23	0.13	12.92	5.21	N.d.	0.09	16.22	17.68	0.81	0.60	0.04	3.13	99.55
Sh-6318*	Wehrlite	40.40	0.05	0.75	4.30	6.70	0.02	35.15	2.80	0.04	N.d.	N.d.	9.90	101.51
148-1a	Plagiowehrlite	40.88	0.03	13.37	5.88	N.d.	0.11	16.31	16.52	0.30	N.d.	0.04	6.29	99.74
155-1	Plagiowehrlite	44.17	0.13	11.21	11.95	N.d.	0.19	22.87	8.47	0.61	N.d.	0.05	0.34	100.01
136	Plagiowehrlite	36.54	0.01	2.28	8.54	N.d.	0.16	37.67	2.37	0.02	N.d.	0.04	12.11	99.75
135	Schriesheimite (cortlandites)	38.86	0.11	3.74	8.92	N.d.	0.16	34.04	4.49	0.41	N.d.	0.05	9.02	99.80
Sh-4186*	Schriesheimite (cortlandites)	41.41	0.01	7.78	3.46	6.31	0.15	27.06	9.13	0.13	0.06	0.07	4.40	100.65
140	Olivine websterite	47.68	0.04	2.77	8.46	N.d.	0.14	20.66	18.71	0.10	N.d.	0.04	0.77	99.36
155	Olivine websterite	49.80	0.04	1.51	13.84	N.d.	0.25	30.84	2.08	0.09	N.d.	0.04	1.23	99.71
152	Websterite	55.30	0.01	1.99	8.02	N.d.	0.17	30.69	3.22	0.10	N.d.	0.05	0.01	99.54
143	Websterite	50.43	0.08	1.56	7.72	N.d.	0.19	16.84	21.91	0.26	N.d.	0.04	0.31	99.33
1593	Websterite	47.83	0.05	2.64	6.12	N.d.	0.14	21.62	17.67	0.06	0.01	0.02	3.74	99.90
1603-1	Websterite	52.05	0.05	2.63	8.57	N.d.	0.19	24.69	9.06	0.23	0.01	0.02	2.44	99.95
155-5	Plagiowebsterite	43.74	0.15	8.17	11.62	N.d.	0.22	17.26	15.50	0.33	N.d.	0.06	2.46	99.50
Sh-4194*	Orthopyroxenite	53.30	0.10	2.90	5.20	4.80	0.14	27.25	6.20	1.41	0.03	N.d.	0.61	102.06
Sh-4263-e*	Orthopyroxenite	57.70	0.15	3.20	3.70	6.20	0.25	22.00	4.01	2.51	0.06	0.07	0.36	100.43
Sh-184-b*	Serpentinized orthopyroxenite	57.00	N.d.	1.02	1.69	5.32	0.24	31.00	1.40	<0.5	<0.1	<0.5	14.10	99.36
G-1606-1*	Olivine clinopyroxenite	47.90	0.03	1.89	2.09	2.67	0.13	22.90	17.30	<0.5	<0.1	<0.5	3.98	98.89
G-159-1*	Olivine clinopyroxenite	41.30	0.03	0.67	6.20	2.89	0.16	34.20	3.57	<0.5	<0.1	<0.5	10.02	99.72
G-1595*	Olivine clinopyroxenite	46.90	0.05	1.74	3.02	3.25	0.16	25.80	13.00	<0.5	0.02	<0.5	5.10	99.04
G-1597*	Olivine clinopyroxenite	38.30	0.68	10.10	1.87	11.90	0.20	13.80	15.70	0.06	0.08	1.26	4.54	89.49
1606-1	Clinopyroxenite	48.67	0.03	2.90	4.64	N.d.	0.13	19.90	20.70	0.07	N.d.	0.02	2.78	99.85

(Continued)

Table 5.1 Continued.

Sample	Rock	SiO$_2$	TiO$_2$	Al$_2$O$_3$	Fe$_2$O$_3$	FeO	MnO	MgO	CaO	Na$_2$O	K$_2$O	P$_2$O$_5$	LOI	Total
1	2	3	4	5	6	7	8	9	10	11	12	13	14	15
Berezovka massif														
Skh-2	Plagiowehrlite	43.73	0.02	21.35	3.31	N.d.	0.07	10.47	16.13	0.97	0.14	0.01	3.57	99.78
Sh-165-P*	Pyroxenite	52.53	0.18	3.36	3.02	1.11	0.11	20.50	15.40	2.16	0.72	N.d.	0.86	100.62
G-1604*	Gabbronorite	47.80	0.08	12.60	1.24	5.42	0.14	13.60	13.40	1.32	0.15	<0.5	3.50	99.25
130	Gabbronorite	43.50	0.02	16.13	6.80	N.d.	0.12	14.79	15.12	0.71	0.08	0.04	2.38	99.69
131	Gabbronorite	41.47	0.02	19.98	3.87	N.d.	0.07	11.50	18.93	0.67	N.d.	0.4	3.14	99.78
131a	Gabbronorite	42.18	0.03	20.66	3.68	N.d.	0.06	10.35	19.48	0.65	N.d.	0.04	2.51	99.64
1607	Gabbronorite	42.96	0.02	20.39	3.49	N.d.	0.08	9.90	17.36	0.65	0.08	0.02	4.99	99.92
Sh-4195-1*	Gabbronorite	49.50	0.20	10.80	0.50	6.50	0.12	12.30	14.99	0.69	0.07	0.27	3.61	95.75
429-5-p	Gabbronorite	49.07	0.05	15.50	4.82	7.00	0.19	6.13	8.40	3.22	0.13	0.81	4.69	100.41
Sh-173*	Gabbronorite	51.28	0.05	15.03	1.74	2.40	0.14	12.80	13.20	0.59	1.59	N.d.	1.36	100.22
129	Gabbronorite	40.50	0.03	18.99	4.74	N.d.	0.08	13.52	14.34	1.44	0.89	0.04	5.14	99.71
Skh-1	Gabbronorite	46.96	0.03	2.47	4.70	N.d.	0.12	20.20	19.27	0.18	N.d.	0.02	5.97	99.92
1605	Gabbronorite	45.60	0.04	17.10	4.93	N.d.	0.09	11.81	17.18	0.56	0.01	0.02	2.44	99.79
1615	Gabbronorite	43.66	0.04	17.17	4.31	N.d.	0.09	11.49	18.16	0.50	0.03	0.01	4.40	99.86
Sh-143*	Gabbronorite	49.96	0.29	16.08	2.35	3.22	0.13	11.30	12.10	2.90	0.52	0.05	1.72	100.58
Sh-6220*	Gabbronorite	48.10	0.05	22.85	0.50	2.40	0.04	5.00	14.28	1.18	1.30	0.26	5.09	101.25
Sh-4192*	Gabbro-pegmatite	42.84	0.32	11.70	2.37	1.58	0.04	4.66	17.10	N.d.	N.d.	N.d.	5.09	N.d.
148-1b	Anorthosite	38.46	N.d.	30.30	0.23	N.d.	N.d.	0.51	22.32	1.09	0.39	0.03	6.53	99.87
702**	Apoharzburgite serpentinite	38.39	N.f.	2.24	4.39	1.43	0.11	37.64	0.62	0.04	0.06	0.01	12.92	100.01
717**	Serpentinized lherzolite	37.12	N.f.	3.76	4.92	2.71	0.18	36.51	2.46	0.04	0.08	0.02	10.09	99.92
709**	Orthopyroxene wehrlite	47.21	0.12	2.00	3.98	6.67	0.18	27.98	8.49	0.08	0.09	0.01	3.05	100.04
700**	Olivine websterite	52.54	0.12	2.38	1.37	4.39	0.14	21.56	16.84	0.17	0.06	0.10	N.d.	99.93
716**	Clinopyroxenite	52.50	0.13	1.40	2.80	4.71	0.31	16.10	21.68	0.42	0.07	0.01	0.11	100.40
703**	Gabbro	45.83	0.32	11.90	1.53	5.89	0.11	19.93	10.35	0.76	0.22	0.02	2.68	99.64
698**	Gabbro	47.05	0.08	11.12	1.88	3.86	0.11	16.55	16.07	0.47	0.18	0.02	1.82	99.61
711**	Gabbro-anorthosite	42.26	0.05	22.56	2.59	2.49	0.09	8.98	15.38	1.18	0.17	N.f.	2.99	100.09
713**	Troctolite (allivalite)	38.00	0.05	5.58	7.83	4.78	3.54	31.16	0.25	0.10	0.07	0.01	8.02	100.32
Shel'ting massif														
174	Olivine orthopyroxenite	56.47	0.01	0.53	11.86	N.d.	0.27	28.95	1.23	0.08	N.d.	0.05	0.16	99.61
176	Olivine orthopyroxenite	53.44	0.07	1.46	20.87	N.d.	0.34	22.48	2.13	0.09	N.d.	0.04	1.13	99.79
182	Gabbronorite	46.51	0.04	14.64	6.97	N.d.	0.13	19.04	10.37	0.73	N.d.	0.04	1.64	100.10
183	Gabbronorite	43.58	0.03	19.85	4.99	N.d.	0.11	14.57	14.38	0.68	N.d.	0.04	1.92	100.14
663***	Serpentinized harzburgite	43.75	0.07	1.48	5.80	4.47	0.14	34.66	1.36	0.04	0.03	0.02	6.70	99.58

Sample	Rock type												Total
661**	Serpentinized harzburgite	39.97	0.07	0.75	8.59	3.76	0.17	34.79	1.11	0.08	0.03	0.01	9.40	99.58
655a**	Serpentinized harzburgite	40.95	0.06	0.41	6.04	3.62	0.17	36.64	0.62	0.04	0.03	0.01	9.96	99.61
404b***	Harzburgite	43.76	0.05	0.52	5.59	2.15	0.13	37.60	0.28	0.06	0.06	N.d.	9.14	99.67
10k***	Harzburgite	40.17	0.03	0.50	4.94	3.56	0.16	36.75	0.64	0.06	0.06	0.04	12.56	99.82
11a***	Harzburgite	42.46	0.03	0.37	4.84	4.55	0.13	33.80	0.70	0.08	0.05	0.03	12.38	100.38
421g***	Dunite	40.42	0.01	0.45	5.05	2.48	0.07	38.77	0.42	0.13	0.06	N.d.	12.99	100.57
12***	Dunite	36.49	0.03	1.53	1.45	8.90	0.20	35.77	0.35	0.05	0.05	0.04	14.74	100.13
651**	Wehrlite	39.69	0.07	2.54	5.63	3.80	0.18	34.96	2.53	0.08	0.03	0.01	9.28	99.85
666**	Enstatitite	55.83	0.06	0.41	2.12	5.07	0.16	34.45	0.99	0.04	0.01	0.01	0.16	99.69
664**	Enstatitite	54.62	0.09	0.90	2.47	8.57	0.20	30.46	2.23	0.08	0.03	0.01	N.d.	99.78
668**	Olivine enstatitite	53.16	0.21	2.07	2.16	9.80	0.23	31.05	1.61	0.04	0.01	0.01	N.d.	100.51
657**	Enstatitite with olivine and Cr-spinel	54.92	0.07	0.41	1.81	6.94	0.18	34.0	1.05	0.08	0.05	0.01	N.d.	99.77
650***	Clinopyroxene enstatitite	53.48	0.09	1.58	2.52	10.29	0.27	29.56	2.17	0.08	0.03	0.01	N.d.	100.22
416b***	Orthopyroxenite	54.53	0.05	0.50	3.43	7.90	0.17	30.20	2.24	0.96	0.06	N.d.	1.16	100.16
12zh***	Orthopyroxenite	54.53	0.07	1.37	1.50	10.17	0.27	27.38	3.66	0.11	0.04	0.05	2.63	99.91
655**	Troctolite	40.05	0.14	8.26	4.17	6.09	0.17	23.57	8.92	0.51	0.07	N.d.	7.08	99.61
667**	Gabbronorite	49.25	0.07	15.63	1.56	5.73	0.14	13.89	11.14	0.85	0.33	0.01	0.65	99.59
13a***	Gabbronorite	48.95	0.07	15.15	3.57	3.84	0.13	11.74	10.28	2.76	0.06	0.03	2.63	99.60
417a***	Norite	48.96	0.05	13.15	3.55	5.00	0.12	11.90	13.90	0.96	0.36	N.d.	1.24	99.59
Komsomol'sk massif														
191	Olivine gabbronorite	39.08	0.10	12.01	7.84	N.d.	0.11	20.89	14.36	0.76	0.08	0.04	4.71	99.99
189	Gabbronorite	43.75	0.05	12.99	7.35	N.d.	0.12	14.11	20.17	0.43	N.d.	0.04	0.52	99.51
South Schmidt massif														
161	Apolherzolitic serpentinite	37.68	N.d.	0.45	7.92	N.d.	0.14	39.07	0.65	0.05	N.d.	0.04	13.90	99.88
163	Apolherzolitic serpentinite	39.43	0.01	1.74	7.84	N.d.	0.15	36.31	1.75	0.04	N.d.	0.04	12.61	99.91
162	Websterite	55.54	0.01	1.75	6.65	N.d.	0.15	32.65	2.37	0.08	<0.01	0.04	1.30	99.50
164	Websterite	52.36	0.01	1.19	7.65	N.d.	0.15	33.22	1.34	0.08	0.01	0.04	3.47	99.51

Note: (*) – data given after [Explanatory note …, 2009]. (**) – data given after [Raznitsyn, 1982]. (***) – data given after [Bekhtol'd and Semenov, 1978]. Other samples were analyzed using the author's collection. Original rock nomenclature is followed. The following components were included in the totals (wt.%): sample 702** – H_2O^- (1.36), CO_2 (0.70); sample 717** – H_2O^- (0.62), CO_2 (0.45); sample 709** – H_2O^- (0.18); sample 700** – H_2O^- (0.21); sample 716** – H_2O^- (0.16); sample 703** – H_2O^- (0.10); sample 698** – H_2O^- (0.50), CO_2 (0.15); sample 711** – H_2O^- (0.35); sample 713** – H_2O^- (0.92); sample 663** – H_2O^- (0.96), CO_2 (0.10); sample 661** – H_2O^- (0.80), CO_2 (0.05); sample 655a** – H_2O^- (1.08); sample 651** – H_2O^- (0.90), CO_2 (0.15); sample 666** – H_2O^- (0.38); sample 664** – H_2O^- (0.12); sample 668** – H_2O^- (0.16); sample 657** – H_2O^- (0.18); sample 650** – H_2O^- (0.58); sample 709** – H_2O^- (0.65); sample 709** – SO_3 (traces); sample 10k*** – Cr_2O_3 (0.14), SO_3 (0.09), CO_2 (0.12); sample 11a*** – Cr_2O_3 (0.02), SO_3 (0.15), CO_2 (0.84), H_2O^- (0.78); sample 412g*** – SO_3 (0.01), CO_2 (0.38); sample 12e*** – Cr_2O_3 (0.34), SO_3 (0.31), H_2O^- (0.87), CO_2 (0.18); sample 416b*** – SO_3 (0.01), CO_2 (traces); 12zh*** – Cr_2O_3 (0.16), SO_3 (0.03), H_2O^- (0.80), CO_2 (0.62); sample 417a*** – SO_3 (traces), CO_2 (0.40). Hereafter, n.d. – no data, n.f. – not found. sample 13a*** – Cr_2O_3 (0.05), SO_3 (0.03), H_2O^- (0.86), CO_2 (0.86).

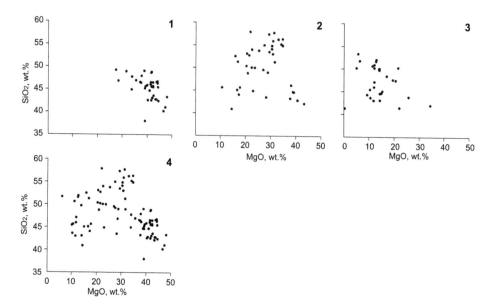

Figure 5.1 Relationship between MgO and SiO$_2$ contents for rocks from the Berezovka, Shel'ting, Komsomol'sk, and South Schmidt massifs (data from Table 5.1).
1 – restitic ultramafic rocks (lherzolites, harzburgites, and dunites) and their serpentinized varieties; *2* – paramagmatic ultramafic and mafic rocks (wehrlites, heterogeneous pyroxenites, other varieties of paramagmatic ultramafic rocks, and olivine gabbro); *3* – orthomagmatic gabbroid rocks (gabbronorites and gabbro); *4* – pooled data on ultramafic and mafic rocks (lherzolites, harzburgites, dunites, wehrlites, pyroxenites, and gabbroids).

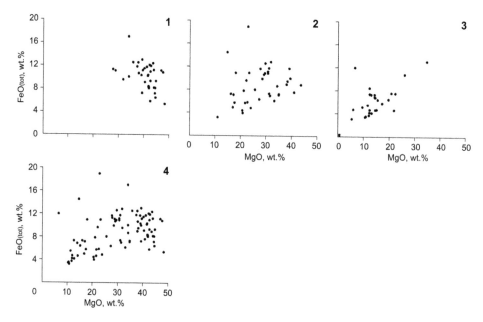

Figure 5.2 Relationship between MgO and FeO$_{tot}$ contents for the rocks from the Berezovka, Shel'ting, Komsomol'sk, and South Schmidt massifs.
1–*4* – see Fig. 5.1 caption for explanation.

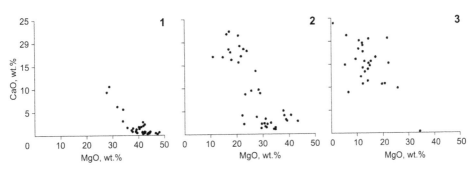

Figure 5.3 Relationship between MgO and CaO contents for the rocks from the Berezovka, Shel'ting, Komsomol'sk, and South Schmidt massifs.
1–3 – see Fig. 5.1 caption for explanation.

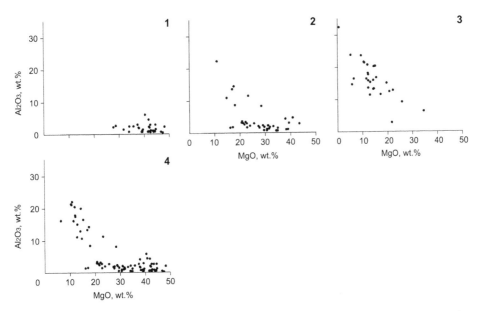

Figure 5.4 Relationship between MgO and Al₂O₃ contents for the rocks from the Berezovka, Shel'ting, Komsomol'sk, and South Schmidt massifs.
1–4 – see Fig. 5.1 caption for explanation.

contain >0.20 wt.% Ti (Fig. 5.5). Chemical composition comparison of ultramafic rocks from the Berezovka, Shel'ting, and South Schmidt massifs for MgO, SiO_2, FeO_{tot}, and Al_2O_3 showed that the fields of the compositional points on the diagrams are nearly coincident, yet the scatter of the points for the Berezovka massif samples was greater than for the samples from the other two massifs (Fig. 5.6). The points for the restitic ultramafic rocks (harzburgites) (the Semail massif, the ophiolite association of Oman) plot more compactly than those for the restitic ultramafic rocks from the Sakhalin

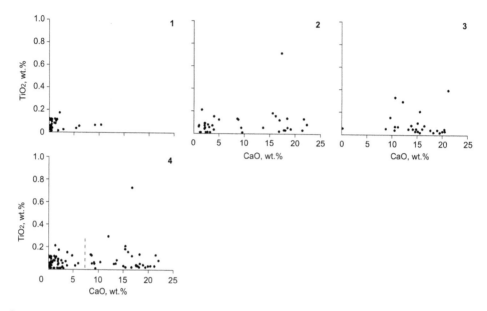

Figure 5.5 Relationship between CaO and TiO$_2$ contents for the rocks from the Berezovka, Shel'ting, Komsomol'sk, and South Schmidt massifs.
1–4 – see Fig. 5.1 caption for explanation.

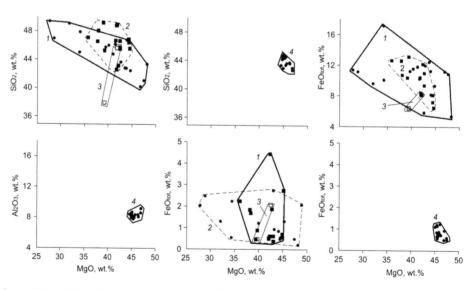

Figure 5.6 Relationship among the contents of MgO and SiO$_2$, MgO and FeO$_{tot}$, and MgO and Al$_2$O$_3$ in ultramafic rocks from mafic-ultramafic massifs of the East Sakhalin ophiolite association. *1* – Berezovka massif (harzburgites, lherzolites, and dunites); *2* – Shel'ting massif (harzburgites, lherzolites, and dunites); *3* – South Schmidt massif (lherzolites) (data from Table 5.1); *4* – Semail massif, Oman (harzburgites) (after [Godard *et al.*, 2000]).

Island massifs. Besides, the ultramafic rocks from the Semail massif are considerably more homogeneous in composition and have greater MgO content.

The data on hand indicate that same-type rock varieties from different massifs of the Sakhalin Island ophiolite association differ in chemical compositions in a various degree. Thus, the contents of the major chemical components in the Berezovka massif ultramafic rocks vary as follows (wt.%): SiO_2 (34–47), Al_2O_3 (0.2–2), FeO_{tot} (1–17), MgO (26–40), and CaO (0.2–10). In paramagmatic ultramafic rocks the contents of the components vary in a wide range (wt.%): SiO_2 (37–58), Al_2O_3 (0.7–21), FeO_{tot} (3–14), MgO (10-38), and CaO (1–22). The Berezovka massif ultramafic rocks are characterized by an overall increased CaO content, compared to similar rocks from the Shel'ting and Schmidt massifs and harzburgites from the Semail massif. The contents of the chemical components in the Berezovka massif orthomagmatic gabbroid rocks vary as follows (wt.%): SiO_2 (41–51), Al_2O_3 (2–23), FeO_{tot} (3–11), MgO (5–20), and CaO (8–19). Note that all the gabbroid rocks from the studied massifs have very low TiO_2 (0.087 wt.% on average) and total alkali (1.2 wt.% on average) contents.

5.2 COMPARATIVE DESCRIPTION OF PETROGENETIC GROUPS OF ROCKS BASED ON THE VALUES OF THEIR PETROCHEMICAL PARAMETERS

We used chemical data obtained for the ultramafic rocks and gabbroids, divided into petrogenetic groups, from the Sakhalin Island mafic-ultramafic massifs to calculate the following chemical parameters: MgO/Fe_2O_{3tot}, MgO/SiO_2, CaO/SiO_2, MgO/Al_2O_3, CaO/Al_2O_3, MgO/TiO_2, and others (Table 5.2). Note that the ratios are not affected by weight loss on ignition (LOI) error, unlike component contents in weight percentage.

The location of the compositional points on the petrochemical diagrams reveals some distinctive features of the chemistry of the rocks under comparison as well as the rock populations. The widest variations were found for the ratios MgO/Fe_2O_{3tot}, MgO/SiO_2, CaO/SiO_2, MgO/Al_2O_3, CaO/Al_2O_3, and MgO/TiO_2 (Fig. 5.7). The location of the compositional points suggests direct relationship between MgO/Fe_2O_{3tot} and MgO/SiO_2, as well as MgO/Fe_2O_{3tot} and MgO/TiO_2. The CaO/Al_2O_3 ratio was ≤ 2 for most of the ultramafic and gabbroid samples, varying from 2 to 14 for only some samples. At the same time, ultramafic rocks are characterized by considerably greater CaO/Al_2O_3 ratio compared to pyrolite (Fig. 5.8, a). In turn, nearly all gabbronorite samples have higher MgO/Fe_2O_{3tot} ratio compared to N-MORB (Fig. 5.8, b). The MgO/SiO_2–CaO/TiO_2 diagram shows that the ratios differ slightly for restitic and hybrid ultramafic rocks, the latter having a greater variation of the CaO/TiO_2 ratio than the former (Fig. 5.9, a). Orthomagmatic gabbroid rocks (gabbronorites) are also characterized by greater variations in CaO/TiO_2, exceeding those for N-MORB (Fig. 5.9, b).

The data for the Sakhalin Island massifs, as well as our earlier observations [Lesnov, 1986a], therefore, indicate that comparison of some petrochemical ratios reveals differences in chemical compositions of separate ultramafic and gabbroid varieties as well as of their petrogenetic groups. The ratios can be used as heterogeneity criteria for the rocks in question and rock populations.

Table 5.2 Petrochemical parameters for the rocks from the Berezovka, Shel'ting, Komsomol'sk, and South Schmidt massifs.

			Dun	Srpt olvt			Hrzt		Srpt hrzt	Lrzt
			Sh-	G-	G-	G-				
Parameter	Pyrolite	MORB	4193	1596-6	1610-2	1596-4	133	1610-2	1610-3	145
MgO/SiO_2	0.840	0.167	0.88	1.16	1.03	1.16	0.76	0.98	1.02	0.56
MgO/Al_2O_3	8.494	0.526	–	104	98.5	268	66.5	47.8	53.7	14.2
MgO/Fe_2O_{3tot}	4.228	0.909	3.85	3.79	3.53	3.96	1.80	3.24	3.17	2.22
MgO/CaO	10.65	0.694	22.98	107	128	212	11.09	62.36	176	2.99
MgO/TiO_2	188.1	7.026	370	1980	1920	402	1531	3679	–	444
CaO/SiO_2	0.079	0.241	0.04	0.01	0.01	0.01	0.07	0.02	0.01	0.19
CaO/Al_2O_3	0.798	0.757	–	0.97	0.77	1.27	6.00	0.77	0.30	4.74
CaO/Fe_2O_{3tot}	0.397	1.310	0.17	0.04	0.03	0.02	0.16	0.05	0.02	0.74
CaO/TiO_2	17.66	10.12	16.1	18.5	15.0	1.90	138.0	59.00	–	149
Al_2O_3/Fe_2O_{3tot}	0.498	1.730	–	0.04	0.04	0.01	0.03	0.07	0.06	0.16
Al_2O_3/TiO_2	22.14	13.37	–	19.00	19.50	1.50	23.00	77.00	–	31.3
Fe_2O_3/TiO_2	44.48	7.726	96.0	522	544	101	851	1137	–	200
Na_2O/K_2O	12.41	12	1.82	–	–	–	–	–	–	–
Na_2O/P_2O_5	17.14	20	2.50	–	–	–	1.75	–	–	3.00

	Lrzt					Pl lrzt	Wrt	Srpt wrt	Pl wrt	Pl wbt
	138	1556	Sh-6320	161	163	142	Sh-6318	139	148-1a	155-5
MgO/SiO_2	0.61	0.72	0.65	1.04	0.92	0.95	0.87	0.38	0.40	0.39
MgO/Al_2O_3	12.1	15.8	25.7	86.8	20.9	9.13	46.9	1.26	1.22	2.11
MgO/Fe_2O_{3tot}	2.36	3.08	3.79	4.93	4.63	4.86	2.99	3.11	2.77	1.49
MgO/CaO	2.72	6.11	5.16	60.1	20.7	10.1	12.6	0.92	0.99	1.11
MgO/TiO_2	376	1020	592	–	3631	–	703	135	544	115
CaO/SiO_2	0.23	0.12	0.13	0.02	0.04	0.09	0.07	0.41	0.40	0.35
CaO/Al_2O_3	4.44	2.60	4.99	1.44	1.01	0.91	3.73	1.37	1.24	1.90
CaO/Fe_2O_{3tot}	0.87	0.50	0.74	0.08	0.22	0.48	0.24	3.39	2.81	1.33
CaO/TiO_2	138	167	115	–	175	–	56.0	147	551	103
Al_2O_3/Fe_2O_{3tot}	0.20	0.19	0.15	0.06	0.22	0.53	0.06	2.48	2.27	0.70
Al_2O_3/TiO_2	31.1	64.3	23.0	–	174	–	15.0	108	446	54.47
Fe_2O_3/TiO_2	160	331	156	–	784	–	235	43.4	196	77.5
Na_2O/K_2O	–	2.67	11.4	–	–	–	0.67	7.27	–	–
Na_2O/P_2O_5	2.50	4.00	42.75	1.25	1.00	2.00	–	20.0	7.50	5.50

	Pl wrt	Schr	Wbt							
Parameter	155-1	135	140	155	1603-1	1593	143	152	162	164
MgO/SiO_2	0.52	0.88	0.43	0.62	0.47	0.45	0.33	0.55	0.59	0.63
MgO/Al_2O_3	2.04	9.10	7.46	20.4	9.39	8.19	10.8	15.4	18.7	27.9
MgO/Fe_2O_{3tot}	1.91	3.82	2.44	2.23	2.88	3.53	2.18	3.83	4.91	4.34
MgO/CaO	2.70	7.58	1.10	14.83	2.73	1.22	0.77	9.53	24.4	25.0
MgO/TiO_2	152	309	517	771	494	432	211	3069	3266	3322
CaO/SiO_2	0.19	0.12	0.39	0.04	0.17	0.37	0.43	0.06	0.02	0.03
CaO/Al_2O_3	0.76	1.20	6.75	1.38	3.44	6.69	14.0	1.62	0.77	1.12
CaO/Fe_2O_{3tot}	0.71	0.50	2.21	0.15	1.06	2.89	2.84	0.40	0.20	0.17
CaO/TiO_2	56.5	40.8	468	52.0	181	353	274	322	134	133
Al_2O_3/Fe_2O_{3tot}	0.94	0.42	0.33	0.11	0.31	0.43	0.20	0.25	0.26	0.16
Al_2O_3/TiO_2	74.7	34.0	69.3	37.8	52.6	52.8	19.5	199	175	119
Fe_2O_3/TiO_2	79.7	81.1	212	346	171	122	96.5	802	665	765
Na_2O/K_2O	–	–	–	–	23.0	6.00	–	–	–	–
Na_2O/P_2O_5	12.2	8.20	2.50	2.25	11.5	3.00	6.50	2.00	2.00	2.00

Table 5.2 Continued.

Parameter	Ortpyr				Ol clnpyr				Clnpyr	
	176	184	174	Sh-4263-e	G-1606-1	G-1596-1	G-1595	G-1597	1606-1	Skh-2
MgO/SiO_2	0.42	0.27	0.51	0.38	0.48	0.83	0.55	0.36	0.41	0.24
MgO/Al_2O_3	15.4	0.62	54.6	6.88	12.1	51.0	14.8	1.37	6.86	0.49
MgO/Fe_2O_{3tot}	1.08	2.27	2.44	2.08	4.52	3.63	3.89	0.91	4.29	3.16
MgO/CaO	10.6	0.68	23.5	5.49	1.32	9.58	1.98	0.88	0.96	0.65
MgO/TiO_2	321	58.0	2895	147	763	1140	516	20.3	663	524
CaO/SiO_2	0.04	0.40	0.02	0.07	0.36	0.09	0.28	0.41	0.43	0.37
CaO/Al_2O_3	1.46	0.91	2.32	1.25	9.15	5.33	7.47	1.55	7.14	0.76
CaO/Fe_2O_{3tot}	0.10	3.33	0.10	0.38	3.41	0.38	1.96	1.04	4.46	4.87
CaO/TiO_2	30.43	85.4	123	26.7	577	119.0	260.0	23.1	690	807
Al_2O_3/Fe_2O_{3tot}	0.07	3.67	0.04	0.30	0.37	0.07	0.26	0.67	0.63	6.45
Al_2O_3/TiO_2	20.9	94.0	53.0	21.3	63.0	22.3	34.8	14.9	96.7	1068
Fe_2O_3/TiO_2	298	25.6	1186	70.6	169	314	133	22.2	155	166
Na_2O/K_2O	–	5.56	–	41.8	–	–	–	0.75	–	6.93
Na_2O/P_2O_5	2.25	34.8	1.60	35.9	–	–	–	0.05	3.50	97.0

Parameter	Hrnb	Gbrt									
	G-4186-1	131	131a	130	1607	191	189	182	183	Sh-4195-1
MgO/SiO_2	0.27	0.28	0.25	0.34	0.23	0.53	0.32	0.41	0.33	0.25
MgO/Al_2O_3	2.07	0.58	0.50	0.92	0.49	1.74	1.09	1.30	0.73	1.14
MgO/Fe_2O_{3tot}	1.19	2.90	2.81	2.18	2.84	2.66	1.92	2.73	2.92	1.59
MgO/CaO	0.86	0.61	0.53	0.98	0.57	1.45	0.70	1.84	1.01	0.82
MgO/TiO_2	35.1	575	345	740	495	209	282	476	486	61.5
CaO/SiO_2	0.32	0.46	0.46	0.35	0.40	0.37	0.46	0.22	0.33	0.30
CaO/Al_2O_3	2.41	0.95	0.94	0.94	0.85	1.20	1.55	0.71	0.72	1.39
CaO/Fe_2O_{3tot}	1.39	4.77	5.29	2.22	4.97	1.83	2.74	1.49	2.88	1.94
CaO/TiO_2	41.0	947	649	756	868	144	403	259	479	75.0
Al_2O_3/Fe_2O_{3tot}	0.58	5.03	5.61	2.37	5.84	1.53	1.77	2.10	3.98	1.40
Al_2O_3/TiO_2	17.0	999	689	807	1020	120	260	366	662	54.0
Fe_2O_3/TiO_2	29.5	199	123	340	175	78.4	147	174	166	38.6
Na_2O/K_2O	–	–	–	8.88	8.13	9.50	–	–	–	9.86
Na_2O/P_2O_5	110	16.8	16.3	17.8	32.5	19.0	10.8	18.3	17.0	2.56

Parameter	Gbrt		Ol gbrt	Gab						Ol gab	Anrt	Drt
	429-5-p	Sh-173	G-1604	129	Skh-1	1605	1612	Sh-143	Sh-6220	G-1607	148-1b	G-197-1
MgO/SiO_2	0.12	0.25	0.28	0.33	0.43	0.26	0.26	0.23	0.10	0.29	0.01	0.05
MgO/Al_2O_3	0.40	0.85	1.08	0.71	8.18	0.69	0.67	0.70	0.22	0.74	0.02	0.20
MgO/Fe_2O_{3tot}	0.49	2.90	1.87	2.85	4.30	2.40	2.67	1.91	1.58	3.19	2.22	0.48
MgO/CaO	0.73	0.97	1.01	0.94	1.05	0.69	0.63	0.93	0.35	0.75	0.02	0.47
MgO/TiO_2	123	256	170	451	673	295	287	39.0	100	1270	–	7.56
CaO/SiO_2	0.17	0.26	0.28	0.35	0.41	0.38	0.42	0.24	0.30	0.39	0.58	0.11
CaO/Al_2O_3	0.54	0.88	1.06	0.76	7.80	1.00	1.06	0.75	0.62	0.99	0.74	0.42
CaO/Fe_2O_{3tot}	0.67	2.99	1.85	3.03	4.10	3.48	4.21	2.04	4.50	4.25	97.0	1.03
CaO/TiO_2	168	264	168	478	642	430	454	41.7	286	1690	–	16.0
Al_2O_3/Fe_2O_{3tot}	1.23	3.41	1.74	4.01	0.53	3.47	3.98	2.71	7.21	4.30	132	2.47
Al_2O_3/TiO_2	310	301	158	633	82.3	428	429	55.4	457	1710	–	38.5
Fe_2O_3/TiO_2	252	88.2	90.8	158	157	123	108	20.4	63.4	398	–	15.6
Na_2O/K_2O	24.8	0.37	8.80	1.62	–	56.0	16.7	5.58	0.91	4.33	2.79	16.2
Na_2O/P_2O_5	3.98	–	–	36.0	9.00	28.0	50.0	58.0	4.54	–	36.3	41.6

Note: Pyrolite – after [McDonough and Sun, 1995], MORB – after [Wedepohl, 1981]. Dun – dunite; Srpt olvt – serpentinized olivinite; Hrzt – harzburgite; Srpt hrzt – serpentinized harzburgite; Lrzt – lherzolite; Pl lrzt – plagiolherzolite; Wrt – wehrlite; Srpt wrt – serpentinized wehrlite; Pl wrt – plagiowehrlite; Schr – schriesheimite; Wbt – websterite; Pl wbt – plagiowebsterite; Ortpyr – orthopyroxenite; Ol clnpyr – olivine clinopyroxenite; Clnpyr – clinopyroxenite; Hrnb – hornblendite; Gbrt – gabbronorite; Ol gbrt – olivine gabbronorite; Gab – gabbro; Ol gab – olivine gabbro; Anrt – anorthosite; Drt – diorite (according to Table 5.1).

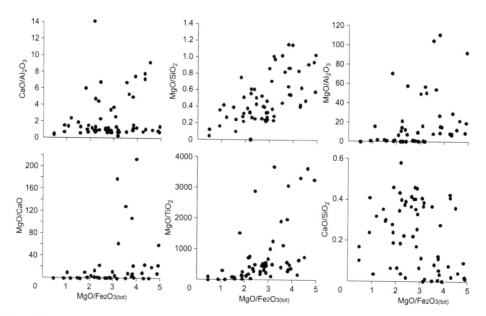

Figure 5.7 Relationship among the values of the petrochemical parameters MgO/Fe_2O_{3tot} and CaO/Al_2O_3, MgO/Fe_2O_{3tot} and MgO/SiO_2, MgO/Fe_2O_{3tot} and MgO/Al_2O_3, MgO/Fe_2O_{3tot} and MgO/CaO, MgO/Fe_2O_{3tot} and MgO/TiO_2, and MgO/Fe_2O_{3tot} and CaO/SiO_2 in ultramafic and mafic rocks from the Berezovka, Shel'ting, Komsomol'sk, and South Schmidt massifs (data from Table 5.2).

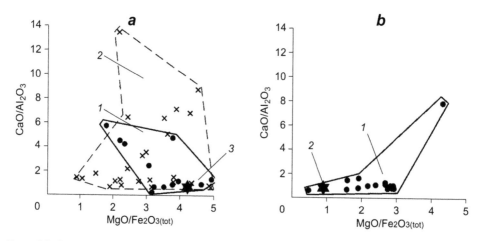

Figure 5.8 Relationship between the values of the petrochemical parameters MgO/Fe_2O_{3tot} and CaO/Al_2O_3 for ultramafic and mafic rocks from the Berezovka, Shel'ting, Komsomol'sk, and South Schmidt massifs (data from Table 5.2).
a: 1 – dunites, harzburgites, and lherzolites; *2* – plagiolherzolites, wehrlites, and olivine and olivine-free websterites, enstatitites, and clinopyroxenites; *3* – pyrolite (after [Ringwood, 1972]); *b: 1* – gabbronorites; *2* – N-MORB (after [Wedepohl, 1981]).

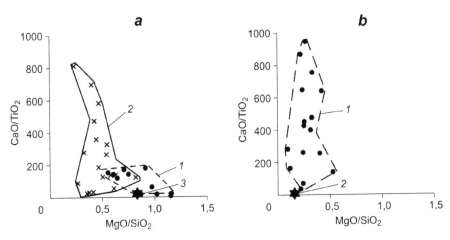

Figure 5.9 Relationship between the values of the petrochemical parameters MgO/SiO$_2$ and CaO/TiO$_2$ for ultramafic and mafic rocks from the Berezovka, Shel'ting, Komsomol'sk, and South Schmidt massifs (data from Table 5.2).
See Fig. 5.8 caption for explanations.

5.3 ON SYSTEMATIZATION OF ULTRAMAFIC AND GABBROID ROCKS BASED ON THE NORMALIZED VALUES OF THEIR PETROCHEMICAL PARAMETERS

To minimize LOI error in petrochemical data, especially those for highly serpentinized ultramafic rocks, comparison and graphical interpretation of the data involved the mentioned petrochemical ratios normalized to those for N-MORB and pyrolite [Lesnov, 2007c]. At an early stage of results processing, the following ratios were calculated for each rock type: MgO/SiO$_2$, Al$_2$O$_3$/MgO, CaO/Al$_2$O$_3$, CaO/TiO$_2$, FeO$_{tot}$/Al$_2$O$_3$, FeO$_{tot}$/MgO, MgO/CaO, CaO/Na$_2$O, Na$_2$O/K$_2$O, and Na$_2$O/P$_2$O$_5$. Ratios obtained for gabbroid rocks were normalized to those for N-MORB [Wedepohl, 1981]; ratios obtained for ultramafic rocks were normalized to those for pyrolite [Ringwood, 1972] as follows:

$$\text{Normalized parameter} = [\text{MgO/CaO}_{(\text{rock type})}]/[\text{MgO/CaO}_{(\text{N-MORB or pyrolite})}]. \quad (5.1)$$

All normalized ratios were plotted in the same fashion as chondrite-normalized REE patterns. The resultant diagrams of ratio variations are shown in Fig. 5.10. Despite slight variations in the ratios, the diagrams for petrographically similar rocks, as well as those for their petrogenetic groups, have similar configuration albeit different from those for other petrographic varieties and petrogenetic groups. This is especially clear in websterite and gabbronorite diagrams and less clear in harzburgite ones. The dunite graph configuration shows that most of its normalized petrochemical ratios are similar to those for pyrolite, while the diagram configuration for harzburgites and lherzolites reveals their compositional differences from pyrolite. Comparison of gabbronorite and websterite diagrams with those for N-MORB brings us to a similar conclusion.

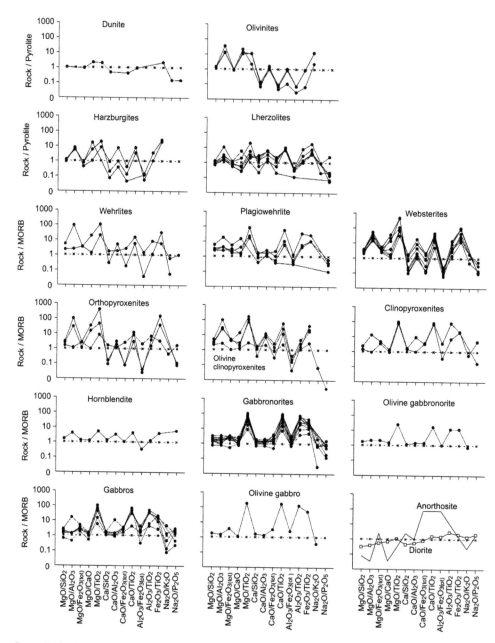

Figure 5.10 Graphs of variations in pyrolite-normalized (for restitic ultramafic rocks) and N-MORB-normalized (for hybrid ultramafic and mafic rocks) values of petrochemical parameters of rocks from the Berezovka, Shel'ting, Komsomol'sk, and South Schmidt massifs (data from Table 5.2).

As the examples suggest, normalized petrochemical ratios for ultramafic and gabbroid rocks and diagrams plotted for them can be used as auxiliary criteria for petrochemical classification of ultramafic and gabbroid rocks.

5.4 THE DEPENDENCE OF THE CHEMICAL COMPOSITION OF ULTRAMAFIC RESTITES ON THE DEGREE OF PARTIAL MELTING OF UPPER-MANTLE SOURCES

Considering possible models for generation of basaltic melts during partial melting of upper-mantle substrates, we came to a conclusion that chemical compositions of basaltic melts, as well as resultant refractory residues (restites), were largely determined by the degree of partial melting of the substrates [Green and Ringwood, 1968]. The conclusion served as basis for modeling processes that could account for the inhomogeneity of the chemical and mineral compositions of restitic ultramafic rocks that compose the mafic-ultramafic massifs of the ophiolite associations. Contemporary studies focus on distribution patterns of chemical components, such as MgO, FeO, CaO, SiO_2, and Al_2O_3, in restitic ultramafic rocks that differ in their characteristics and variation patterns in some petrochemical ratios, such as MgO/SiO_2, CaO/SiO_2, and Al_2O_3/SiO_2 [Glebovitsky et al., 2009]. The available petrochemical and experimental data led the authors to the conclusion that the Mg/Si and Al/Si ratios should be used as more reliable criteria to evaluate the degree of melting of a matter compositionally similar to primitive mantle. Physical experiments demonstrated that the ratios consistently changed with the increase in the degree of partial melting of ultramafic samples. For example, V.A. Glebovitsky et al. showed that the Mg/Si ratio increased and Al/Si decreased with the increase in pressure from 1.0 to 7.0 GPa and in temperature from 1250 to 1950°C during the experiment on partial melting of ultramafic rock.

With these considerations in mind, we calculated Mg/Si, Al/Si, and Ca/Al ratios for restitic harzburgites and lherzolites from the Berezovka, Shel'ting, and South Schmidt massifs and plotted the values (Table 5.3, Fig. 5.11). For comparison, the

Table 5.3 Petrochemical parameters for the restitic ultramafic rocks from the Berezovka, Shel'ting, and South Schmidt massifs.

Sample	Rock	SiO_2	Al_2O_3	MgO	CaO	Si	Al	Mg	Ca	Mg/Si	Al/Si	Ca/Al
Berezovka massif												
133	Harzburgite	40.37	0.46	30.61	2.76	18.85	0.122	18.37	1.97	0.975	0.006	16.18
6320	Lherzolite	45.40	1.15	29.60	5.74	21.20	0.305	17.76	4.10	0.838	0.014	13.46
138	-"-	42.81	2.18	26.30	9.97	19.99	0.578	15.78	6.91	0.789	0.029	11.96
1596	-"-	42.77	1.93	30.59	5.10	19.97	0.511	18.35	3.58	0.919	0.026	7.01
145	-"-	47.25	1.88	26.65	8.92	22.07	0.498	15.99	6.38	0.725	0.023	12.81
Shel'ting massif												
404b	Harzburgite	43.76	0.52	37.60	0.28	20.44	0.138	22.56	0.20	1.104	0.007	1.45
10k	-"-	40.17	0.50	36.75	0.64	18.76	0.133	22.05	0.46	1.175	0.007	3.44
11a	-"-	42.46	0.37	33.80	0.70	19.83	0.098	20.28	0.50	1.023	0.005	5.11
South Schmidt massif												
161	Lherzolite	37.68	0.45	39.07	0.65	17.60	0.119	23.44	0.46	1.332	0.007	3.91
163	-"-	39.43	1.74	36.31	1.75	18.41	0.461	21.79	1.25	1.184	0.025	2.71
	Pyrolite	45.10	3.30	38.10	3.21	21.06	0.875	22.86	2.30	1.085	0.042	2.62

Note: Pyrolite – after [Ringwood, 1981].

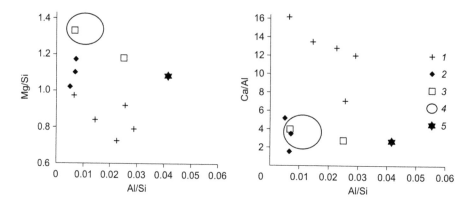

Figure 5.11 Relationship between the values of the petrochemical parameters Mg/Si–Al/Si and Ca/Al–Al/Si for restitic ultramafic rocks (data from Table 5.3).
1 – Berezovka massif (harzburgites and lherzolites); *2* – Shel'ting massif (harzburgite); *3* – South Schmidt massif (lherzolites); *4* – field of compositional points for harzburgites of the Semail massif, Oman (after [Godard *et al.*, 2000]); *5* – pyrolite (after [Ringwood, 1972]).

compositional point for pyrolite as a model of primitive mantle and the group of points representing the chemical composition of harzburgites from the Semail massif were plotted. Analysis of the diagrams leads us to conclude that (1) ultramafic restites from the Berezovka massif have lower Mg/Si and higher Ca/Al ratios than ultramafic rocks from the Shel'ting and South Schmidt massifs, similar to pyrolite; the ratios also have greater variations; (2) ultramafic rocks from the three massifs under consideration, as well as harzburgites from the Semail massif, have lower Al/Si ratio than pyrolite; (3) ultramafic rocks from the Shel'ting and South Schmidt massifs, as well as harzburgites from the Semail massif, have low Ca/Al ratio, being in this way similar to pyrolite; (4) experimental data from [Glebovitsky *et al.*, 2009] suggest that the increased Mg/Si ratio and the decreased Al/Si ratio for the ultramafic restites from the Shel'ting and South Schmidt massifs compared to pyrolite, as well as those in harzburgites from the Semail massif, should be regarded as evidence of considerable depletion of the ultramafic rocks. This fact points to a very high degree of partial melting of upper-mantle sources during the formation of the ultramafic restites.

***

The available petrochemical data suggest discreteness of chemical compositions of restitic ultramafic rocks and orthomagmatic gabbroid rocks from the Sakhalin Island massifs. Chemical compositions of hybrid ultramafic and gabbroid rocks are intermediate between the two aforementioned petrogenetic rock groups. Same-type varieties of ultramafic and gabbroid rocks from different massifs differ in chemical compositions in a various degree. All the gabbroid rocks from the massifs have very low TiO_2 and alkali contents. The ratios of major-component contents (petrochemical parameters) and diagrams showing variations in the normalized ratios can be used as auxiliary criteria for classification of ultramafic and gabbroid rocks and for revealing their heterogeneity. Higher Mg/Si and lower Al/Si ratios for ultramafic restites from the South Schmidt and Shel'ting massifs compared to pyrolite indicate high degree of partial melting of upper-mantle sources during their formation.

Chapter 6

Geochemistry of rocks of mafic-ultramafic massifs of the East Sakhalin ophiolite association

According to Zharikov and Yaroshevskii [2003], geochemistry is the science that studies the occurrence and distribution of chemical elements and their isotopes in the Earth's substance. Geochemical studies of igneous rocks and massifs built up by them are the most important area in modern petrology. The introduction of new analytical methods, ensuring very low limits of element detection along with the use of very small batches of analyzed substance, has largely contributed to such importance. Analytical studies that have been carried out in the last decades have revealed new patterns in the distribution of many indicator impurity elements in a variety of rocks, which has contributed to the progress in the solution of some problems related to rock, mineral, and ore formation. To perform the discriminatory treatments using geochemical data and to carry out petrogenetic reconstructions, a number of diagrams, plots, and algorithms have been developed [Rollinson, 1993; Sklyarov, 2001; etc.].

As Zharikov and Yaroshevskii [2003] stated, available geochemical data on the distribution of elements in igneous rocks allow interpreting them reasonably enough within the concept of crystallization differentiation of mafic magmas. At the same time, they stressed that many geochemical and petrological issues remain unresolved and alternative genetic models for the same rocks and massifs built up by them are used increasingly. As shown in the following chapters, the last point is highly relevant in the study of the genesis of mafic-ultramafic massifs, because the results of their more detailed study by geochemical and other methods often contradict the model of crystallization differentiation of mafic magmas. They point to the discrete geochemical parameters of rocks in the studied massifs at different times of their formation and suggest active influence of mafic melts on earlier ultramafic rocks. As Coleman [1979] emphasized, petrologists possess limited information on trace elements, isotope ratios, and the ophiolite genesis. The current situation in this field of petrology has significantly improved; nevertheless, for the majority of mafic-ultramafic massifs, especially those that are common in the folded structures of Eastern Russia, geochemical data obtained with the use of modern analytical methods are very scarce. We hope that the below results of geochemical studies of ultramafic and mafic rocks from the Berezovka, Shel'ting, Komsomol'sk, and South Schmidt massifs by XRF-SR, INAA, RNAA, ICP-MS, and LA ICP-MS will help to fill this gap.

6.1 DISTRIBUTION OF CA, K, AND SOME IMPURITY ELEMENTS IN ROCKS OF THE BEREZOVKA MASSIF (FROM XRF-SR DATA)

Element distribution patterns in rocks were obtained by the X-ray fluorescence method with synchrotron radiation (XRF-SR) for a collection of samples of gabbroid and ultramafic rocks from the Berezovka massif. The following elements were determined: Cr, Ni, Cu, Zn, Ga, Rb, Sr, Y, Zr, Nb, Mo, Pb, Ti, Mn, Fe, V, as well as Ca and K. Thus, the element detection limits were in the range from 0.1 to 0.5 ppm, error being no higher than 15–30% [Baryshev *et al.*, 1986]. This collection included samples of rocks belonging to three petrogenetic groups: ultramafic restites (harzburgite and lherzolites), hybrid ultramafic rocks (wehrlites, websterites, and clinopyroxenites), and orthomagmatic gabbroids (gabbronorites and gabbro). Preliminary results from these studies were published in [Lesnov and Kolmogorov, 2009].

Contents of the overwhelming majority of detected elements in rocks from this collection vary over a wide range (Tables 6.1, 6.2, Fig. 6.1). It was found that ultramafic

Table 6.1 Petrographic characteristics of rock samples from the Berezovka massif, analyzed by the XRF-SR method.

Sample	Rock	Petrographic characteristics
1610-2	Harzburgite	Moderately serpentinized. Orthopyroxene forms large porphyroblasts.
1610-3	Harzburgite	Significantly serpentinized. Orthopyroxene is almost completely replaced by bastite. Rare thin veinlets of chrysotile asbestos are present.
1596	Lherzolite with websterite veinlet	Medium- to coarse-grained texture. The content of orthopyroxene is insignificant.
1593	Wehrlite	The content of olivine is low, and it is almost completely replaced by serpentine.
Skh-2	Plagiowehrlite	Coarse-grained texture. Olivine is almost completely replaced by serpentine. The amount of plagioclase is very small; it is completely replaced by secondary products.
1603-1	Olivine websterite	Coarse-grained texture. Orthopyroxene and clinopyroxene almost did not undergo secondary alterations.
1606-1	Olivine clinopyroxenite	Clinopyroxene dominates over olivine, which is completely replaced by serpentine.
1607	Gabbronorite	Pyroxenes and plagioclase are slightly altered. There is an admixture of late-magmatic amphibole, and an accessory ore mineral is absent.
1605	Gabbro	Medium-grained texture; clinopyroxene and plagioclase are slightly altered. Amphibole forms kelyphitic rims around clinopyroxene grains.
Skh-1	Gabbro	Rock is significantly altered. Clinopyroxene and plagioclase are preserved as relics. Secondary amphibole is present.
1612	Gabbro	Fine-grained texture, moderately altered rock.

restites differ from their hybrid varieties and orthomagmatic gabbroids in much higher contents of refractory and medium-refractory elements, such as Cr, Ni, Cu, Zn, Rb, V, and Ti. Elevated Ca, K, and Sr contents were identified in some varieties of hybrid ultramafic rocks and orthomagmatic gabbroids. Almost all analyzed rocks are depleted in Y and Zr. A direct relationship was found between Ca and K contents as well as between V and Ti. Based on the Rb and Sr content ratio, the rocks under consideration can be divided into two geochemically distinct groups: orthomagmatic gabbroids, in which an increase in Sr content is accompanied by an increase in Rb content; restitic and hybrid ultramafic rocks, in which, along with low Sr contents, significant variations in Rb contents were observed.

Several geochemical differences between the rocks were observed when comparing their chondrite-normalized multielement spectra (Fig. 6.2). According to the configuration of their spectra, harzburgites are markedly depleted in elements, such as K, Ti, Ni, Cu, Zn, Ga, Zr, Mo, and Pb, but enriched in Cr compared to chondrite. According to the location and configuration of the spectra, the analyzed lherzolite samples are identical in trace-element composition. Thus, they are depleted in Ni, Cu, and Zn, while being enriched in Ca, Rb, and Sr as compared with chondrite. Plagiolherzolite

Table 6.2 Contents of chemical elements and their ratios in rocks from the Berezovka mafic-ultramafic massif, ppm.

Element	Sample										
	133 1	1610-2 2	1610-3 3	138 4	1596 5	142 6	139 7	1593 8	Skh-2 9	140 10	1603-1 11
K	70	40	10	550	590	190	2070	960	1130	1130	630
Ca	19200	4200	1700	63900	70800	24400	110000	103000	106000	120000	56400
Ti	190	260	210	380	470	100	660	240	110	220	310
Mn	2150	1900	1800	1280	1890	1190	690	940	730	1100	1280
Fe	141000	101000	97900	81500	102000	63900	33000	37400	23900	52600	57900
V	64	41.1	35	61	92	46	60	60	29.2	76	52.7
Cr	3113	7359	7662	1918	2515	5279	1051	4752	5485	2155	2571
Ni	1751	1286	1271	1233	865	2619	279	378	282	341	451
Cu	26	18.8	25.2	18.8	6.93	11.9	20.7	10.5	7.72	12.6	13.6
Zn	107	81	77	74	85	72	30.5	35.9	19	42.3	78
Ga	0.99	2.04	1.69	2.88	2.89	3.3	10.6	3.52	2.78	4.05	3.3
Rb	4.18	3.78	3.94	4.83	2.94	1.37	1.61	10.3	0.8	1.06	2.26
Sr	3.86	3.39	2.92	21.5	11.2	29.1	210	10.1	9.44	10.8	19.1
Y	<0.1	<0.1	<0.1	0.78	1.24	<0.1	4.73	1.83	0.68	3.04	1.66
Zr	1.32	<0.1	<0.1	8	94	59	7.83	0.63	0.4	18.8	1.63
Nb	0.26	<0.1	0.22	0.28	0.22	<0.1	<0.1	<0.1	<0.1	0.15	<0.1
Mo	0.25	<0.1	<0.1	1.13	0.48	0.22	0.79	1.01	0.37	3.15	1.21
Pb	0.38	2.7	<0.3	1.22	1.92	0.63	1.28	1.7	1.3	1.78	1.4
Fe/Ni	80.5	78.5	77.0	66.1	118	24.4	118	98.9	84.8	154	128
Fe/Cr	45.3	13.7	12.8	42.5	40.6	12.1	31.4	7.87	4.36	24.4	22.5
Cr/V	48.6	179	219	31.4	27.3	115	17.5	79.2	188	28.4	48.8
Cr/Mn	1.45	3.87	4.26	1.50	1.33	4.44	1.52	5.06	7.51	1.96	2.01
Ti/V	2.97	6.33	6.00	6.23	5.11	2.17	11.0	4.00	3.77	2.89	5.88
Ca/Ti	101	16.2	8.10	168	51	244	167	429	964	546	182
Ca/K	274	105	170	116	120	128	53.1	107	93.8	106	89.5
Sr/Rb	0.92	0.90	0.74	4.45	3.81	21.2	130	0.98	11.8	10.2	8.45
Zn/Cu	4.12	4.31	3.06	3.94	12.3	6.05	1.47	3.42	2.46	3.36	5.74
Zn/Pb	282	30.0	–	50.7	44.3	114	23.8	21.1	14.6	23.8	55.7

Table 6.2 Continued.

Element	Sample 143 12	1606-1 13	130 14	131 15	131a 16	1607 17	Skh-1 18	1605 19	1612 20	Detection limits 21	Cl chondrite 22
K	1300	1080	1830	1430	1610	1550	1750	1050	1310	–	558
Ca	142000	107000	90700	111000	124000	99300	79400	87000	96200	–	9280
Ti	480	110	160	190	190	50	50	150	100	–	436
Mn	1280	690	840	510	520	430	360	480	470	0.5	1990
Fe	45400	20700	44100	24500	24900	20700	17300	26600	23300	–	190400
V	54	37.7	49	34	39	13.4	12.7	23.9	23.5	0.5	56.5
Cr	893	5298	258	308	327	398	403	530	1192	0.5	2660
Ni	401	229	202	167	159	130	131	151	182	0.5	110
Cu	3.45	10.4	10.5	3.52	3.85	7.56	8.86	10.4	11.9	0.2	126
Zn	66	23.5	42.9	29.2	31.2	27.9	17.3	29.4	25.4	0.2	312
Ga	4.21	3.08	11.5	13.9	14.8	10.1	9.49	8.25	7.88	0.2	10
Rb	1.02	0.59	2.98	1.31	1.37	2.22	3.88	0.96	1.65	0.1	2.3
Sr	15.1	17.3	295	292	323	471	661	175	410	0.1	7.80
Y	13.3	0.81	1.15	1.9	2.02	0.62	0.8	1.21	1.51	0.1	1.56
Zr	71	0.74	8.13	53.2	85	3.04	3.96	1.82	3.71	0.1	3.94
Nb	0.08	<0.1	<0.1	<0.1	<0.1	<0.1	<0.1	<0.1	<0.1	0.1	0.245
Mo	0.41	1.1	1.5	0.54	0.78	0.88	<0.1	0.37	<0.1	0.1	0.928
Pb	2.24	2.1	2.36	2.59	1.78	2.5	1.3	0.97	1.4	0.3	2.470
Fe/Ni	113	90.4	218	147	157	159	132	176	128	–	17.8
Fe/Cr	50.8	3.91	171	79.5	76.1	52.0	42.9	50.2	19.5	–	71.6
Cr/V	16.5	141	5.27	9.06	8.38	29.7	31.7	22.2	50.7	–	47.1
Cr/Mn	0.698	7.68	0.307	0.604	0.629	0.926	1.12	1.10	2.54	–	1.34
Ti/V	8.98	2.92	3.27	5.59	4.87	3.73	3.94	6.28	4.26	–	7.72
Ca/Ti	297	973	567	584	652	1986	1588	580	962	–	21.3
Ca/K	109	99.1	49.6	77.6	77.0	64.1	45.4	82.9	73.4	–	16.6
Sr/Rb	14.8	29.3	99.0	223	236	212	170	182	248	–	3.39
Zn/Cu	19.1	2.26	4.09	8.30	8.10	3.69	1.95	2.93	2.13	–	2.48
Zn/Pb	29.5	11.2	18.2	11.3	17.5	11.2	13.3	30.3	18.1	–	126

Note: 1–3 – harzburgites; 4, 5 – lherzolites; 6 – plagiolherzolite; 7, 8 – wehrlites; 9 – plagiowehrlite; 10, 11 – olivine websterites; 12 – websterite; 13 – olivine clinopyroxenite; 14–17 – gabbronorites; 18–20 – gabbro; 21 – lower limits of element detection; 22 – contents of elements in Cl chondrite [Anders, Grevesse, 1989]. Analyses were carried out by the XRF-SR method at the Institute of Geology and Mineralogy (analyst Yu.P. Kolmogorov).

is enriched in Zr and depleted in Cu compared to chondrite, whereas wehrlites, websterites, clinopyroxenites, gabbronorites, and gabbro are depleted in Ni, Cu, and Zn compared to chondrite. Intense Zr maxima are present in websterite spectra, while Sr maxima occur in the spectra of gabbronorites and gabbro.

Let us discuss some of the patterns in the distribution of elements in the analyzed rock varieties (their content variations are given in ppm).

Chromium is present in increased amounts in harzburgites (3113–7662), lherzolites (1918–2515), plagiolherzolites (5279), wehrlites (1051–5485), and olivine clinopyroxenites (5298). Its somewhat lower content and a narrower range of variations are observed in olivine websterites (2155–2571).

Harzburgites (1271–1751), lherzolites (865–1233), and plagiolherzolites (2619) are enriched in *nickel*. Its contents are much lower (130–451) in wehrlites, olivine websterites, olivine clinopyroxenites, and some gabbroids. For comparison, note that in harzburgites of the Papua New Guinea complex, Ni is present in an amount of

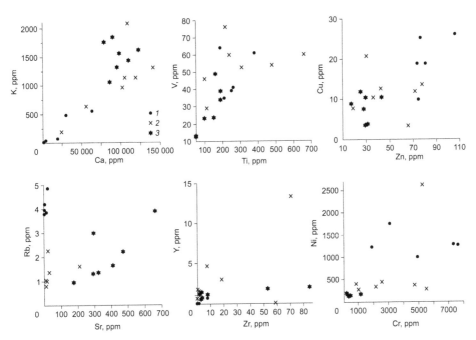

Figure 6.1 Variations in the contents of Ca, K, and some impurity elements in ultramafic and mafic rocks from the Berezovka massif (data from Table 6.2).
1 – restitic ultramafic rocks (harzburgites and lherzolites); *2* – hybrid ultramafic rocks (plagiolherzolites, wehrlites, plagiowehrlites, olivine and olivine-free websterites, olivine clinopyroxenites); *3* – orthomagmatic mafic rocks (gabbronorites, gabbro).

about 1100 ppm. In harzburgites of the Kroton massif (Kumroch Ridge, Kamchatka Peninsula) Ni content is 2800 ppm, in those from the Ust'-Bel'sk massif (Chukchi Peninsula) – 1500 ppm, and in those from the midocean ridges – 2500 ppm [Lutts, 1980].

Iron is the most abundant in harzburgite (97,000–141,000), lherzolites (81,500–102,000), olivine websterites (52,600–57,900), and wehrlites (23,900–37,400).

According to *manganese* content, harzburgites (1800–2150), lherzolites (1280–1890), and olivine websterites (1100–1280) differ slightly.

In all rocks under study *vanadium* is present in low and roughly equal amounts: restitic ultramafic rocks (35–92), hybrid ultramafic rocks (29–76), and gabbroids (13–49). In average content of V harzburgites from the Berezovka massif (47) are comparable with harzburgites from the Papua New Guinea complex (30), Kroton massif (50), and the midocean ridges (45) [Lutts, 1980].

Titanium in the largest amounts is found in wehrlites (110–660) and websterites (220–480).

According to *copper* content, the rocks from the Berezovka massif differ slightly: harzburgites (19–25), lherzolites (7–19), and wehrlites (8–21). Copper is present in gabbroids (3–12) in somewhat smaller amounts.

The distribution of *zinc* demonstrates the same trend as that of copper: Its elevated contents are defined in harzburgite (77–107), lherzolites (74–85), and plagiolherzolites

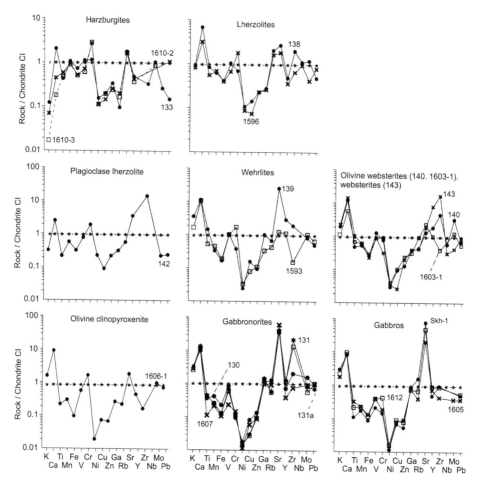

Figure 6.2 Distribution patterns of chondrite-normalized contents of impurity elements in rocks of the Berezovka massif (data from Table 6.2).
CI chondrite normalization (after [McDonough and Sun, 1995]).

(72); they are slightly lower in wehrlites (19–36), olivine websterites (42.3–78), clinopyroxenites (24), and gabbroids (17–43).

The largest amount of *calcium* is defined in wehrlites (103,000–110,000), websterites (110,000–140,000), and gabbronorites (99,300–124,000).

The highest *potassium* content is identified in gabbronorites (1430–1830) and gabbro (1050–1750); somewhat lower K content is detected in wehrlites (960–2070), websterites (630–1300), and clinopyroxenites (1080); harzburgites (10–70) are the most depleted. Approximately the same amounts of K were determined from the Kroton harzburgite massif (50). Abnormally high K content is detected in harzburgites of the Papua New Guinea complex (400) [Lutts, 1980].

Rubidium is present in very small amounts in all rock varieties under study: harzburgites (3.8–4.2), lherzolites (2.9–4.8), plagiolherzolites (1.4), wehrlites (0.8–10), olivine websterites (1.1–2.3), websterites (1.0), olivine clinopyroxenites (0.6), and gabbroids (1.0–3.0). Note that harzburgites of the Berezovka massif are richer in Rb compared to harzburgites of the Papua New Guinea complex (0.8) [Lutts, 1980].

Gallium content insignificantly drops in the series gabbroids (7.9–14.8) → wehrlites (2.8–11) → websterites (4.2) → olivine websterites (3.3–4.1) → plagiolherzolites (3.3) → olivine clinopyroxenites (3.1) → lherzolites (2.9) → harzburgites (1–2).

Strontium is present in rocks in the following amounts: harzburgites (2.9–3.9), lherzolites (11.2–21.5), websterites (10.8–19.1), olivine clinopyroxenites (17.3), plagiolherzolites (29.1), wehrlites (9.44–210), and olivine gabbro (175–661). Also, Sr content is slightly higher in harzburgites from the Kumroch Ridge massif (10) and from the midocean ridges (13.6) [Lutts, 1980] than in such rocks from the Berezovka massif.

The highest *yttrium* content is defined in one of websterite samples (13.3); it is lower in wehrlites (4.78), olivine websterites (1.66–3.04), and gabbroids (0.62–2.02); and it is minimal in harzburgites and plagiolherzolites (<0.1).

Approximately similar *zirconium* content is found in lherzolites (94), websterites (71), plagiowehrlites (59), and gabbronorites (85); it is minimal in harzburgites (<0.1). For comparison, Zr content of harzburgites in the Papua New Guinea complex (20) and in the midocean ridges (30) is much higher [Lutts, 1980].

Minimum *molybdenum* contents are detected in harzburgites (<0.1), and they are much higher in the olivine websterites (1.21–3.15).

Lead is found in comparable amounts in all rock varieties under study: harzburgites (0.38–2.7), lherzolites (1.22–1.92), wehrlites (1.28–1.70), websterites (2.24), olivine websterites (1.40–1.78), gabbronorites (1.78–2.59), and gabbro (0.97–1.4).

The above-mentioned data indicate that elements determined by XRF-SR are unevenly distributed among petrographic and genetically different rocks of the Berezovka massif. It is assumed that the identified differences are related to the fractionation of elements in the formation of ultramafic restites and generation of mafic melts as well as in interaction between them. This process is accompanied by contamination of mafic melts with ultramafic substance and addition of elements from the melts into hybrid ultramafic rocks during the ultramafic restite transformation.

6.2 GEOCHEMICAL PARAMETERS OF ROCKS OF THE BEREZOVKA MASSIF (FROM XRF-SR DATA)

Some parameters of distribution of impurity elements whose contents were determined by XRF-SR were used as additional indicators to identify geochemical differences between the petrographic varieties of rocks. For this purpose the ratios between the maximum and the minimum contents of each element was calculated, which reflect the extent of heterogeneity in the element distribution in the rocks of the massif. According to calculations, by the degree of dispersion elements form the descending series

Zr (235) → Sr (226) → K (207) → Cr (30) → Y (22) → Ni (20) → Rb (18) → Ga (15) → Mo (14) → Ti (13) → Fe (8.2) → Cu (8) → V (7.2) → Pb (7) → Zn (6.2) → Mn (6). These data suggest that Zr, Sr, and K underwent the most intense fractionation during the formation of rocks of the Berezovka massif, while Fe, Cu, V, Pb, Zn, and Mn underwent the least intense fractionation.

In addition, the values of some geochemical parameters (ratios) between the contents of elements (Table 6.2) were calculated. Depending on the petrographic attribution of rocks and quantitative mineral composition, these parameters change at different intervals.

Fe/Ni varies in the range of 24–218 with elevated values inherent mainly to gabbroids (128–176).

Fe/Cr. From 4 in olivine clinopyroxenites to 171 in gabbronorites.

Cr/V. From 5 in gabbronorites to 219 in harzburgites.

Cr/Mn. From 0.31 in gabbronorites to 7.68 in olivine clinopyroxenites.

Ti/V. From 2.2 in plagiolherzolites to 11 in wehrlites.

Ca/Ti. From 8.1 in harzburgites to 1986 in gabbronorites. Elevated values of the parameter are observed mainly in the gabbroids and some hybrid ultramafic rocks (websterites and wehrlites).

Ca/K. In gabbroids – 45–99, in harzburgites, lherzolites, and plagiolherzolites – 105–274.

Sr/Rb. In gabbroids – 99–248, in websterites – 8.5–11.0, in harzburgites and lherzolites – 0.74–4.45.

Zn/Cu. In gabbroids – 1.95–8.30, in harzburgites and lherzolites – 3.1–12.3.

Zn/Pb. In gabbroids – 11–30, in restitic and hybrid ultramafic rocks – 14–282.

K/Rb. Elevated values were defined in gabbroids (451–1175), wehrlites (93–1413), olivine websterites (279–1066), websterites (1275), and olivine clinopyroxenites (1831). Much lower values were observed in harzburgites (2.5–16.7) and lherzolites (114–201), while in the latter the values are lower than in harzburgites of the Papua New Guinea complex (570) and the Ust'-Bel'sk massif (333) [Lutts, 1980].

The geochemical features of the rocks of the Berezovka massif are also described based on the values of the above parameters, normalized to the corresponding values of these parameters for the CI chondrite. Variation plots of normalized values of parameters of different rocks vary in their position on the diagrams and in the intensity of the maxima and minima of some parameters (Fig. 6.3). For example, plots of harzburgites are complicated by Fe/Ni, Ca/K, and Cr/V maxima and Fe/Cr and Sr/Rb minima. Plots for lherzolites have a different configuration, with Ca/Ti and Ca/K maxima. Fe/Cr minima are present on plots for plagiolherzolites, wehrlites, plagiowehrlites, and olivine and olivine-free websterites, whereas Ca/Ti maxima are on plots of all varieties of rocks under study. Furthermore, intense Zn/Pb minima are observed on the plots of all rocks except plagiolherzolite.

Thus, the results of XRF-SR analyses of the Berezovka massif rocks showed that a significant part of elements under study is distributed in rocks very unevenly; refractory elements are mainly concentrated in restitic and hybrid ultramafic rocks, whereas medium-refractory and fusible elements are concentrated in orthomagmatic gabbro and some hybrid ultramafic rocks. The observed differences in the distribution of impurity elements in the rocks of the Berezovka massif are generally consistent with the model of its polygenic formation.

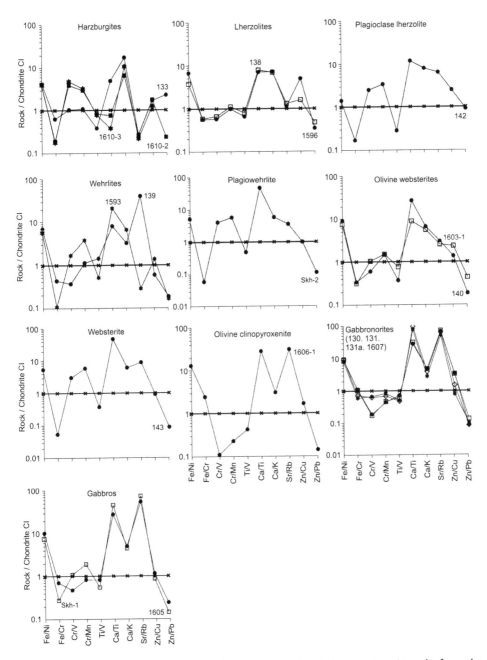

Figure 6.3 Variations of chondrite-normalized values of geochemical parameters in rocks from the Berezovka massif (data from Table 6.2).
CI chondrite normalization (after [Evensen *et al.*, 1978]).

6.3 DISTRIBUTION OF RARE EARTH ELEMENTS IN ROCKS OF MAFIC-ULTRAMAFIC MASSIFS OF THE EAST SAKHALIN OPHIOLITE ASSOCIATION

Rare earth elements (REE) are considered important indicators used for the taxonomy of igneous rocks of different compositions and for solution of problems of their genesis [Lesnov, 2005, 2007b]. Laws of distribution of these elements in ultramafic rocks and gabbroids of Sakhalin Island were first studied on the basis of collections of their samples from the Berezovka, Shel'ting, Komsomol'sk, and South Schmidt massifs. These studies were carried out by four analytical methods: instrumental neutron activation analysis (INAA), radiochemical neutron activation analysis (RNAA), inductively coupled plasma (from solutions) mass spectrometry (ICP-MS), and laser-ablation inductively coupled plasma mass spectrometry (LA ICP-MS) with the use of glass-fused powder rock samples.

6.3.1 Methods for REE analysis of rocks

Studies by *the INAA method* were made at the Analytical Center of the Institute of Geology and Geophysics, Siberian Branch of the USSR Academy of Sciences (Novosibirsk), at an early stage of research into the REE distribution in the Berezovka massif rocks [Lesnov *et al.*, 1998a]. Powder samples weighing 100 mg were used which were preirradiated at the Tomsk reactor by the flow with intensity 1×10^{13} neutrons cm^{-2} s^{-1} for 12 h. After cooling the sample activity was measured; then, based on the intensity of the characteristic lines, contents of ten REE were determined. Measurements were taken on coaxial and planar detectors with the resolution of 2 keV at line 1332 keV (Co-60) and 0.4 keV at line 59 keV (Am-241), respectively. Geochemical standards ST-1A, ST-2, and SG-3, as well as BIL-1 and ZUK-1, were used as control samples. The following lower limits of element detection (ppm) were achieved during the tests: La (0.5), Ce (0.8), Nd (5.0), Sm (0.05), Eu (0.01), Gd (0.2), Tb (0.1), Tm (0.1), Yb (0.1), and Lu (0.1).

The RNAA method, being a perfect option of the INAA method, allowed significantly reducing limits of the REE detection. Analyses were also performed at the Analytical Center of the Institute of Geology and Geophysics, Siberian Branch of the USSR Academy of Sciences. Reduction in limits of the REE detection was achieved using an additional procedure for the radiochemical separation of REE from radionuclides disturbing their detection (Sc, Cr, Fe, and Co), based on ion exchange chromatography effect. In this case, similar to the INAA, powder samples weighing 100 mg were preirradiated at the Tomsk reactor by the flow with intensity 1×10^{13} neutrons cm^{-2} s^{-1} for 20 h. Metered amounts of a solution of a mixture of elements being defined, spread on ashless filter paper along with the geochemical standards ST-1A, JP-1, and DTS-1, were used as standards. The irradiated sample was cooled for 5 days and then subjected to several successive chemical procedures. The first of them was to separate REE from interfering radionuclides. It was performed with the use of chromatographic columns 0.9×18 cm in size, filled with Dowex AG 50WX8 resin (100–200 mesh), followed by elution with acid solutions of 0.1 M $H_2C_2O_4$, 0.5 M HCl, and 2 M HNO_3. Then REE fraction was eluted from the cationite with 6 M and 8 M HNO_3. Geochemical standards underwent the same radiochemical procedures.

The latter procedure was to measure the activity of samples and to determine REE contents similarly to the INAA method. This method of analysis allowed achieving the following lower detection limits of REE (ppm): La (0.02), Ce (0.05), Nd (0.05), Sm (0.005), Eu (0.001), Gd (0.03), Tb (0.005), Tm (0.02), Yb (0.02), and Lu (0.02).

Analyses by *the ICP-MS method* were carried out in the laboratory of the Analytical Center of the Sobolev Institute of Geology and Mineralogy (Novosibirsk) and in the laboratory of the Analytical Center of the Institute of Tectonics and Geophysics (Khabarovsk).

In the laboratory of the Analytical Center of the Sobolev Institute of Geology and Mineralogy, sample preparation and determination of REE contents were performed as follows. Initially, samples were opened in sealed Teflon autoclaves of a MARS-5 microwave digestion system using different mixtures of nitric, hydrofluoric, and hydrochloric acids, prepurified by subboiling. After the decomposition and transfer of samples into solutions, REE contents were determined in them on an Element mass spectrometer (Finnigan MAT, Germany) by external calibration using standard geological samples for comparison and with the addition of indium as an internal standard. The following lower detection limits of REE (ppm) were achieved: La (0.02), Ce (0.03), Pr (0.003), Nd (0.004), Sm (0.001), Eu (0.0006), Gd (0.0005), Tb (0.0005), Dy (0.0006), Ho (0.0005), Er (0.0005), Tm (0.0002), Yb (0.0003), and Lu (0.0002).

In the laboratory of the Analytical Center of the Institute of Tectonics and Geophysics, the scheme of sample preparation and determination of the REE contents was different in the following: (1) a sample weighing 0.05 g was placed into a glassy-carbon crucible and treated with 5 mL HNO_3 and HF in a ratio of 1 : 1; (2) solution was evaporated on a hot plate to moist salts; (3) 1 mL concentrated HNO_3 and 0.5 mL hydrogen peroxide were added; (4) the resulting solution was again evaporated to dryness; (5) the dried deposit was dissolved in 10 mL 10% HNO_3; (6) the resulting solution was heated till complete dissolution of salts. When undecomposed deposit was detected in the sample, it was again evaporated and the whole process was repeated. After complete dissolution the sample was transferred into a 50 mL measuring tube and the solution was brought up to the mark with the addition of deionized water. Determination of REE contents in the resulting solution was carried out on an ELAN DRC II mass spectrometer (Perkin Elmer, USA). Analyses by *the LA ICP-MS method* were performed in the laboratory of the Analytical Center of the Sobolev Institute of Geology and Mineralogy. For this purpose, those samples of fused glass powder rocks were used for which the overall chemical analysis by the X-ray fluorescence (XRF) method had been performed. These samples were analyzed on an Element mass spectrometer (Finnigan MAT, Germany) in combination with a UV Laser Probe laser detachable device (Nd:YAG laser, $\lambda = 266$ nm, Finnigan MAT, Germany). As a calibration standard (reference sample) glass NIST-612 (USG) was used. The lower limits of element detection were as follows (ppm): La (0.001), Ce (0.001), Pr (0.0005), Nd (0.0008), Sm (0.0004), Eu (0.0003), Gd (0.0003), Tb (0.0002), Dy (0.0002), Ho (0.0002), Er (0.0002), Tm (0.0001), Yb (0.0002), and Lu (0.0001).

REE detection in individual samples of gabbroids and ultramafic rocks of the Sakhalin Island massifs at an early research stage was performed by INAA and RNAA [Lesnov et al., 1998a]. As a result, it was found that lherzolites and websterites are significantly depleted in all REE and the total contents of elements in them are lower than those in similar rocks from some massifs of other ophiolite associations. In a few samples of gabbronorites total REE contents exceeded 2 ppm. In most samples of gabbronorites and websterites chondrite-normalized contents of heavy REE were slightly higher than the contents of light elements. Later the REE detection in the rocks of the Sakhalin Island massifs was performed by ICP-MS and LA ICP-MS. In total, more than 60 samples were analyzed: harzburgites, lherzolites, and their serpentinized varieties, as well as wehrlites, different types of pyroxenites, gabbronorites, gabbro, olivine gabbro, and anorthosites.

6.3.2 Distribution of REE in restitic ultramafic rocks

The analyzed restitic ultramafic rocks are harzburgites and lherzolites from the Berezovka and South Schmidt massifs (Table 6.3). The total REE contents are defined in the range 0.32–3.6 ppm. The total REE content in harzburgites from the Berezovka massif is somewhat higher (0.97–1.97 ppm) than that in harzburgites from the Voikar–Syn'in (Polar Urals) (0.21 ppm) and Kempirsay (South Urals) (0.67 ppm) massifs but much lower than that in such rocks from the Khutul (Mongolia) (22.8 ppm) and Harzburg (Germany) (11.4 ppm) massifs [Lesnov, 2007b].

Judging from distribution spectra of chondrite-normalized element contents, abnormally high contents of light REE and a slight deficit of Eu (Fig. 6.4, 1, 3, 4) are observed in some samples. The contents of heavy REE in these rocks are more stable compared to the contents of light elements, as indicated by the ratios between

Table 6.3 Contents of REE in harzburgites and lherzolites of the Berezovka and South Schmidt massifs, ppm.

Element	Sample							
	133 1	1610-2 2	1610-3 3	138 4	145 5	161 6	163 7	1596 8
La	0.176	0.26	0.097	0.345	0.16	0.063	0.020	0.80
Ce	0.433	0.9	0.36	0.918	0.29	0.12	0.045	2.0
Pr	0.031	0.096	0.025	0.129	0.065	0.010	0.005	0.22
Nd	0.175	0.22	0.12	0.805	0.40	0.025	0.023	0.97
Sm	0.025	0.072	0.048	0.226	0.20	0.006	0.012	0.21
Eu	N.d.	<0.001	<0.001	0.056	0.058	0.001	0.005	0.034
Gd	0.040	0.078	0.057	0.298	0.25	0.007	0.036	0.24
Tb	N.d.	0.017	0.013	0.031	0.040	0.002	0.007	0.042
Dy	0.046	0.11	0.081	0.327	0.29	0.019	0.054	0.24
Ho	N.d.	0.030	0.019	0.057	0.071	0.005	0.015	0.056
Er	0.030	0.091	0.059	0.207	0.18	0.016	0.059	0.21
Tm	N.d.	0.012	0.011	0.016	0.036	0.004	0.013	0.034
Yb	0.044	0.077	0.069	0.214	0.23	0.033	0.088	0.24
Lu	N.d.	0.008	0.007	0.018	0.034	0.008	0.015	0.041
Σ REE	N.d.	1.972	0.967	3.647	2.304	0.319	0.397	5.337
$(La/Yb)_n$	2.70	2.28	0.95	1.09	0.47	1.29	0.17	2.25
$(Eu/Eu^*)_n$	N.d.	N.d.	N.d.	0.66	0.79	0.47	0.68	0.46
$(La/Sm)_n$	4.43	2.27	1.27	0.96	0.50	6.61	1.05	2.40

Note: 1–3 – harzburgites; 4–7 – lherzolites; 8 – lherzolite with websterite veinlet. Massifs: 1–5 – Berezovka; 6, 7 – South Schmidt. Methods: 1, 4–7 – ICP-MS, from solutions; 2, 3, 8 – LA ICP-MS. Analyst D.V. Avdeev, Institute of Tectonics and Geophysics (Khabarovsk); analysts I.V. Nikolaeva and S.V. Palesskii, Institute of Geology and Mineralogy (Novosibirsk). N.d. – No data.

their maximum and minimum values: Yb (7.3), Er (13), Ce (44), and La (40). Available data suggest that the frequently observed anomalous enrichment of ultramafic restites in light REE and sometimes in other REE is due to the presence of varying amounts of their unstructured impurities in the analyzed samples; they are concentrated in the intergranular and intragranular microcracks during the infiltration of epigenetic fluids. This assumption is confirmed by the results of geochemical studies of ultramafic restites from other ophiolite associations and from deep-seated xenoliths in alkali basalts [Lesnov, 2007b; Lesnov et al., 2009a,b].

6.3.3 Distribution of REE in hybrid ultramafic rocks

As it was noted, the hybrid ultramafic rocks are regarded as products of magmatic metasomatic transformations of different nature and intensity occurring with restitic ultramafic rocks under the influence of mafic melts and their fluids. This explains the higher heterogeneity of hybrid ultramafic rocks with respect to their overall chemical and quantitative mineral compositions and distribution of impurity elements, including REE.

The REE composition of hybrid ultramafic rocks, which are plagiolherzolites, wehrlites, plagiowehrlites, olivine websterites, websterites, olivine clinopyroxenites, amphibole peridotites (schriesheimites), and olivine and olivine-free orthopyroxenites from the Berezovka, Shel'ting, and South Schmidt massifs, was investigated on a more representative collection compared to restitic ultramafic rocks (Table 6.4). In samples in which all REE were identified, their total contents vary in the range of 0.40–15.0 ppm. In some of them chondrite-normalized contents of heavy REE exceed contents of light elements; therefore, $(La/Yb)_n < 1$, and the REE distribution patterns

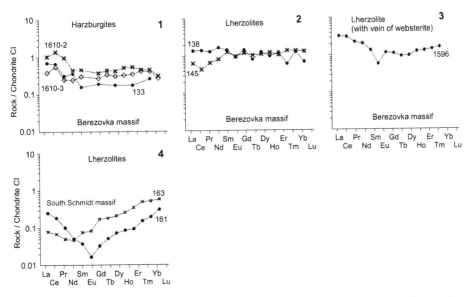

Figure 6.4 Chondrite-normalized REE patterns in restitic ultramafic rocks of the Berezovka and South Schmidt massifs (data from Table 6.3).
CI chondrite normalization (after [Evensen et al., 1978]).

Table 6.4 Contents of REE in plagiolherzolites, wehrlites, olivine plagiowehrlites, and olivine-free websterites and enstatitites from the Berezovka, Shel'ting, and South Schmidt massifs, ppm.

Element	Sample											
	142 1	139 2	136 3	140 4	137 5	155 6	134a 7	134 8	1593 9	1603-1 10	143 11	143a 12
La	0.034	0.168	0.13	0.138	0.180	0.103	0.07	0.390	0.11	0.23	0.60	0.405
Ce	0.079	0.530	0.28	0.332	0.250	0.262	0.84	0.240	0.25	0.62	2.20	1.690
Pr	0.011	0.086	0.015	0.033	N.d.	0.003	N.d.	N.d.	0.045	0.069	0.50	0.360
Nd	0.046	0.661	0.12	0.259	0.200	0.119	<1.9	0.180	0.23	0.49	3.00	2.486
Sm	0.016	0.273	0.011	0.090	0.047	0.003	0.04	0.056	0.086	0.19	1.16	0.943
Eu	0.005	0.140	N.d.	0.025	0.019	N.d.	0.024	0.024	0.015	0.051	0.32	0.239
Gd	0.029	0.475	0.023	0.192	0.083	0.018	N.d.	0.098	0.15	0.30	1.55	1.375
Tb	0.007	0.072	N.d.	0.025	0.016	N.d.	<0.06	0.019	0.039	0.063	0.29	0.232
Dy	0.043	0.633	0.027	0.327	N.d.	0.043	N.d.	N.d.	0.26	0.43	1.97	1.720
Ho	0.015	0.123	N.d.	0.066	N.d.	N.d.	N.d.	N.d.	0.064	0.11	0.42	0.369
Er	0.045	0.374	0.018	0.251	N.d.	0.044	N.d.	N.d.	0.22	0.36	1.29	1.115
Tm	0.009	0.039	N.d.	0.023	0.021	N.d.	N.d.	0.026	0.033	0.056	0.21	0.144
Yb	0.063	0.344	0.028	0.260	0.130	0.084	0.19	0.180	0.23	0.36	1.30	1.095
Lu	0.010	0.035	N.d.	0.026	0.019	N.d.	<0.05	0.029	0.038	0.054	0.20	0.146
ΣREE	0.41	3.95	N.d.	2.05	N.d.	N.d.	N.d.	N.d.	1.57	3.38	15.0	12.32
(La/Yb)$_n$	0.36	0.33	3.13	0.36	0.94	0.83	0.25	1.46	0.32	0.43	0.31	0.25
(Eu/Eu*)$_n$	0.70	1.18	N.d.	0.57	0.92	N.d.	3.19	0.98	0.40	0.65	0.73	0.64
(La/Sm)$_n$	1.34	0.39	7.44	0.97	2.41	21.6	1.10	4.39	0.81	0.76	0.33	0.27

Elements	Sample							
	1606-1 13	Skh-2 14	135 15	164 16	162 17	174 18	184 19	176 20
La	0.22	0.051	0.271	0.131	0.051	0.116	0.310	0.165
Ce	0.38	0.13	0.699	0.364	0.10	0.20	0.917	0.401
Pr	0.044	0.018	0.094	0.008	0.009	0.014	0.150	0.018
Nd	0.20	0.068	0.583	0.121	0.045	0.061	1.085	0.140
Sm	0.074	0.025	0.185	N.d.	0.021	0.031	0.437	0.010
Eu	<0.001	<0.001	0.063	N.d.	0.007	0.007	0.213	N.d.
Gd	0.12	0.045	0.307	0.009	0.026	0.033	0.753	0.017
Tb	0.024	0.015	0.042	N.d.	0.004	0.005	0.120	N.d.
Dy	0.15	0.060	0.431	0.011	0.028	0.032	0.992	0.042
Ho	0.042	0.014	0.086	N.d.	0.010	0.006	0.200	N.d.
Er	0.13	0.041	0.299	0.005	0.038	0.022	0.600	0.040
Tm	0.024	0.008	0.032	N.d.	0.009	0.004	0.062	N.d.
Yb	0.15	0.041	0.031	0.019	0.059	0.030	0.547	0.070
Lu	0.021	0.004	0.037	N.d.	0.011	0.005	0.062	N.d.
ΣREE	1.58	0.52	3.16	0.67	0.42	0.57	6.45	N.d.
(La/Yb)$_n$	0.99	0.84	5.90	4.65	0.58	2.61	0.38	1.59
(Eu/Eu*)$_n$	N.d.	N.d.	0.80	N.d.	0.92	0.92	1.13	N.d.
(La/Sm)$_n$	1.87	1.28	0.92	N.d.	1.53	0.67	0.45	10.4

Note: 1 – plagiolherzolite; 2, 9 – wehrlites; 3, 14 – plagiowehrlite; 4, 5, 8, 10 – olivine websterites; 6, 11, 12, 16, 17 – websterites; 13 – olivine clinopyroxenite; 15 – cortlandite (schriesheimite); 18, 20 – olivine enstatitites; 19 – enstatitite. Massifs: 1–15 – Berezovka; 16, 17 – South Schmidt; 18–20 – Shel'ting. Methods: 1–4, 6, 11, 12, 15–20 – ICP-MS, from solutions; 9, 10, 13, 14 – LA ICP-MS, from glass-fused samples; 7 – INAA; 5, 8 – RNAA. Performed: 1, 17, 18 – I.V. Nikolaeva and S.V. Palesskii (Institute of Geology and Mineralogy, Novosibirsk); 9, 10, 13, 14 – S.V. Palesskii (Institute of Geology and Mineralogy, Novosibirsk); 7 – V.A. Bobrov (Institute of Geology and Geophysics, Khabarovsk)); 5, 8 – V.A. Kovaleva (Institute of Geology and Mineralogy, Novosibirsk); 2–4, 6, 11, 12, 15, 16, 19, 20 – D.V. Avdeev (Institute of Tectonics and Geophysics, Khabarovsk). N.d. – No data.

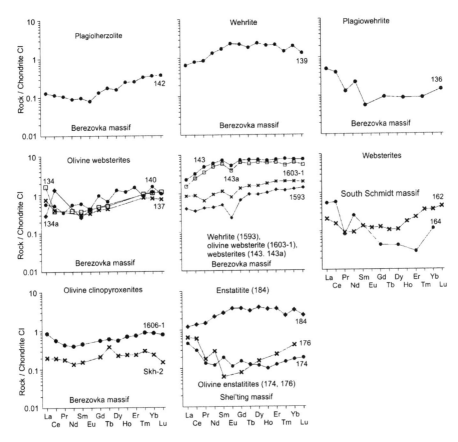

Figure 6.5 Chondrite-normalized REE patterns in hybrid ultramafic rocks from the Berezovka, Shel'ting, and South Schmidt massifs (data from Table 6.4).
CI chondrite normalization (after [Evensen *et al.*, 1978]).

have a common positive slope (Fig. 6.5, 1, 4, 5). Another group of hybrid ultramafic rock samples is made up of their varieties enriched in light REE in varying degrees. In them $(La/Yb)_n > 1$ and, accordingly, REE patterns have a negative slope (Fig. 6.5, 3, 6, 8). In hybrid varieties, unlike restitic ultramafic rocks, heavy REE are distributed less evenly compared to light ones, as indicated by the correlation between the maximum and minimum average contents of Yb (58), Er (223), Ce (28), and La (18).

Comparison of REE patterns in plagiolherzolites from sample 142 (Fig. 6.5, 1) and in wehrlite from sample 139 (Fig. 6.5, 2) according to their position on the plot indicates how much the content of these impurities depends on the modal clinopyroxene content in these rocks. However, the configuration of the spectrum of plagiowehrlite from sample 136 (Fig. 6.5, 3), whose left flank is raised, is most likely due to the presence of a significant amount of plagioclase as the major concentrator of light REE in this hybrid ultramafic rock. Fig. 6.5, 5 is also quite indicative, reflecting the differences between REE contents in olivine-free websterites, on the one hand, and the olivine-rich wehrlites and olivine websterites, on the other.

Table 6.5 Contents of REE in olivine websterites, gabbronorites, gabbro, olivine gabbro, and anorthosites from the Berezovka, Shel'ting, and Komsomol'sk massifs, ppm.

Element	Sample 130 / 1	129 / 2	131 / 3	132a / 4	138a / 5	141 / 6	144 / 7	147 / 8	132ak / 9	182 / 10	183 / 11	155-5 / 12	191 / 13	189 / 14
La	0.126	0.155	0.28	N.d.	N.d.	0.21	0.25	0.28	N.d.	0.077	0.070	0.94	0.64	0.134
Ce	0.313	0.454	0.44	<0.1	0.58	1.10	1.7	3.0	0.230	0.21	0.18	2.00	1.48	0.456
Pr	0.026	0.056	N.d.	N.d.	N.d.	N.d.	N.d.	N.d.	N.d.	0.039	0.025	0.32	0.23	0.059
Nd	0.204	0.399	0.48	0.18	<0.9	1.20	N.d.	N.d.	0.170	0.20	0.18	1.57	1.06	0.546
Sm	0.058	0.120	0.16	N.d.	0.067	0.076	0.23	0.103	0.060	0.095	0.098	0.54	0.34	0.184
Eu	0.025	0.053	0.075	0.038	0.030	0.053	0.18	0.05	0.038	0.051	0.057	0.19	0.13	0.062
Gd	0.101	0.178	N.d.	N.d.	N.d.	N.d.	N.d.	N.d.	0.092	0.13	0.16	0.77	0.48	0.288
Tb	N.d.	0.014	0.045	0.033	0.025	0.10	N.d.	N.d.	0.017	0.027	0.028	0.13	0.095	0.033
Dy	0.153	0.216	N.d.	N.d.	N.d.	N.d.	N.d.	N.d.	N.d.	0.17	0.20	0.86	0.67	0.380
Ho	0.022	0.033	N.d.	N.d.	N.d.	N.d.	N.d.	N.d.	N.d.	0.039	0.045	0.17	0.15	0.063
Er	0.107	0.133	N.d.	N.d.	N.d.	N.d.	N.d.	N.d.	N.d.	0.14	0.13	0.49	0.46	0.246
Tm	N.d.	0.003	N.d.	N.d.	N.d.	N.d.	N.d.	N.d.	N.d.	0.024	0.027	0.074	0.069	0.012
Yb	0.124	0.131	0.245	0.104	0.158	0.20	0.197	0.17	0.110	0.16	0.17	0.47	0.45	0.239
Lu	0.005	0.006	0.03	N.d.	N.d.	0.017	0.047	0.022	0.020	0.027	0.021	0.061	0.069	0.013
ΣREE	1.27	1.95	N.d.	N.d.	N.d.	N.d.	N.d.	N.d.	N.d.	1.39	1.39	8.59	6.32	2.72
(La/Yb)$_n$	0.69	0.80	2.49	N.d.	N.d.	0.71	0.86	1.11	N.d.	0.33	0.28	1.35	0.96	0.38
(Eu/Eu*)$_n$	0.99	1.11	N.d.	N.d.	N.d.	N.d.	N.d.	N.d.	1.56	1.40	1.38	0.90	0.98	0.82
(La/Sm)$_n$	1.37	0.81	1.11	N.d.	N.d.	1.74	0.68	1.71	N.d.	0.51	0.45	1.10	1.19	0.46

Element	Sample 1607 / 15	1612 / 16	1612a / 17	Skh-1 / 18	1605 / 19	148-1b / 20
La	0.16	0.19	0.19	0.058	0.073	0.084
Ce	0.34	0.56	0.40	0.12	0.25	0.19
Pr	0.026	0.11	0.076	0.017	0.032	0.023
Nd	0.13	0.22	0.22	0.080	0.18	0.092
Sm	0.036	0.088	0.080	0.033	0.074	0.039
Eu	<0.001	0.020	0.015	<0.001	0.025	0.047
Gd	0.048	0.15	0.13	0.067	0.11	0.035
Tb	0.009	0.027	0.027	0.018	0.021	0.005
Dy	0.055	0.19	0.18	0.085	0.16	0.021
Ho	0.015	0.040	0.042	0.017	0.042	0.005
Er	0.037	0.13	0.12	0.065	0.12	0.010
Tm	0.007	0.018	0.020	0.011	0.019	0.002
Yb	0.037	0.13	0.11	0.067	0.12	0.010
Lu	N.d.	0.014	0.013	0.007	0.015	0.002
ΣREE	0.90	1.89	1.62	0.65	1.24	0.57
(La/Yb)$_n$	2.92	0.99	1.17	0.58	0.41	5.67
(Eu/Eu*)$_n$	N.d.	0.53	0.45	N.d.	0.85	3.82
(La/Sm)$_n$	2.80	1.36	1.50	1.11	0.62	1.36

Note: 1, 3–11, 14, 15 – gabbronorites; 2, 16–19 – gabbro; 12 – olivine websterite; 13 – olivine gabbro; 20 – anorthosite. Massifs: Berezovka (1–9, 12, 15–20); Shel'ting (10, 11); Komsomol'sk (13, 14). Methods: ICP-MS, from solutions (1, 2, 10–14, 15, 20); LA ICP-MS, in the glass-fused samples (15–19); INAA (3–8); RNAA (9). 10, 11, 12, 13, 15, 20 (analysts I.V. Nikolaeva and S.V. Palesskii, Institute of Geology and Mineralogy, Novosibirsk); 1, 2, 14 (analyst D.V. Avdeev, Institute of Tectonics and Geophysics, Khabarovsk); 3–8 (analyst V.A. Bobrov, Institute of Geology and Mineralogy, Novosibirsk); 9 (analyst V.A. Kovaleva, Institute of Geology and Mineralogy, Novosibirsk). N.d. – No data.

6.3.4   Distribution of REE in orthomagmatic gabbroids

REE distribution in orthomagmatic gabbroids was investigated on gabbronorite and gabbro samples from the Berezovka, Shel'ting, and Komsomol'sk massifs (Table 6.5). In those samples in which all REE were detected, their total content ranged from 0.65 to 2.72 ppm. According to the general REE content, these rocks are close to, or poorer than, CI chondrite; at the same time, a portion of samples has positively inclined spectra of chondrite-normalized element contents (Fig. 6.6). Contents of La vary in the range of 0.059–0.28 ppm with an average of 0.16 ppm; contents of Ce are in the range of 0.12–1.70 ppm with an average of 0.49 ppm; the content of Yb is in the range of 0.037–0.245 ppm with an average of 0.145 ppm. These three elements are not significantly different in ratio between the maximum and minimum contents: La (4.7), Ce (14.2), and Yb (6.6). Overall, orthomagmatic gabbroids of the studied massifs are depleted in REE, as indicated by their comparison with N-MORB: La – 2.50 ppm, Ce – 7.50 ppm, and Yb – 3.05 ppm. The near-chondrite content of REE and the lack of evidence for their intense fractionation with enrichment with light elements

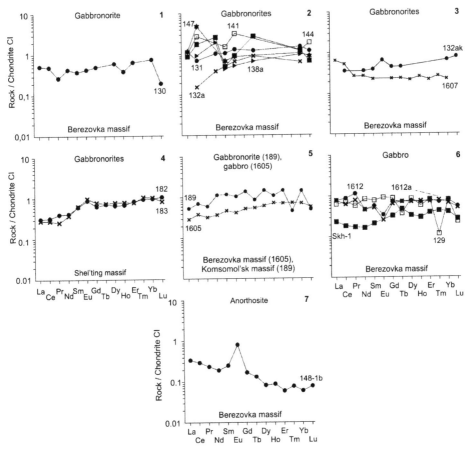

Figure 6.6   Chondrite-normalized REE patterns in mafic rocks from the Berezovka, Shel'ting, and Komsomol'sk massifs (data from Table 6.5).
CI chondrite normalization (after [Evensen et al., 1978]).

suggest that the parental melts of these gabbroids formed from a previously depleted upper-mantle source.

6.3.5 Distribution of REE in hybrid gabbroids

The REE composition of hybrid gabbroids is investigated on a limited number of samples of olivine gabbro and anorthosites (Table 6.5). Olivine gabbro differs from orthomagmatic gabbro in the higher general content of REE, the total contents of which are in the range of 6.32–8.59 ppm, and their near-horizontal spectra indicate that elements were hardly subjected to fractionation (Fig. 6.6, 7). Probably this is due to the overall anomalous trace-element composition of contaminated mafic melts from which hybrid gabbroids crystallized. As for the anorthosite, as it can be seen in its spectrum, it has features of almost monomineral plagioclase rock, i.e., low total content of REE (0.57 ppm), their significant fractionation, and an intense positive Eu anomaly (Fig. 6.6, 8).

**

The conducted geochemical studies of rocks of mafic-ultramafic massifs of Sakhalin Island have greatly expanded knowledge of their trace-element composition. However, the data are only the first approximation to the real evaluation of the contents of various elements in the rocks of these massifs.

Analyses of the rocks from the Berezovka massif by XRF-SR have demonstrated that the majority of impurity elements in them is very unevenly distributed. Refractory elements mainly accumulated in restitic and hybrid ultramafic rocks, but fusible elements – in orthomagmatic gabbroids and some hybrid ultramafic rocks and gabbroids. These data are consistent with the model of partial melting in the formation of ultramafic restites and the mechanism of contamination of parental mafic melts with ultramafic substance during their interreacting.

Almost all rocks under study are depleted in REE impurity. In gabbroids and ultramafic rocks these elements are mainly concentrated in the form of structural (isomorphous) impurities in clinopyroxenes and amphiboles, to a much lesser extent – in orthopyroxenes, plagioclases, and accessory phases (apatite, zircon, titanite, and others). Uneven distribution of REE in rocks is often associated with the presence of varying amounts of unstructured impurity of light REE, concentrated in the intergranular and intragranular microcracks as submicroscopic particles of epigenetic minerals, such as apatite [Lesnov et al., 2009]. Analytical errors arising during the performance of analyses using different methods in different laboratories can serve as another reason for the heterogeneity of REE content estimates [Itoh et al., 1993]. As we can assume, the overall low REE content in the ultramafic restites and orthomagmatic gabbroids under study is due to the depletion of their upper-mantle source in these elements and the high degree of its partial melting.

Chapter 7

Rb–Sr isotope systems in rocks of mafic-ultramafic massifs of the East Sakhalin ophiolite association

Results of study of Rb–Sr isotope systems are increasingly used in geochemical studies and systematics of igneous rocks differing in composition and genesis. The first data on these systems in rocks from mafic-ultramafic massifs of the East Sakhalin ophiolite association were obtained for samples of ultramafic rocks and gabbroids from the Berezovka and, partly, other massifs [Kiseleva *et al.*, 2010; Lesnov *et al.*, 2010b]. These studies were carried out in the laboratory of the Analytical Center of the Sobolev Institute of Geology and Mineralogy by the standard technique, including (1) acid digestion of weighted rock specimens, (2) separation of Rb and Sr by ion exchange chromatography on Dowex AG 50WX8 resin (200–400 mesh) with 2 N HCl as an eluent, and (3) measurement of Rb and Sr isotope ratios with an MI 1201T solid-phase thermal-ionization mass spectrometer in double-ribbon mode with one-collector recording. The accuracy of measurement of the $^{87}Sr/^{86}Sr$ ratios was controlled by parallel measurements of the VNIIM (Mendeleyev Institute for Metrology) (0.70800 ± 10, specification; 0.70810 ± 10, measurement) and ISG-1 (0.71731 ± 10, specification; 0.71727 ± 4, measurement) geochemical standards.

According to synchrotron-radiation XRF data (Table 6.2), Rb and Sr contents in ultramafic and gabbroid rocks from the studied massifs vary within a narrow range of values. Harzburgites from the Berezovka massif contain 3.78–4.18 ppm Rb; lherzolites, 2.94–4.83 ppm. Similar contents of this element were determined in plagiolherzolites (1.37 ppm), wehrlites (0.8–10.3 ppm), olivine websterites (1.06–2.26 ppm), websterites (1.02 ppm), olivine clinopyroxenites (0.59 ppm), and gabbroids (0.96–2.96 ppm). The same samples have the following Sr contents (ppm): harzburgites (2.92–3.86), lherzolites (11.2–21.5), olivine websterites (10.8–19.1), websterites (15.1), olivine clinopyroxenites (17.3), plagiolherzolites (29.1), wehrlites (9.44–210), and gabbroids (175–661). Refined estimates of Rb and Sr contents were obtained by analysis of samples of these rocks with an MI 1201T mass spectrometer (Table 7.1). It was found that lherzolite and websterite are characterized by low Rb and Sr contents and elevated Rb/Sr ratios compared to other rocks from the analyzed collection. Higher Rb and Sr contents were determined in wehrlites, gabbronorites, gabbro, and anorthosites. Gabbroids show low Rb/Sr ratios (0.0017–0.0107). According to the obtained estimates, the values of the coefficient 1/Sr increase from anorthosites (0.0008) to lherzolites (0.0585) and websterites (0.1302), whereas $^{87}Rb/^{86}Sr$ increases from 0.00492 (gabbro, sample 1615) to 0.08745 (websterite, sample 140).

Table 7.1 Rb and Sr contents (ppm) and their isotope ratios in rocks from the Berezovka, Shel'ting, and Komsomol'sk massifs.

Sample	Rock	Rb	Sr	Rb/Sr	1/Sr	Rb^{87}/Sr^{86}	Sr^{87}/Sr^{86} (*)	Sr^{87}/Sr^{86} (**)
Berezovka massif								
138	Lherzolite	0.230	17.1	0.0134	0.0585	0.03887	0.70382 ± 15	0.70373 ± 4
139	Wehrlite	1.907	158	0.0121	0.0063	0.03492	0.70360 ± 6	0.70352 ± 9
140	Websterite	0.232	7.68	0.0302	0.1302	0.08745	0.70430 ± 12	0.70410 ± 2
130	Gabbronorite	3.497	231	0.0152	0.0043	0.04382	0.70364 ± 11	0.70354 ± 8
131a	Gabbronorite	1.296	246	0.0053	0.0041	0.01522	0.70353 ± 5	0.70350 ± 10
1607	Gabbronorite	1.427	422.12	0.0034	0.0024	0.00978	0.70351 ± 9	0.70349 ± 11
131	Gabbronorite	1.561	255.02	0.0061	0.0039	0.01770	0.70343 ± 8	0.70339 ± 13
1612	Fine-grained gabbro	1.10	387	0.0029	0.0026	0.00824	0.70344 ± 8	0.70342 ± 12
1605	Gabbro	0.278	164	0.0017	0.0061	0.00492	0.70344 ± 8	0.70335 ± 14
Skh-1	Gabbro	6.26	584	0.0107	0.0017	0.03098	0.70317 ± 11	0.70310 ± 15
148-1b	Anorthosite	8.91	1189	0.0075	0.0008	0.02167	0.70375 ± 6	0.70370 ± 5
155-5	Olivine gabbro from a vein in plagiowehrlite	1.33	128	0.0104	0.0078	0.02993	0.70408 ± 12	0.70401 ± 3
Shel'ting massif								
182	Gabbronorite	0.32	150	0.0022	0.0067	0.00627	0.70372 ± 11	0.70371 ± 6
183	Gabbronorite	0.87	169	0.0051	0.0059	0.01480	0.70363 ± 7	0.70360 ± 7
Komsomol'sk massif								
191	Olivine gabbronorite	1.36	807	0.0017	0.0012	0.00489	0.70429 ± 13	0.70428 ± 1

Note: The determination of Rb and Sr contents and the estimation of their isotope ratios were carried out with an MI 1201T mass-spectrometer (TIMS) at the Sobolev Institute of Geology and Mineralogy (Novosibirsk). The accuracy of measurement of the $^{87}Sr/^{86}Sr$ ratio was controlled by parallel measurements of the VNIIM and ISG-1 standards (analyst V.Yu. Kiseleva). (*) – measured Sr isotope ratios. (**) – primary Sr isotope ratios, calculated based on the estimated isotope age of rock (158 Ma, Late Jurassic), obtained on analysis of zircon from sample 1607.

In lherzolites, wehrlites, and samples of some other rocks they vary within the range of values for the above two samples.

Based on the measured $^{87}Sr/^{86}Sr$ ratio and estimated isotope age of zircon from gabbronorite (sample 1607), equal to 158 ± 3.1 Ma (see Chapter 10), we calculated primary Sr isotope ratios in rocks from the studied collection. All the analyzed rocks, from lherzolites (earliest rocks in the considered massifs) to gabbronorites and gabbro (latest rocks), are characterized by quite low $^{87}Sr/^{86}Sr$ ratios (0.70343–0.70430) (Figs. 7.1, 7.2). Diagrams show the hypothetical evolution trends of Rb and Sr isotope ratios in ultramafic and gabbroid rocks from the Berezovka and other massifs. As one can see, the $^{87}Sr/^{86}Sr$ ratio for most of the samples of the studied rocks has values corresponding to its lower limit determined for ultramafic rocks and island-arc basalts [Cox et al., 1982]. On the other hand, the obtained estimates are very close to the values for N-MORB (0.7025–0.7035) [Faure, 1989]. Samples of websterites, plagiowehrlites, and olivine gabbronorites, which are regarded as hybrid rocks, as stated above, show the $^{87}Sr/^{86}Sr$ ratios of >0.704.

On the diagram ($^{87}Sr/^{86}Sr$)–1/Sr (Fig. 7.2, (2)), the compositional points of the studied rocks form two trends: near-vertical and near-diagonal. The near-diagonal

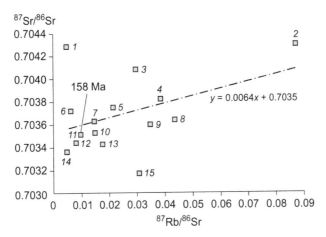

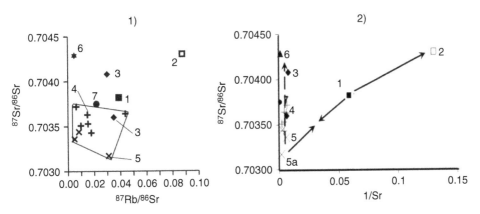

Figure 7.1 Diagram $^{87}Rb/^{86}Sr$ – $^{87}Sr/^{86}Sr$ for rocks from the Berezovka, Shel'ting, and Komsomol'sk massifs.
1 – olivine gabbronorite (sample 191); *2* – websterite (sample 140); *3* – olivine gabbro from a vein in plagiowehrlite (sample 155-5); *4* – lherzolite (sample 138); *5* – anorthosite (sample 148-1b); *6–8* – gabbronorites (samples 182, 183, 130); *9* – wehrlite (sample 139); *10, 11, 13* – gabbronorites (samples 131a, 1607, 131); *12, 14, 15* – gabbro (samples 1612, 1605, Skh-1) (data from Table 7.1). The age of zircon from sample 1607 (point 11) is 158 ± 3.1 Ma (see Chapter 10).

Figure 7.2 Diagrams $^{87}Sr/^{86}Sr$ – $^{87}Rb/^{86}Sr$ (*1*) and $^{87}Sr/^{86}Sr$ – 1/Sr (*2*) for rocks from the Berezovka, Shel'ting, and Komsomol'sk massifs.
1 – lherzolite (black square); *2* – websterite (white square); *3* – wehrlite and plagiowehrlite (rhombuses); *4* – gabbronorites (straight cross); *5, 5a* – gabbro (oblique cross); *6* – olivine gabbronorite (star on plot *1* and triangle – on plot *2*); *7* – anorthosite (dot) (data from Table 7.1).

trend suggests isotope mixing during interaction between the mafic melt from which orthomagmatic gabbroids crystallized (point *5a*) and older ultramafic restite (point 1). That process produced hybrid ultramafic rocks (websterites) (point 2). Hybrid gabbroids (olivine gabbronorite) are characterized by the lowest ^{87}Rb/^{86}Sr ratio; hybrid ultramafic rocks (websterite), by the highest one.

The extremely low ^{87}Rb/^{86}Sr ratios in the ultramafic and gabbroid rocks are similar to those for the primitive mantle [Faure, 1989]. This suggests that the restitic ultramafic rocks and orthomagmatic gabbroids were minimally contaminated with crustal material, like the hybrid ultramafic and gabbroid rocks formed by their interaction, which make up massifs on Sakhalin Island.

Data on Rb–Sr systems for rocks from mafic-ultramafic massifs of Sakhalin Island were compared with such data for similar complexes. Having studied Sr isotope ratios in rocks from mafic-ultramafic massifs of the ophiolite association of the Koryakia Mountains and the Chukchi Peninsula, Mikhalev and Popeko [1986] determined extremely low values in ultramafic rocks. For example, the ^{87}Sr/^{86}Sr ratio is 0.7058–0.7073 in rocks of one massif of the Khatyrka zone, 0.7005–0.7075 in rocks from the Ust'-Bel'sk massif, and 0.7004 in rocks from the North Pekul'nei and Krasnogorsk massifs; the Rb/Sr ratio in rocks from massifs of this region is 0.0018–0.0060. Based on a simplified one-stage model for evolution of mantle material, the authors presumed that the minimum age of ultramafic rocks from the studied massifs is ∼3 Ga.

A study of Rb–Sr isotope systems in rocks from the Dugda mafic-ultramafic massif (northeastern Tuva) showed ^{87}Sr/^{86}Sr = 0.7048–0.7055 in wehrlites, 0.7053 in pyroxenites, and 0.7050–0.7051 in gabbro and anorthosites [Mekhonoshin *et al.*, 1986]. According to the authors, parental mafic melts did not experience significant contamination with crustal material during the formation of the Dugda massif.

Results of studies of Rb–Sr isotope systems in gabbroids from the Dovyren mafic-ultramafic massif (North Baikal region) were presented by E.V. Kislov *et al.* [1989]. The ^{87}Sr/^{86}Sr ratio in these rocks is quite high (0.709–0.715), which testifies to the important role of crustal rocks in the contamination of the parental melts of gabbroids of the Dovyren massif. Also, the ^{87}Sr/^{86}Sr ratios for gabbroids from the nearby Chaya mafic-ultramafic massif were determined as very low (0.7022–0.7037). It might be that during the formation of the Chaya massif gabbroids, their parental melts were not contaminated with crustal material.

* * *

The first studies of Rb–Sr isotope systems in ultramafic and gabbroid rocks from massifs of the East Sakhalin ophiolite association have shown that all the analyzed samples are depleted in Rb and Sr and are characterized by the low ^{87}Sr/^{86}Sr and ^{87}Rb/^{86}Sr ratios. It has been found that the ^{87}Sr/^{86}Sr and 1/Sr values decrease in the series from hybrid ultramafic rocks (websterites) to restitic lherzolites and then to orthomagmatic gabbronorites. The established trend of variation in the above parameters might be due to the mixing of the isotope systems of restitic ultramafic rocks and orthomagmatic mafic melts during the formation of mafic-ultramafic massifs on Sakhalin Island. This hypothesis confirms the model of polygenic formation of these massifs. The extremely low ^{87}Sr/^{86}Sr ratios suggest that the upper-mantle mafic melts from which gabbroids of massifs of Sakhalin Island crystallized were only slightly contaminated with crustal rocks.

Chapter 8

Chemical composition of rock-forming and accessory minerals from rocks of mafic-ultramafic massifs of the East Sakhalin ophiolite association

This chapter presents results of the first systematic studies of the chemical composition of rock-forming and accessory minerals from the most widespread ultramafic rocks and gabbroids of the Berezovka, Shel'ting, Komsomol'sk, and South Schmidt massifs (Table 8.1). Attention was focused not only on the research into the regularities of concentration of chemical components in minerals but also on their distribution between coexisting phases. Most of mineral analyses were carried out in transparent rock plates by the X-ray spectral method, and the other were made in monomineral fractions, using a wet chemical technique. More than 300 chemical analyses of olivines, orthopyroxenes, clinopyroxenes, plagioclases, amphiboles, zircons, Cr-spinels, magnetites, ilmenites, and sulfides from 52 samples of mafic rocks were performed.

8.1 ROCK-FORMING MINERALS

8.1.1 Olivines

The composition of olivines from harzburgites, lherzolites, plagiolherzolites, wehrlites, plagiowehrlites, olivine websterites, olivine orthopyroxenites, olivine gabbronorites, and olivine gabbro was studied (Table 8.2). The contents of FeO and MgO in all olivines vary over a wide range of values, showing a clear inverse relationship due to isomorphous mutual substitution of these components (Fig. 8.1). The maximum MgO and, correspondingly, minimum FeO contents were found in olivines from lherzolites of the South Schmidt massif. Slightly lower contents of MgO are observed in the mineral from harzburgites, plagiowehrlites, and olivine websterites of the Berezovka massif and from olivine orthopyroxenites of the Shel'ting massif. The minimum MgO and maximum FeO contents were detected in olivines from olivine gabbronorites and olivine gabbro.

The analyzed olivine samples are characterized by $Mg^{\#} = Mg/(Mg+Fe)$ varying from 0.91 (ferran forsterite from lherzolites) to 0.73 (chrysolite from olivine gabbro and olivine-containing orthopyroxenites). Chemical analysis of small series of olivine grains from several rock samples showed their constant or negligibly varying $Mg^{\#}$ value. For example, different olivine grains from harzburgite and plagiolherzolite show a constant $Mg^{\#}$ value, 0.86 and 0.91, respectively. In different olivine grains from olivine websterite, olivine orthopyroxenite, and olivine gabbro its values differ by hundredths of unit: 0.83–0.84, 0.73–0.74, and 0.73–0.74, respectively.

Table 8.1 Chemical composition of coexisting minerals from rocks of the Berezovka, Shel'ting, Komsomol'sk, and South Schmidt massifs, wt.%.

Sample	Rock	Mineral	SiO$_2$	TiO$_2$	Al$_2$O$_3$	Cr$_2$O$_3$	Fe$_2$O$_3$	FeO	MnO	MgO	CaO	Na$_2$O	K$_2$O	NiO	H$_2$O	Total	Mg# (An, %, Cr#)
1	2	3	4	5	6	7	8	9	10	11	12	13	14	15	16	17	18
Berezovka massif																	
133	Cpx Harz	Ol*	39.55	N.d.	N.d.	N.d.	N.d.	17.46	0.25	42.25	0.02	N.d.	N.d.	0.31	N.d.	99.84	0.81
"	"	Ol*	39.61	N.d.	N.d.	N.d.	N.d.	18.02	0.27	42.82	0.01	N.d.	N.d.	0.25	N.d.	100.98	0.81
"	"	Ol*	38.78	N.d.	N.d.	N.d.	N.d.	17.68	0.23	43.06	0.01	N.d.	N.d.	0.27	N.d.	100.03	0.81
"	"	Opx*	55.65	0.02	0.86	0.16	N.d.	11.84	0.30	30.68	0.41	N.d.	N.d.	N.d.	N.d.	99.92	0.82
"	"	Cpx*	54.29	0.01	0.91	0.25	N.d.	3.52	0.07	17.50	24.04	0.07	N.d.	N.d.	N.d.	100.66	0.89
"	"	Cpx*	53.69	0.04	1.03	0.15	N.d.	4.19	0.12	16.23	23.44	0.13	N.d.	N.d.	N.d.	99.02	0.87
"	"	Cpx*	54.00	0.02	1.03	0.13	N.d.	4.25	0.11	17.47	23.56	0.12	N.d.	N.d.	N.d.	100.71	0.88
"	"	Srp*	40.67	N.d.	0.01	0.01	N.d.	10.22	0.11	36.14	0.07	N.d.	0.01	N.d.	12.00	99.24	0.86
"	"	Srp*	38.86	0.02	0.49	0.07	N.d.	14.31	0.31	34.57	0.16	N.d.	N.d.	N.d.	12.00	100.90	0.81
133-2a	Harz	Ol	39.27	N.d.	N.d.	N.d.	N.d.	17.81	0.23	42.64	0.02	N.d.	N.d.	0.26	N.d.	100.23	0.81
1610-2	Harz	Ol*	39.85	0.026	0.001	0.001	N.d.	13.45	0.216	46.96	0.005	N.f.	0.002	N.d.	N.d.	100.51	0.86
"	"	Ol*	39.88	0.011	0.003	N.f.	N.d.	13.34	0.194	46.92	0.016	0.032	N.f.	N.d.	N.d.	100.39	0.86
"	"	Ol*	39.57	N.f.	0.002	0.286	N.d.	13.30	0.203	46.23	0.014	0.003	0.004	N.d.	N.d.	99.61	0.86
"	"	Ol*	39.76	0.015	0.003	0.063	N.d.	13.42	0.226	46.46	0.007	0.028	N.f.	N.d.	N.d.	99.97	0.86
"	"	Opx*	55.62	0.030	1.69	0.512	N.d.	8.71	0.208	31.55	2.06	0.003	N.f.	N.d.	N.d.	100.38	0.87
"	"	Opx*	55.25	0.030	1.50	0.634	N.d.	8.67	0.239	31.16	1.77	N.f.	N.f.	N.d.	N.d.	99.25	0.87
"	"	Opx*	55.97	0.021	1.53	0.437	N.d.	8.96	0.220	32.42	0.377	0.025	N.f.	N.d.	N.d.	99.96	0.87
"	"	Opx*	55.56	0.037	1.65	0.493	N.d.	9.04	0.223	32.14	0.548	N.f.	0.006	N.d.	N.d.	99.68	0.86
"	"	CrSp*	0.005	0.149	18.27	37.18	N.d.	35.65	0.325	8.08	0.007	0.006	N.f.	N.d.	N.d.	99.68	Cr#-0.58
"	"	CrSp*	N.f.	0.187	17.24	32.25	N.d.	42.92	0.433	6.09	N.f.	0.015	N.f.	N.d.	N.d.	99.16	Cr#-0.56
1596	Harz	Ol*	40.36	N.d.	N.f.	0.01	N.d.	13.49	0.19	46.58	0.01	0.02	N.d.	0.24	N.d.	100.90	0.86
"	"	Ol*	40.06	N.d.	N.f.	0.02	N.d.	13.78	0.24	45.75	N.f.	N.f.	N.d.	0.25	N.d.	100.66	0.86
"	"	Ol*	40.30	N.d.	N.f.	0.01	N.d.	13.77	0.22	46.81	0.02	0.02	N.d.	0.25	N.d.	100.31	0.86
"	"	Ol*	40.67	N.d.	N.f.	0.01	N.d.	13.59	0.23	46.56	0.01	0.05	N.d.	0.26	N.d.	101.37	0.86
"	"	Ol*	39.63	N.d.	0.01	N.f.	N.d.	13.51	0.23	47.53	N.f.	N.f.	N.d.	0.26	N.d.	101.18	0.86
"	"	Ol*	40.34	N.d.	0.02	0.01	N.d.	13.54	0.23	46.28	0.02	N.f.	N.d.	0.24	N.d.	100.69	0.86
"	"	Cpx*	54.06	0.01	0.89	0.14	N.d.	2.76	0.13	16.76	24.37	0.02	N.f.	N.d.	N.d.	99.14	0.92
"	"	Cpx*	52.23	0.08	2.43	0.88	N.d.	3.40	0.11	16.37	23.60	0.10	0.01	N.d.	N.d.	99.20	0.90

"	"	Cpx*	53.88	0.08	1.21	0.26	N.d.	2.87	0.11	17.13	24.27	0.05	N.f.	N.d.	N.d.	99.87	0.92
"	"	Amph*	46.28	0.46	10.69	1.37	N.d.	5.85	0.08	17.65	12.43	1.28	0.30	N.d.	2.00	98.37	0.84
"	"	Amph*	48.29	0.34	9.28	1.12	N.d.	5.64	0.09	18.44	12.61	1.08	0.19	N.d.	2.00	99.08	0.85
"	"	Amph*	46.71	0.34	8.93	1.06	N.d.	6.22	0.07	20.20	11.37	1.00	0.21	N.d.	2.00	98.11	0.85
"	"	CrSp*	N.f.	0.34	11.22	30.75	N.d.	51.16	0.43	4.68	0.01	0.03	N.f.	N.d.	N.d.	98.62	$Cr^{\#}$−0.65
"	"	Ilm	0.04	46.54	0.49	1.86	N.d.	40.52	0.23	9.10	N.f.	0.02	N.d.	0.11	N.d.	98.90	—
138	Lherz	Ol	39.88	N.d.	N.d.	N.d.	N.d.	16.52	0.25	43.96	0.02	N.d.	N.d.	0.06	0.004	100.53	0.83
"	"	Opx*	50.00	0.08	3.03	0.38	N.d.	12.23	0.14	30.88	1.43	N.d.	N.d.	N.d.	N.d.	98.23	0.91
"	"	Opx*	55.74	0.05	1.91	0.19	N.d.	9.66	0.25	31.65	0.59	0.02	N.d.	N.d.	N.d.	100.21	0.85
"	"	Cpx*	52.64	0.19	2.53	0.29	N.d.	4.22	0.15	16.40	23.46	0.07	N.d.	N.d.	N.d.	99.95	0.87
138	Lherz	Cpx*	48.90	0.21	3.46	0.46	N.d.	1.94	0.11	16.47	23.86	0.29	N.d.	0.04	0.04	98.88	0.86
154	Lherz	Opx*	55.80	N.d.	2.45	N.d.	N.d.	7.53	N.d.	33.93	0.58	N.d.	N.d.	N.d.	N.d.	100.29	0.89
"	"	Cpx*	53.57	0.04	2.79	N.d.	N.d.	2.79	N.d.	17.95	24.58	0.10	N.d.	N.d.	N.d.	101.82	0.92
"	"	Cpx*	50.09	N.d.	3.31	N.d.	N.d.	4.09	0.09	15.43	25.66	0.14	N.d.	N.d.	N.d.	98.81	0.87
142	PI Lherz	Ol*	40.27	0.004	0.045	0.118	N.d.	9.04	0.149	48.97	0.036	0.077	0.018	N.d.	N.d.	98.73	0.91
"	"	Ol*	41.03	N.d.	N.d.	N.d.	N.d.	9.17	0.150	49.24	0.010	N.d.	0.005	N.d.	N.d.	99.60	0.91
"	"	Ol*	40.67	N.d.	0.012	0.008	N.d.	9.03	0.145	49.42	0.009	0.050	0.014	N.d.	N.d.	99.36	0.91
"	"	Opx	51.80	0.12	0.86	N.d.	0.96	14.42	0.43	27.62	1.67	N.d.	N.d.	0.06	N.d.	97.94	0.76
"	"	Opx*	55.01	0.07	0.79	N.d.	N.d.	14.23	N.d.	29.12	0.46	N.d.	N.d.	N.d.	N.d.	99.68	0.78
"	"	Cpx	53.83	0.01	2.40	N.d.	2.53	N.d.	N.d.	17.22	23.64	0.17	N.d.	N.d.	N.d.	99.61	0.93
"	"	Cpx*	51.85	0.059	2.58	0.930	N.d.	2.48	0.091	16.44	22.41	0.098	0.004	N.d.	0.004	96.95	0.92
"	"	Cpx*	52.10	0.049	2.63	0.966	N.d.	2.47	0.090	16.43	22.72	0.144	N.d.	N.d.	N.d.	97.60	0.92
"	"	Cpx*	52.48	0.039	2.30	0.378	N.d.	2.48	0.106	17.02	21.65	0.139	0.022	N.d.	0.022	96.62	0.92
"	"	Cpx*	52.96	0.035	2.18	0.384	N.d.	2.47	0.115	16.59	22.37	0.230	0.008	N.d.	0.008	97.33	0.92
"	"	CrSp*	0.042	0.011	38.10	24.77	N.d.	20.27	0.223	14.20	0.025	0.022	0.018	N.d.	0.018	97.68	$Cr^{\#}$−0.30
"	"	CrSp*	0.043	0.046	37.61	25.49	N.d.	21.06	0.217	13.77	N.d.	0.049	N.d.	N.d.	N.d.	98.49	$Cr^{\#}$−0.31
139	Wehrl	Ol	40.02	N.d.	N.d.	N.d.	N.d.	13.15	0.27	46.34	0.03	0.04	N.d.	N.d.	0.13	99.93	0.86
"	"	Cpx*	54.41	0.03	1.68	0.30	N.d.	2.92	0.10	17.92	22.51	0.04	N.d.	N.d.	N.d.	99.90	0.92
"	"	Cpx*	54.74	0.05	1.94	0.027	N.d.	2.73	0.07	18.79	20.75	0.06	N.d.	N.d.	N.d.	99.40	0.92
"	"	Cpx*	54.41	0.03	1.70	0.26	N.d.	2.79	0.08	18.34	21.63	0.01	N.d.	N.d.	N.d.	99.26	0.93
"	"	Cpx*	54.03	0.04	1.64	0.40	N.d.	3.21	0.12	17.26	23.67	0.05	N.d.	N.d.	N.d.	100.43	0.91
"	"	Cpx*	54.45	N.d.	1.42	0.26	N.d.	2.93	0.13	17.29	23.99	0.04	N.d.	N.d.	N.d.	100.51	0.91
"	"	Cpx*	53.00	0.11	2.37	0.67	N.d.	2.18	0.08	17.37	23.79	0.06	N.d.	N.d.	N.d.	99.63	0.93
136	PI Wehrl	Srp*	37.97	N.d.	N.d.	N.d.	N.d.	6.81	0.12	35.67	0.25	N.d.	N.d.	0.38	14.00	95.20	0.90
"	"	CrSp*	N.d.	0.05	32.17	29.23	N.d.	26.11	0.32	13.23	0.03	0.04	N.d.	N.d.	N.d.	101.19	$Cr^{\#}$−0.38

(Continued)

Table 8.1 Continued.

Sample 1	Rock 2	Mineral 3	SiO2 4	TiO2 5	Al2O3 6	Cr2O3 7	Fe2O3 8	FeO 9	MnO 10	MgO 11	CaO 12	Na2O 13	K2O 14	NiO 15	H2O 16	Total 17	Mg# (An, %, Cr#) 18
145	Ol Web	Opx	55.30	0.15	1.10	0.19	4.02	7.54	0.43	28.57	1.38	N.d.	N.d.	0.03	N.d.	98.71	0.82
-"-	-"-	Opx*	55.69	0.064	1.57	0.266	N.d.	11.13	0.278	30.64	0.646	N.f.	0.011	N.d.	N.d.	100.30	0.89
-"-	-"-	Opx*	55.09	0.047	1.42	0.207	N.d.	11.00	0.283	30.45	0.827	0.029	0.011	N.d.	N.d.	99.36	0.88
-"-	-"-	Cpx	52.90	0.05	1.30	0.28	1.53	3.36	0.14	17.60	21.71	0.23	N.d.	0.01	N.d.	99.11	0.87
-"-	-"-	Cpx*	53.60	0.05	1.63	N.d.	N.d.	4.10	N.d.	16.85	23.65	0.10	N.d.	N.d.	N.d.	99.98	0.88
-"-	-"-	Cpx*	53.09	0.08	1.47	0.29	N.d.	3.70	0.122	16.40	24.25	0.068	0.001	N.d.	N.d.	99.48	0.83
-"-	-"-	Cpx*	52.76	0.072	1.82	0.432	N.d.	4.00	0.15	16.65	23.40	0.080	0.010	N.d.	N.d.	99.42	0.83
-"-	-"-	CrSp*	0.019	1.30	7.69	18.58	N.d.	60.35	0.399	2.68	0.011	0.019	0.004	N.d.	N.d.	91.06	Cr#-0.71
-"-	-"-	CrSp*	0.032	1.48	8.44	21.05	N.d.	58.25	0.390	2.82	0.004	N.d.	0.014	N.d.	N.d.	92.48	Cr#-0.71
155	Ol Web	Ol*	39.82	N.d.	N.d.	N.d.	N.d.	15.74	0.18	44.26	0.011	N.d.	N.d.	0.26	N.d.	100.27	0.83
-"-	-"-	Ol*	39.86	N.d.	N.d.	N.d.	N.d.	15.63	0.20	44.46	0.004	N.d.	N.d.	0.28	N.d.	100.47	0.84
-"-	-"-	Opx	53.50	0.13	0.80	0.30	1.15	8.43	0.50	30.23	0.92	N.d.	N.d.	0.04	N.d.	98.00	0.85
-"-	-"-	Cpx	52.90	N.d.	1.20	0.45	0.11	2.96	0.14	18.27	22.64	0.27	N.d.	0.01	N.d.	98.95	0.91
-"-	-"-	Cpx*	53.40	0.08	1.50	0.45	N.d.	3.67	0.12	16.30	23.33	0.13	N.d.	N.d.	N.d.	98.97	0.89
134	Ol Web	Opx	55.00	N.d.	1.35	0.07	0.42	9.98	0.18	30.56	1.38	N.d.	N.d.	0.04	N.d.	98.98	0.84
-"-	-"-	Cpx	53.83	0.06	1.50	0.19	0.09	4.07	0.13	17.27	21.72	0.14	N.d.	0.02	N.d.	99.02	0.88
-"-	-"-	Cpx*	49.68	N.d.	1.79	0.89	N.d.	2.44	0.04	17.36	24.12	0.24	N.d.	N.d.	N.d.	96.56	0.93
137	Ol Web	Opx	55.00	N.d.	1.75	0.17	0.27	10.05	0.20	30.89	0.92	0.16	N.d.	0.03	N.d.	99.28	0.84
-"-	-"-	Cpx	53.00	0.06	2.10	0.33	0.09	4.07	0.07	17.94	21.25	0.16	N.d.	0.03	N.d.	99.10	0.89
140	Ol Web	Ol	39.74	N.d.	N.d.	N.d.	N.d.	17.17	0.29	43.70	0.020	N.d.	N.d.	N.d.	N.d.	100.49	0.82
-"-	-"-	Opx	52.20	0.07	1.51	0.43	1.71	9.77	0.21	29.89	2.86	N.d.	N.d.	0.02	N.d.	98.67	0.86
-"-	-"-	Opx *	54.42	0.10	2.13	N.d.	N.d.	11.54	0.31	31.30	0.53	N.d.	N.d.	N.d.	N.d.	100.33	0.83
-"-	-"-	Opx *	54.71	0.02	2.09	0.33	N.d.	11.57	0.28	30.82	0.49	0.03	0.01	N.d.	N.d.	100.32	0.83
-"-	-"-	Cpx	50.40	0.10	2.37	0.38	1.22	3.48	0.14	16.98	23.38	0.25	N.d.	0.01	N.d.	98.71	0.87
-"-	-"-	Cpx*	52.28	0.08	2.41	0.49	N.d.	4.30	0.14	16.41	23.58	0.11	N.f.	N.d.	N.d.	99.79	0.87
1603-1	Ol Web	Opx*	55.46	0.039	1.74	0.264	N.d.	10.30	0.260	31.60	0.506	0.010	0.010	N.d.	N.d.	100.17	0.85
-"-	-"-	Opx*	55.98	0.039	1.63	0.249	N.d.	10.37	0.271	31.35	0.567	0.012	0.006	N.d.	N.d.	100.47	0.84
-"-	-"-	Opx*	55.70	0.028	1.53	0.242	N.d.	10.53	0.272	31.20	0.591	0.009	N.f.	N.d.	N.d.	100.11	0.84
-"-	-"-	Opx*	55.94	0.069	1.65	0.220	N.d.	10.65	0.263	31.14	0.425	0.010	0.006	N.d.	N.d.	100.37	0.84
-"-	-"-	Opx *	55.24	0.017	1.62	0.290	N.d.	11.13	0.295	30.94	0.388	N.d.	0.004	N.d.	N.d.	99.93	0.83
-"-	-"-	Opx*	55.25	0.047	1.49	0.280	N.d.	10.78	0.275	31.25	0.528	0.007	0.006	N.d.	N.d.	99.92	0.84
1603-1	Ol Web	Cpx*	52.93	0.107	2.00	0.459	N.d.	4.04	0.123	16.25	23.27	0.089	N.d.	N.d.	N.d.	99.47	0.88
-"-	-"-	Cpx*	52.97	0.100	2.14	0.495	N.d.	3.73	0.154	15.94	24.05	0.042	N.d.	N.d.	N.d.	99.61	0.88

Sample	Rock	Mineral														Total	
1593	Ol Web	Cpx*	52.90	0.07	2.63	0.95	N.d.	3.62	0.12	15.65	23.72	0.13	N.d.	N.d.	N.d.	99.78	0.87
-"-	-"-	Cpx*	52.68	0.06	2.71	0.82	N.d.	3.73	0.11	15.71	23.49	0.12	N.d.	N.d.	N.d.	99.41	0.88
-"-	-"-	Cpx*	52.95	0.07	2.42	0.65	N.d.	3.80	0.12	16.17	23.11	0.13	N.d.	N.d.	N.d.	99.41	0.88
-"-	-"-	Cpx*	52.97	0.07	2.45	0.64	N.d.	3.42	0.11	15.77	23.84	0.10	N.d.	N.d.	N.d.	99.37	0.89
-"-	-"-	Cpx*	52.57	0.09	2.51	0.70	N.d.	3.31	0.10	15.81	23.67	0.06	N.d.	N.d.	N.d.	98.80	0.90
-"-	-"-	Cpx*	52.59	0.07	2.63	0.72	N.d.	3.58	0.12	15.79	23.31	0.12	N.d.	N.d.	N.d.	98.93	0.89
-"-	-"-	Cpx*	53.05	0.10	2.64	0.70	N.d.	3.74	0.12	15.66	23.32	0.13	0.01	N.d.	N.d.	99.47	0.88
143	Web	Opx*	54.86	0.065	0.876	0.089	N.d.	15.06	0.413	27.96	0.821	0.036	0.011	N.d.	N.d.	100.18	0.77
-"-	-"-	Opx*	53.78	0.078	0.812	0.104	N.d.	15.36	0.426	28.33	0.583	N.d.	N.d.	N.d.	N.d.	99.47	0.77
-"-	-"-	Opx*	54.11	0.077	0.719	0.083	N.d.	15.27	0.454	28.21	0.572	0.014	0.003	N.d.	N.d.	99.52	0.77
-"-	-"-	Cpx*	51.85	0.09	1.29	N.d.	N.d.	5.40	N.d.	15.00	23.03	0.18	N.d.	N.d.	N.d.	96.84	0.83
-"-	-"-	Cpx*	53.27	0.117	1.22	0.216	N.d.	5.57	0.179	15.57	22.97	0.225	0.010	N.d.	N.d.	99.36	0.83
-"-	-"-	Cpx*	53.53	0.112	1.14	0.214	N.d.	5.60	0.208	15.46	23.03	0.205	0.002	N.d.	N.d.	99.60	0.83
-"-	-"-	Cpx*	53.37	0.094	1.169	0.121	N.d.	5.43	0.181	15.33	23.16	0.269	N.d.	N.d.	N.d.	99.12	0.83
-"-	-"-	Cpx*	53.20	0.143	1.429	0.198	N.d.	7.13	0.220	16.75	20.30	0.239	N.d.	N.d.	N.d.	99.61	0.81
-"-	-"-	Mt*	0.036	0.438	0.413	2.70	N.d.	90.28	0.166	0.078	0.153	0.105	N.d.	N.d.	N.d.	94.37	—
-"-	-"-	Mt*	0.010	0.371	0.316	1.96	N.d.	91.13	0.200	0.020	0.090	0.108	N.d.	N.d.	N.d.	94.21	—
-"-	-"-	Mt*	0.071	0.425	0.439	2.78	1.72	90.15	0.150	0.011	0.153	0.067	N.d.	N.d.	N.d.	94.25	—
150	Web	Cpx	51.50	N.d.	2.40	0.17	N.d.	4.66	0.21	14.95	23.10	0.24	0.01	N.d.	N.d.	98.96	0.81
152	Web	Opx*	56.58	0.01	1.09	0.33	N.d.	6.63	0.17	34.19	0.22	0.03	N.d.	N.d.	N.d.	99.22	0.90
-"-	-"-	Cpx*	53.87	0.04	1.86	N.d.	N.d.	2.03	N.d.	17.55	24.30	0.20	N.d.	N.d.	N.d.	99.85	0.94
-"-	-"-	Cpx*	53.98	0.02	1.41	0.76	N.d.	2.12	0.08	17.48	24.13	0.14	0.01	0.01	N.d.	100.10	0.94
154-2	Web	Cpx*	51.52	N.d.	1.80	N.d.	N.d.	4.33	0.09	16.20	26.08	0.16	0.011	N.d.	N.d.	100.16	0.87
Cx-2	Ol Cpx	Cpx*	52.90	0.040	2.54	1.05	N.d.	3.01	0.094	17.02	23.25	0.137	0.011	N.d.	N.d.	100.06	0.91
-"-	-"-	Cpx*	52.56	0.031	2.45	0.938	N.d.	3.05	0.079	17.60	23.45	0.099	N.f.	N.d.	N.d.	99.26	0.91
-"-	-"-	Cpx*	53.05	0.033	2.49	0.910	N.d.	3.12	0.087	17.17	23.54	0.178	N.f.	N.d.	N.d.	100.58	0.91
-"-	-"-	Cpx*	53.65	0.045	1.77	1.45	N.d.	2.66	0.92	16.76	23.68	0.065	N.d.	N.d.	N.d.	100.18	0.92
-"-	-"-	Cpx*	51.47	0.037	2.54	3.12	N.d.	3.09	0.066	16.16	22.72	0.124	0.005	N.d.	N.d.	99.33	0.90
-"-	-"-	Cpx*	52.48	0.039	2.28	0.982	N.d.	2.94	0.093	16.74	23.46	0.106	0.005	N.d.	N.d.	99.23	0.91
1606-1	Ol Cpx	Amph	44.66	0.52	12.72	0.64	N.d.	7.14	0.16	17.96	11.32	2.31	0.21	N.d.	2.00	99.64	0.82
135	Cortl	Srp*	51.74	N.d.	N.d.	N.d.	N.d.	2.77	0.04	30.48	0.04	N.d.	N.d.	0.06	14.00	99.13	0.95
155-3	Ol Gb	Ol*	38.38	N.d.	0.007	0.016	N.d.	20.64	0.336	40.61	0.018	0.022	N.d.	N.d.	N.d.	100.02	0.78
-"-	-"-	Ol*	38.60	0.007	0.013	0.010	N.d.	20.87	0.333	40.31	0.013	0.039	N.d.	N.d.	N.d.	100.19	0.77
-"-	-"-	Pl*	45.46	N.d.	34.98	N.d.	N.d.	0.342	0.003	N.d.	18.23	0.647	0.020	N.d.	N.d.	99.68	An-93.9
-"-	-"-	Pl*	44.98	0.010	34.63	0.003	N.d.	0.411	N.d.	N.d.	18.29	0.732	0.012	N.d.	N.d.	99.07	An-93.2
-"-	-"-	Cpx*	51.79	0.047	1.63	0.057	N.d.	5.65	0.156	15.79	21.40	0.179	0.005	N.d.	N.d.	96.69	0.83
-"-	-"-	Mt*	0.051	0.581	0.855	3.35	N.d.	88.82	0.175	0.479	0.056	0.040	0.005	N.d.	N.d.	94.41	—
-"-	-"-	Mt*	0.064	0.249	0.597	2.30	N.d.	89.22	0.171	0.564	0.046	0.027	0.007	N.d.	N.d.	93.24	—

(Continued)

Table 8.1 Continued.

Sample 1	Rock 2	Mineral 3	SiO₂ 4	TiO₂ 5	Al₂O₃ 6	Cr₂O₃ 7	Fe₂O₃ 8	FeO 9	MnO 10	MgO 11	CaO 12	Na₂O 13	K₂O 14	NiO 15	H₂O 16	Total 17	Mg# (An, %, Cr#) 18
155-5	Ol Gb	Ol*	38.01	N.f.	0.006	N.f.	N.d.	24.61	0.416	36.88	0.022	0.035	N.f.	N.d.	N.d.	99.98	0.73
''	''	Ol*	37.99	N.f.	0.012	N.f.	N.d.	23.85	0.371	37.25	0.023	N.f.	N.f.	N.d.	N.d.	99.50	0.74
''	''	Ol*	37.89	N.f.	0.012	N.f.	N.d.	24.36	0.394	36.82	0.023	0.001	N.f.	N.d.	N.d.	99.49	0.73
''	''	Ol*	37.77	N.f.	0.008	0.027	N.d.	23.52	0.387	37.61	0.023	0.036	N.f.	N.d.	N.d.	99.37	0.74
''	''	Cpx*	51.22	0.295	2.69	0.179	N.d.	6.88	0.200	15.51	22.11	0.232	N.f.	N.d.	N.d.	99.31	0.80
''	''	Cpx*	51.53	0.320	2.62	0.226	N.d.	8.64	0.232	16.57	19.42	0.172	0.005	N.d.	N.d.	99.74	0.77
''	''	Cpx*	52.34	0.228	2.23	0.108	N.d.	8.42	0.63	16.94	18.43	0.154	0.005	N.d.	N.d.	99.13	0.78
''	''	Amph*	53.50	0.16	1.87	0.119	N.d.	12.04	0.304	19.79	12.25	0.110	N.f.	N.d.	N.d.	100.15	0.75
''	''	Pl*	45.38	0.019	34.07	0.005	N.d.	0.315	0.003	N.f.	17.93	1.17	0.012	N.d.	N.d.	98.90	An–89.4
''	''	Pl*	45.39	0.016	34.68	0.022	N.d.	0.381	N.f.	N.f.	18.53	1.00	0.018	N.d.	N.d.	100.03	An–91.0
''	''	Pl*	43.63	0.011	35.28	0.019	N.d.	0.218	0.007	N.f.	19.32	0.42	0.016	N.d.	N.d.	98.92	An–96.1
''	''	Pl*	45.64	0.006	34.16	0.015	N.d.	0.353	N.f.	N.f.	18.17	1.20	0.014	N.d.	N.d.	99.56	An–89.3
148-1	Ol Gb	Opx*	54.23	0.052	1.63	0.221	N.d.	15.23	0.318	27.88	0.633	0.051	N.f.	N.d.	N.d.	100.24	0.77
''	''	Opx*	54.16	0.036	1.69	0.235	N.d.	16.20	0.377	26.90	0.685	0.050	0.001	N.d.	N.d.	100.34	0.75
''	''	Cpx*	52.43	0.081	1.97	0.353	N.d.	5.34	0.131	15.87	23.06	0.152	0.010	N.d.	N.d.	99.40	0.84
''	''	Cpx*	52.67	0.024	1.74	0.373	N.d.	5.29	0.142	15.99	23.25	0.207	0.002	N.d.	N.d.	99.68	0.84
''	''	Amph*	48.94	0.080	8.33	0.468	N.d.	8.26	0.131	17.49	12.94	1.26	0.102	N.d.	2.00	99.39	0.84
''	''	Amph*	50.66	0.052	6.83	0.329	N.d.	7.40	0.127	18.45	12.25	1.16	0.083	N.d.	2.00	99.34	0.85
''	''	Pl*	44.72	N.f.	35.00	0.038	N.d.	0.218	0.002	N.f.	18.97	0.792	0.013	N.d.	N.d.	99.75	An–92.9
''	''	Pl*	43.96	N.f.	35.41	0.021	N.d.	0.109	0.007	N.f.	19.40	0.529	N.f.	N.d.	N.d.	99.44	An–95.3
''	''	CrSp*	0.059	1.46	9.54	28.75	N.d.	51.13	0.313	2.25	0.137	N.f.	0.005	N.d.	N.d.	99.64	Cr# = 0.67
''	''	Sulf*	N.f.	N.f.	0.230	0.275	N.d.	40.97	N.f.	N.f.	0.383	0.381	0.063	N.d.	N.d.	42.67	—
''	''	Sulf*	N.f.	0.010	0.018	0.049	N.d.	34.96	N.f.	N.f.	0.154	0.062	0.008	N.d.	N.d.	35.33	—
128	Gbn	Cpx	53.00	0.07	2.62	0.16	0.01	5.32	N.d.	15.45	22.18	0.30	N.d.	0.03	N.d.	99.30	0.84
''	''	Pl	44.10	N.d.	34.00	N.d.	0.07	0.55	N.d.	N.d.	18.53	1.20	0.30	N.d.	1.28	100.03	An–88.2
130	Gbn	Opx*	55.07	0.03	1.35	0.07	N.d.	14.24	0.33	27.42	0.84	0.01	0.01	N.d.	N.d.	99.34	0.77
''	''	Opx*	55.04	0.03	1.39	0.06	N.d.	14.27	0.33	27.25	086	0.02	0.02	N.d.	N.d.	99.24	0.77
''	''	Opx*	55.19	0.03	1.23	0.08	N.d.	14.22	0.31	27.64	0.60	N.f.	N.f.	N.d.	N.d.	99.30	0.77
''	''	Opx*	54.65	0.03	1.47	0.07	N.d.	13.82	0.33	27.19	1.27	0.01	N.f.	N.d.	N.d.	98.85	0.78

Sample	Rock	Mineral														Total	An/Mg#
,,	,,	Opx*	55.39	0.03	1.31	0.08	N.d.	14.63	0.34	27.58	0.63	0.01	0.01	N.d.	N.d.	100.02	0.77
,,	,,	Cpx*	53.14	0.03	1.56	0.08	N.d.	5.19	0.14	15.99	22.99	0.13	N.d.	N.d.	N.d.	99.25	0.85
,,	,,	Cpx	54.03	0.01	2.02	N.d.	0.20	4.81	0.14	15.95	21.72	0.30	N.d.	0.03	N.d.	99.21	0.85
,,	,,	Pl	44.26	N.d.	34.96	N.d.	0.46	0.11	N.d.	0.55	16.21	1.05	0.64	N.d.	1.80	100.04	An–85.7
,,	,,	Amph*	50.55	0.18	7.48	0.10	N.d.	7.63	0.11	17.40	12.15	0.96	0.09	N.d.	2.00	98.65	0.80
,,	,,	Amph*	50.51	0.18	7.42	0.18	N.d.	7.70	0.11	16.96	11.99	0.91	0.06	N.d.	2.00	98.02	0.80
,,	,,	Cpx	54.03	0.05	2.16	0.03	0.21	4.66	0.14	15.45	22.18	0.30	N.d.	0.03	N.d.	99.24	0.85
,,	,,	Pl	43.30	N.d.	33.23	N.d.	N.d.	0.55	N.d.	0.74	14.67	1.56	1.45	N.d.	4.00	99.50	An–75.7
131a	Gbn	Cpx*	54.50	0.06	2.12	N.d.	N.d.	4.53	N.d.	16.91	20.26	0.17	N.d.	N.d.	N.d.	98.55	0.87
,,	,,	Pl	42.78	N.d.	35.17	N.d.	0.65	0.21	N.d.	0.37	17.75	0.76	0.24	N.d.	2.16	100.09	An–91.4
132a	Gbn	Cpx*	52.46	0.14	2.20	N.d.	N.d.	4.60	0.20	16.81	23.33	0.19	N.d.	N.d.	N.d.	99.93	0.87
,,	,,	Pl	42.66	N.d.	35.39	N.d.	0.71	0.15	N.d.	0.37	17.75	0.76	0.42	N.d.	2.32	100.53	An–90.3
138a	Gbn	Opx	54.96	0.06	2.77	N.d.	N.d.	11.20	0.28	30.90	0.81	0.03	N.d.	N.d.	N.d.	101.01	0.85
,,	,,	Cpx*	50.97	0.12	3.02	0.62	N.d.	5.57	0.17	16.51	23.04	0.23	N.d.	N.d.	N.d.	98.81	0.84
141	Gbn	Opx*	55.49	N.d.	2.17	N.d.	N.d.	10.35	N.d.	31.89	0.61	0.02	N.d.	N.d.	N.d.	100.53	0.85
,,	,,	Cpx	49.90	0.12	3.02	0.62	2.96	1.80	0.11	16.81	23.38	0.43	0.20	0.01	N.d.	99.16	0.87
,,	,,	Pl	41.51	N.d.	33.66	N.d.	0.34	0.09	N.d.	0.37	18.57	1.30	N.d.	N.d.	3.76	99.80	An–87.8
144	Gbn	Opx	51.20	0.40	2.59	0.05	1.31	12.58	0.28	28.82	1.43	N.d.	N.d.	N.d.	N.d.	98.70	0.79
,,	,,	Cpx	53.82	0.12	2.46	N.d.	N.d.	4.76	N.d.	16.26	22.17	0.21	N.d.	N.d.	N.d.	99.80	0.86
,,	,,	Pl	43.40	N.d.	34.53	N.d.	0.31	0.24	N.d.	0.74	16.98	1.54	0.64	N.d.	2.12	100.50	An–83.5
146	Gbn	Opx	54.60	N.d.	1.20	N.d.	5.22	9.25	0.43	25.90	0.92	N.d.	N.d.	0.03	N.d.	97.55	0.77
,,	,,	Cpx	53.30	N.d.	2.33	0.15	0.61	4.58	0.21	15.61	22.18	0.35	N.d.	0.01	N.d.	99.33	0.84
,,	,,	Pl	43.50	N.d.	33.90	N.d.	0.19	0.55	N.d.	0.55	16.47	1.55	0.20	N.d.	1.90	98.81	An–84.0
147	Gbn	Opx*	56.04	0.03	0.96	N.d.	1.51	12.97	N.d.	30.30	0.49	0.39	N.d.	N.d.	N.d.	100.79	0.81
,,	,,	Cpx*	50.40	0.12	2.38	0.13	N.d.	4.14	0.11	15.76	23.86	0.39	N.d.	0.01	N.d.	98.81	0.84
,,	,,	Cpx*	54.04	0.08	1.30	N.d.	0.65	4.69	N.d.	16.13	23.33	0.23	N.d.	N.d.	N.d.	100.00	0.86
,,	,,	Pl	43.38	N.d.	34.53	N.d.	N.d.	0.21	N.d.	0.37	16.97	1.34	0.80	0.03	1.68	99.94	An–83.9
149	Gbn	Opx	53.20	N.d.	2.80	0.20	1.79	9.10	0.50	27.90	1.84	N.d.	N.d.	N.d.	N.d.	97.36	0.82
,,	,,	Cpx	52.00	0.26	2.40	0.30	2.47	3.11	N.d.	14.95	23.10	0.32	N.d.	0.01	N.d.	98.92	0.83
151	Gbn	Cpx	53.40	N.d.	1.80	0.18	0.56	3.55	0.21	15.61	2.64	0.34	N.d.	0.01	N.d.	98.30	0.87
,,	,,	Pl	43.22	N.d.	34.52	N.d.	0.27	0.15	N.d.	0.18	17.75	0.84	0.64	N.d.	N.d.	99.55	An–89.3

(Continued)

Table 8.1 Continued.

Sample 1	Rock 2	Mineral 3	SiO₂ 4	TiO₂ 5	Al₂O₃ 6	Cr₂O₃ 7	Fe₂O₃ 8	FeO 9	MnO 10	MgO 11	CaO 12	Na₂O 13	K₂O 14	NiO 15	H₂O 16	Total 17	Mg# (An, %, Cr#) 18
1607	Gbn	Cpx*	54.10	0.024	0.75	0.04	N.d.	2.52	0.164	16.65	24.93	N.f.	N.f.	N.d.	N.d.	99.18	0.92
-"-	-"-	Cpx*	53.93	N.f.	1.84	0.183	N.d.	3.57	0.125	17.02	22.39	0.124	N.f.	N.d.	N.d.	99.18	0.89
-"-	-"-	Cpx*	53.76	0.021	0.73	0.169	N.d.	2.89	0.146	15.75	25.23	0.041	0.006	N.d.	N.d.	98.75	0.91
-"-	-"-	Cpx*	54.22	0.007	1.41	0.248	N.d.	3.18	0.150	16.23	24.77	0.075	N.f.	N.d.	N.d.	100.29	0.90
-"-	-"-	Cpx*	53.48	0.001	1.64	0.158	N.d.	3.37	0.141	16.85	24.60	0.027	N.f.	N.d.	N.d.	100.25	0.90
-"-	-"-	Cpx*	53.88	0.017	0.99	0.137	N.d.	3.09	0.137	15.85	25.34	0.027	0.009	N.d.	N.d.	99.49	0.90
1607	Gbn	Pl*	43.15	0.026	34.16	0.057	N.d.	0.007	N.f.	N.f.	18.99	0.703	0.011	N.d.	N.d.	97.10	An–93.7
-"-	-"-	Pl*	43.32	N.f.	34.85	0.042	N.d.	0.011	N.f.	N.f.	19.70	0.269	0.011	N.d.	N.d.	98.20	An–97.6
-"-	-"-	Pl*	43.05	N.f.	34.92	0.021	N.d.	0.162	0.013	N.f.	20.04	0.192	N.f.	N.d.	N.d.	98.39	An–98.3
-"-	-"-	Pl*	43.60	0.012	35.89	0.073	N.d.	0.107	N.f.	N.f.	19.85	0.251	N.f.	N.d.	N.d.	99.78	An–97.8
-"-	-"-	Pl*	43.47	N.f.	36.27	0.019	N.d.	0.054	N.f.	N.f.	20.04	0.129	N.f.	N.d.	N.d.	99.98	An–98.8
-"-	-"-	Pl*	43.16	N.f.	36.06	0.056	N.d.	0.033	N.f.	N.f.	20.13	0.056	N.f.	N.d.	N.d.	99.49	An–99.5
1605	Gbn	Opx*	55.22	N.f.	1.81	0.078	N.d.	12.25	0.336	30.37	0.441	0.056	0.009	N.d.	N.d.	100.58	0.82
-"-	-"-	Opx*	55.33	0.032	1.80	0.078	N.d.	12.27	0.321	30.19	0.406	0.017	0.006	N.d.	N.d.	100.45	0.81
-"-	-"-	Opx*	55.17	0.011	1.98	0.109	N.d.	11.86	0.287	30.30	0.518	0.017	0.001	N.d.	N.d.	100.22	0.83
-"-	-"-	Opx*	55.14	0.002	1.76	0.103	N.d.	12.03	0.309	9.83	0.682	N.f.	0.002	N.d.	N.d.	99.86	0.82
-"-	-"-	Cpx*	52.97	0.072	2.24	0.171	N.d.	4.37	0.121	16.53	23.62	N.f.	0.104	N.d.	N.d.	100.19	0.87
-"-	-"-	Cpx*	52.80	0.068	2.18	0.360	N.d.	4.42	0.152	16.32	23.13	N.f.	0.163	N.d.	N.d.	99.60	0.87
-"-	-"-	Cpx*	52.58	0.051	2.16	0.209	N.d.	5.11	0.129	15.40	23.59	0.215	0.003	N.d.	N.d.	99.46	0.84
-"-	-"-	Pl*	43.12	0.020	36.01	0.015	N.d.	0.186	N.f.	N.f.	19.58	0.216	N.f.	N.d.	N.d.	99.16	An–98.0
129	Gb	Cpx*	52.96	0.03	2.19	0.17	N.d.	4.52	0.17	16.53	23.59	0.13	0.01	N.d.	N.d.	100.30	0.87
-"-	-"-	Cpx*	53.69	0.10	2.46	0.18	N.d.	4.62	0.14	15.52	22.18	0.14	N.d.	N.d.	N.d.	99.03	0.86
-"-	-"-	Cpx*	52.96	0.03	2.19	0.17	N.d.	4.52	0.17	16.53	23.59	0.13	0.01	N.d.	N.d.	100.30	0.87
-"-	-"-	Amph*	57.18	0.01	1.29	0.03	N.d.	4.07	0.20	22.70	12.94	0.15	N.d.	N.d.	2.00	100.57	0.91
-"-	-"-	Amph*	54.03	0.01	4.84	0.09	N.d.	5.32	0.18	21.54	12.24	0.74	0.02	N.d.	2.00	101.01	0.88
-"-	-"-	Amph*	54.79	0.05	2.86	0.14	N.d.	3.94	0.09	20.60	12.92	0.29	0.01	N.d.	2.00	97.69	0.90
-"-	-"-	Amph*	46.01	0.14	11.86	0.11	N.d.	7.13	0.11	16.74	12.47	1.55	0.06	N.d.	2.00	98.18	0.81

1612	Gb	Cpx*	52.60	0.066	2.01	0.44	N.d.	4.92	0.136	18.60	20.85	0.135	N.d.	N.d.	N.d.	99.74	0.87
,,	,,	Cpx*	53.51	0.070	2.16	0.46	N.d.	4.25	0.147	16.45	23.39	0.154	N.d.	N.d.	N.d.	100.61	0.87
,,	,,	Pl*	44.16	N.f.	35.73	N.f.	N.d.	0.133	0.005	N.f.	19.57	0.329	N.f.	N.d.	N.d.	99.92	An-97.1
,,	,,	Pl*	43.63	N.f.	35.52	0.015	N.d.	0.135	0.013	N.f.	19.49	0.495	N.f.	N.d.	N.d.	99.30	An-95.6
155-4	Leuc Gb	Cpx*	52.96	0.041	1.866	0.338	N.d.	4.17	0.126	15.30	24.02	0.204	0.017	N.d.	N.d.	99.03	0.87
,,	,,	Pl*	45.05	0.014	35.12	N.d.	N.d.	0.234	N.d.	N.d.	19.07	0.672	0.014	N.d.	N.d.	100.17	An-93.9
,,	,,	Pl*	44.80	N.d.	35.42	0.006	N.d.	0.159	N.d.	N.d.	19.11	0.484	0.009	N.d.	N.d.	99.99	An-95.8
,,	,,	Amph*	46.26	0.135	11.83	0.248	N.d.	8.20	0.106	16.88	12.36	1.869	0.122	N.d.	2.00	100.00	0.78
,,	,,	Amph*	45.58	0.176	12.48	0.500	N.d.	7.72	0.108	16.91	12.32	1.886	0.125	N.d.	2.00	99.81	0.78
,,	,,	Amph*	45.29	0.136	13.50	0.411	N.d.	6.66	0.102	17.37	12.06	1.909	0.142	N.d.	2.00	99.58	0.77
155-1	Leuc Ol Gb	Ol*	38.89	N.d.	N.d.	N.d.	N.d.	20.08	0.32	41.03	0.007	N.d.	N.d.	0.19	N.d.	100.51	0.78
,,	,,	Ol*	39.26	N.d.	N.d.	N.d.	N.d.	19.37	0.29	41.84	0.011	N.d.	N.d.	0.19	N.d.	100.96	0.79
,,	,,	Ol*	39.14	N.d.	N.d.	N.d.	N.d.	18.90	0.30	41.74	0.010	N.f.	N.d.	0.20	N.d.	100.29	0.80
,,	,,	Opx*	53.85	0.143	2.58	0.335	N.d.	13.45	0.312	28.83	0.691	0.243	0.001	N.d.	N.d.	100.19	0.79
,,	,,	Cpx*	52.30	0.474	3.07	0.33	N.d.	4.93	0.14	16.26	21.89	1.96	0.007	N.d.	N.d.	99.64	0.85
,,	,,	Pl*	47.56	0.031	33.01	0.002	N.d.	0.158	0.037	N.d.	16.54	0.798	N.d.	N.d.	N.d.	99.30	An-82.3
,,	,,	Pl*	45.19	0.003	34.72	N.d.	N.d.	0.030	0.003	N.d.	18.29	N.d.	0.005	N.d.	N.d.	99.05	An-92.7
Shel'ting massif																	
174	Ol Enst	Ol*	40.46	0.002	N.d.	0.007	N.d.	13.54	0.183	46.16	0.024	0.044	N.d.	N.d.	N.d.	100.42	0.86
,,	,,	Ol*	40.41	N.d.	0.006	0.015	N.d.	13.77	0.210	46.21	0.013	N.d.	0.005	N.d.	N.d.	100.63	0.96
,,	,,	Opx*	56.07	0.008	0.45	0.29	N.d.	8.92	0.210	32.53	0.887	0.015	0.003	N.d.	N.d.	99.38	0.87
,,	,,	Opx*	57.02	0.018	0.44	0.29	N.d.	8.93	0.201	32.32	0.982	0.011	0.005	N.d.	N.d.	100.22	0.87
,,	,,	Opx*	56.65	0.016	0.50	0.24	N.d.	8.99	0.218	32.10	0.929	0.028	N.d.	N.d.	N.d.	99.68	0.86
,,	,,	Opx*	56.76	0.032	0.47	0.28	N.d.	8.91	0.236	32.08	1.031	N.d.	N.d.	N.d.	N.d.	99.81	0.87
,,	,,	CrSp*	0.014	0.105	7.65	50.56	N.d.	35.10	0.356	5.83	0.006	N.d.	N.d.	N.d.	N.d.	99.61	Cr#-0.82
,,	,,	CrSp*	0.039	0.241	11.33	42.90	N.d.	37.96	0.357	6.58	N.d.	N.d.	N.d.	N.d.	N.d.	99.40	Cr#-0.72
176	Cpx-Ol Enst	Ol*	38.44	N.d.	N.d.	N.d.	N.d.	23.83	0.34	37.42	0.016	N.d.	N.d.	0.18	N.d.	100.22	0.74
,,	,,	Ol*	38.34	N.d.	N.d.	N.d.	N.d.	24.26	0.31	37.61	0.016	N.d.	N.d.	0.19	N.d.	100.73	0.73
,,	,,	Ol*	38.54	N.d.	N.d.	N.d.	N.d.	24.09	0.35	37.45	0.014	N.d.	N.d.	0.19	N.d.	100.63	0.74
,,	,,	Opx*	54.21	0.029	1.22	0.116	N.d.	16.43	0.36	27.02	0.92	N.d.	N.d.	N.d.	N.d.	100.30	0.75
,,	,,	Opx*	54.88	0.050	1.28	0.161	N.d.	16.25	0.34	27.01	1.07	0.04	N.d.	N.d.	N.d.	101.08	0.75
,,	,,	Opx*	54.16	0.032	1.36	0.181	N.d.	16.10	0.34	27.03	1.00	N.d.	N.d.	N.d.	N.d.	100.22	0.75
,,	,,	Opx*	54.68	0.053	1.32	0.174	N.d.	16.12	0.37	26.97	1.14	N.d.	N.d.	N.d.	N.d.	100.82	0.75
,,	,,	Opx*	54.61	0.047	1.28	0.152	N.d.	16.18	0.32	27.01	1.25	0.04	N.d.	N.d.	N.d.	100.88	0.75
,,	,,	Opx*	54.69	0.057	1.26	1.160	N.d.	16.30	0.36	26.97	1.15	0.03	N.d.	N.d.	N.d.	100.98	0.75
,,	,,	Opx*	55.13	0.037	1.21	0.188	N.d.	16.15	0.34	28.80	1.01	0.03	N.d.	N.d.	N.d.	102.90	0.76
,,	,,	Cpx*	51.84	0.08	1.91	0.32	N.d.	6.75	0.17	15.34	21.99	0.15	0.15	N.d.	N.d.	98.55	0.84
,,	,,	CrSp*	0.023	1.15	5.84	12.66	N.d.	77.68	0.25	2.09	N.d.	0.13	0.008	N.d.	N.d.	99.81	Cr#-0.59

(Continued)

Table 8.1 Continued.

Sample 1	Rock 2	Mineral 3	SiO₂ 4	TiO₂ 5	Al₂O₃ 6	Cr₂O₃ 7	Fe₂O₃ 8	FeO 9	MnO 10	MgO 11	CaO 12	Na₂O 13	K₂O 14	NiO 15	H₂O 16	Total 17	Mg# (An, %, Cr#) 18
182	Gbn	Opx*	53.76	0.069	1.39	0.264	N.d.	11.98	0.233	31.22	0.775	0.005	0.002	N.d.	N.d.	99.71	0.82
-,-	-,-	Opx*	53.91	0.035	1.49	0.248	N.d.	12.04	0.237	31.16	0.810	0.038	N.d.	N.d.	N.d.	99.96	0.82
-,-	-,-	Cpx*	52.59	0.099	1.95	0.487	N.d.	4.77	0.134	16.65	19.91	0.227	0.007	N.d.	N.d.	96.83	0.86
-,-	-,-	Cpx*	52.18	0.120	2.40	0.461	N.d.	4.86	0.125	16.22	20.16	0.351	0.010	N.d.	N.d.	96.88	0.86
-,-	-,-	Cpx*	52.35	0.092	2.11	0.499	N.d.	4.66	0.112	16.56	20.56	0.316	N.d.	N.d.	N.d.	97.29	0.86
-,-	-,-	Cpx*	52.24	0.125	2.17	0.471	N.d.	5.11	0.116	16.25	20.07	0.264	0.019	N.d.	N.d.	96.83	0.85
-,-	-,-	Pl*	47.25	0.011	32.87	0.014	N.d.	0.293	N.d.	N.d.	16.95	1.74	0.021	N.d.	N.d.	99.15	An–84.2
-,-	-,-	Pl*	47.40	0.015	33.17	N.d.	N.d.	0.312	0.011	N.d.	16.88	1.91	0.025	N.d.	N.d.	99.73	An–82.9
183	Gbn	Opx*	54.54	0.06	1.34	0.08	N.d.	13.23	0.26	28.85	1.02	N.d.	N.d.	N.d.	N.d.	99.39	0.80
-,-	-,-	Opx*	54.30	0.07	1.13	0.11	N.d.	13.24	0.27	29.32	0.70	0.02	0.004	N.d.	N.d.	99.16	0.80
-,-	-,-	Opx*	54.27	0.07	1.40	0.11	N.d.	13.24	0.28	28.90	0.02	0.03	0.003	N.d.	N.d.	99.31	0.80
-,-	-,-	Opx*	55.14	0.06	1.28	0.07	N.d.	12.68	0.30	28.82	0.73	0.01	0.004	N.d.	N.d.	99.10	0.80
-,-	-,-	Cpx*	52.63	0.11	2.11	0.24	N.d.	5.65	0.16	16.02	22.78	0.24	0.004	N.d.	N.d.	99.94	0.83
-,-	-,-	Cpx*	52.71	0.10	1.97	0.20	N.d.	5.38	0.16	16.02	23.03	0.24	0.009	N.d.	N.d.	99.82	0.83
-,-	-,-	Cpx*	52.28	0.12	2.17	0.21	N.d.	5.46	0.16	16.40	22.98	0.22	N.d.	N.d.	N.d.	100.00	0.84
-,-	-,-	Pl*	45.28	N.d.	33.27	0.013	N.d.	0.495	0.009	N.d.	17.99	1.29	0.011	N.d.	N.d.	98.35	An–88.5
-,-	-,-	Pl*	45.87	0.032	34.12	N.d.	N.d.	0.487	N.d.	N.d.	17.94	1.22	0.009	N.d.	N.d.	99.97	An–89.0
-,-	-,-	Pl*	45.83	0.028	34.18	0.004	N.d.	0.451	0.004	N.d.	18.21	1.29	0.019	N.d.	N.d.	100.01	An–88.6
-,-	-,-	Pl*	45.49	N.d.	33.98	0.018	N.d.	0.370	0.013	N.d.	17.77	1.34	0.005	N.d.	N.d.	98.99	An–88.0
-,-	-,-	Pl*	46.16	N.d.	33.74	0.007	N.d.	0.328	0.006	N.d.	17.62	1.63	0.003	N.d.	N.d.	99.40	An–86.4
-,-	-,-	Pl*	45.75	0.006	34.05	N.d.	N.d.	0.376	N.d.	N.d.	18.16	1.181	0.020	N.d.	N.d.	99.54	An–89.4
-,-	-,-	Pl*	45.29	0.031	34.17	0.012	N.d.	0.325	0.018	N.d.	18.09	1.038	0.011	N.d.	N.d.	98.98	An–90.6

Komsomol'sk massif

Sample 1	Rock 2	Mineral 3	SiO₂ 4	TiO₂ 5	Al₂O₃ 6	Cr₂O₃ 7	Fe₂O₃ 8	FeO 9	MnO 10	MgO 11	CaO 12	Na₂O 13	K₂O 14	NiO 15	H₂O 16	Total 17	Mg# (An, %, Cr#) 18
191	Ol Gbn	Ol*	39.80	N.d.	N.d.	0.016	N.d.	16.43	0.326	43.59	0.010	0.057	0.005	N.d.	N.d.	100.23	0.83
-,-	-,-	Ol*	40.25	N.d.	0.015	N.d.	N.d.	16.68	0.285	43.70	0.043	N.d.	N.d.	N.d.	N.d.	100.98	0.82
-,-	-,-	Ol*	40.09	0.013	N.d.	0.016	N.d.	16.54	0.313	43.51	0.019	0.002	N.d.	N.d.	N.d.	100.51	0.82
-,-	-,-	Cpx*	51.91	0.087	3.57	0.323	N.d.	4.46	0.150	15.16	23.49	0.139	0.003	N.d.	N.d.	99.29	0.86
-,-	-,-	Cpx*	53.14	0.058	2.63	0.182	N.d.	4.10	0.141	15.47	23.75	0.100	0.011	N.d.	N.d.	99.59	0.87
-,-	-,-	Amph*	45.12	0.325	13.04	0.838	N.d.	6.91	0.108	16.23	12.58	1.54	0.35	N.d.	2.00	99.04	0.81

Sample	Rock	Mineral															
,,	,,	Srp*	41.76	N.d.	0.070	0.045	N.d.	5.41	0.023	39.02	0.122	N.d.	0.005	N.d.	14.00	100.46	0.93
,,	,,	Chl*	40.32	0.003	31.54	0.024	N.d.	0.036	0.010	N.d.	12.32	2.89	0.006	N.d.	12.00	99.14	1.00
189	Gbn	Opx*	53.09	0.039	2.30	0.073	N.d.	15.71	N.d.	28.53	0.66	N.d.	N.d.	N.d.	N.d.	100.74	0.76
,,	,,	Cpx*	52.33	0.11	2.46	0.15	N.d.	5.80	0.20	15.73	22.43	0.15	N.d.	N.d.	N.d.	99.35	0.83
,,	,,	Cpx*	52.07	0.11	2.88	0.16	N.d.	5.91	0.17	15.17	22.37	0.21	N.d.	N.d.	N.d.	99.05	0.82
,,	,,	Cpx*	52.45	0.35	2.49	0.50	N.d.	6.25	0.16	15.91	20.64	0.29	N.d.	N.d.	N.d.	99.04	0.82
,,	,,	Cpx*	52.15	0.13	2.93	0.18	N.d.	6.15	0.19	14.86	22.63	0.28	0.004	N.d.	N.d.	99.51	0.81
,,	,,	Cpx*	52.03	0.11	2.93	0.14	N.d.	6.07	0.16	14.99	23.30	0.23	0.01	N.d.	N.d.	99.98	0.81
,,	,,	Pl*	44.61	0.019	35.39	N.d.	N.d.	0.322	N.d.	N.d.	19.93	0.62	0.003	N.d.	N.d.	99.90	An–94.4
,,	,,	Pl*	44.13	0.018	35.97	N.d.	N.d.	0.204	N.d.	N.d.	19.25	0.39	0.005	N.d.	N.d.	99.96	An–96.4
,,	,,	Amph*	44.48	0.37	12.14	0.27	N.d.	9.37	0.08	15.58	11.77	1.95	0.06	N.d.	2.00	98.07	0.75
,,	,,	Amph*	44.35	0.40	12.40	0.276	N.d.	9.91	0.096	15.24	.15	2.05	0.075	N.d.	2.00	98.94	0.73
192	Gb-Pegm	Pl*	45.51	N.d.	35.23	0.016	N.d.	0.285	N.d.	N.d.	17.21	0.653	0.016	N.d.	N.d.	98.92	An–93.5
,,	,,	Amph*	45.84	0.663	11.81	0.021	N.d.	11.88	0.153	15.30	9.99	1.73	0.052	N.d.	2.00	99.44	0.76
,,	,,	Amph*	45.61	0.682	11.77	0.100	N.d.	11.74	0.156	15.44	10.22	1.75	0.057	N.d.	2.00	99.52	0.76
,,	,,	Amph*	45.91	0.722	12.23	N.d.	N.d.	11.93	0.159	15.11	10.05	1.70	0.055	N.d.	2.00	99.86	0.76
,,	,,	Ti-Mt*	0.009	2.21	4.17	0.319	N.d.	86.50	0.232	1.62	0.001	0.029	N.d.	N.d.	N.d.	95.09	—
,,	,,	Ti-Mt*	1.15	2.55	3.04	0.068	N.d.	88.56	0.144	1.85	0.001	0.016	0.006	N.d.	N.d.	97.37	—
,,	,,	Ti-Mt*	0.029	2.42	1.68	0.072	N.d.	88.86	0.264	0.556	0.002	N.d.	N.d.	N.d.	N.d.	93.88	—

South Schmidt massif

Sample	Rock	Mineral															
161	Lherz	Ol*	40.17	0.004	0.017	N.d.	N.d.	9.33	0.118	49.45	0.036	N.d.	0.005	N.d.	N.d.	99.14	0.90
,,	,,	Opx*	55.18	0.008	1.79	0.629	N.d.	5.93	0.151	34.88	0.775	N.d.	0.017	N.d.	N.d.	99.35	0.91
,,	,,	Opx*	55.34	0.029	1.73	0.614	N.d.	5.83	0.136	34.16	1.567	0.046	N.d.	N.d.	N.d.	99.46	0.91
,,	,,	Cpx*	54.24	0.004	1.39	0.501	N.d.	2.03	0.092	17.07	23.64	0.200	0.006	N.d.	N.d.	99.18	0.94
,,	,,	Cpx*	53.68	0.024	1.90	0.948	N.d.	1.98	0.088	16.52	23.77	0.181	N.d.	N.d.	N.d.	99.09	0.94
,,	,,	Cpx*	53.36	0.022	1.75	0.889	N.d.	2.55	0.089	16.97	23.27	0.196	0.006	N.d.	N.d.	99.09	0.92
,,	,,	CrSp*	0.022	0.033	22.67	43.58	N.d.	19.85	0.291	11.76	0.005	0.044	0.004	N.d.	N.d.	100.26	Cr#–0.56
,,	,,	CrSp*	0.028	0.030	24.15	41.38	N.d.	20.50	0.269	11.79	0.012	N.d.	N.d.	N.d.	N.d.	100.16	Cr#–0.53
,,	,,	CrSp*	0.022	0.054	22.51	44.22	N.d.	17.41	0.226	13.22	N.d.	N.d.	N.d.	N.d.	N.d.	99.67	Cr#–0.57
,,	,,	CrSp*	0.026	0.052	23.16	42.48	N.d.	19.34	0.278	11.96	0.011	0.045	0.009	N.d.	N.d.	99.34	Cr#–0.55
,,	,,	CrSp*	0.043	0.038	23.29	45.87	N.d.	19.59	0.266	11.77	0.016	0.033	N.d.	N.d.	N.d.	100.92	Cr#–0.57
,,	,,	CrSp*	0.033	0.034	22.95	45.62	N.d.	19.57	0.265	12.11	0.011	0.018	0.006	N.d.	N.d.	100.61	Cr#–0.57
,,	,,	CrSp*	0.103	0.048	23.03	43.59	N.d.	20.16	0.284	12.25	0.027	0.065	0.020	N.d.	N.d.	99.57	Cr#–0.56
,,	,,	CrSp*	0.035	0.048	23.05	46.41	N.d.	17.09	0.258	13.06	N.d.	0.011	0.005	N.d.	N.d.	99.96	Cr#–0.57
,,	,,	CrSp*	0.041	0.042	24.10	45.16	N.d.	19.19	0.254	11.70	0.018	0.016	0.003	N.d.	N.d.	100.52	Cr#–0.56

(Continued)

Table 8.1 Continued.

Sample 1	Rock 2	Mineral 3	SiO₂ 4	TiO₂ 5	Al₂O₃ 6	Cr₂O₃ 7	Fe₂O₃ 8	FeO 9	MnO 10	MgO 11	CaO 12	Na₂O 13	K₂O 14	NiO 15	H₂O 16	Total 17	Mg# (An, %, Cr#) 18
163	Lherz	Ol*	40.61	0.008	0.007	0.012	N.d.	9.77	0.157	49.03	0.014	0.006	N.f.	N.d.	N.d.	99.62	0.90
,,	,,	Ol*	40.48	N.f.	0.019	0.001	N.d.	9.57	0.137	49.22	0.011	0.003	0.012	N.d.	N.d.	99.46	0.90
,,	,,	Opx*	55.64	0.018	3.05	0.658	N.d.	6.38	0.154	32.71	0.648	0.032	0.003	N.d.	N.d.	99.30	0.90
,,	,,	Opx*	54.92	0.036	3.36	0.855	N.d.	6.29	0.126	32.05	0.881	0.002	0.011	N.d.	N.d.	100.55	0.90
,,	,,	Opx*	57.15	0.018	2.39	0.575	N.d.	6.18	0.158	33.73	0.200	N.f.	N.f.	N.d.	N.d.	100.39	0.91
,,	,,	Opx*	54.83	0.047	3.35	1.17	N.d.	6.36	0.145	32.50	0.564	0.075	0.047	N.d.	N.d.	99.07	0.90
,,	,,	Opx*	54.93	0.035	3.27	1.17	N.d.	6.44	0.161	32.46	0.560	0.070	0.035	N.d.	N.d.	99.12	0.90
,,	,,	Cpx*	51.99	0.092	3.44	1.05	N.d.	2.31	0.057	15.86	23.95	0.081	N.f.	N.d.	N.d.	98.83	0.92
,,	,,	Cpx*	53.09	0.092	2.59	0.724	N.d.	2.11	0.081	16.17	24.07	0.102	0.003	N.d.	N.d.	99.00	0.93
,,	,,	Cpx*	52.51	0.062	3.04	0.950	N.d.	2.13	0.082	16.27	24.13	0.101	N.f.	N.d.	N.d.	99.28	0.93
,,	,,	Cpx*	52.04	0.092	3.59	1.166	N.d.	2.16	0.070	15.81	24.35	0.102	0.009	N.d.	N.d.	99.38	0.93
,,	,,	Cpx*	52.36	0.075	3.44	1.007	N.d.	2.28	0.082	15.74	23.98	0.096	N.f.	N.d.	N.d.	99.06	0.92
,,	,,	CrSp*	0.014	0.039	44.82	23.25	N.d.	15.03	0.144	16.85	0.010	0.037	0.002	N.d.	N.d.	100.20	Cr#−0.26
,,	,,	CrSp*	N.f.	0.055	42.28	25.85	N.d.	16.32	0.182	15.64	0.015	0.006	0.003	N.d.	N.d.	100.34	Cr#−0.29
164	Ol Web	Ol*	40.98	N.d.	N.d.	N.d.	N.d.	9.50	0.13	49.23	0.019	N.d.	N.d.	0.34	N.d.	100.19	0.90
,,	,,	Opx*	56.98	0.012	0.927	0.480	N.d.	6.03	0.151	34.00	1.03	0.069	0.007	N.d.	N.d.	99.69	0.91
,,	,,	Opx*	56.59	0.017	0.812	0.534	N.d.	6.07	0.127	34.30	0.901	0.011	0.010	N.d.	N.d.	99.47	0.91
,,	,,	Opx*	56.92	0.006	0.830	0.501	N.d.	6.11	0.155	34.26	0.681	0.020	0.001	N.d.	N.d.	99.48	0.91
162	Web	Opx*	57.04	0.027	0.93	0.45	N.d.	5.64	0.149	33.30	1.14	0.221	0.021	N.d.	N.d.	98.91	0.91
,,	,,	Opx*	57.71	N.d.	0.46	0.12	N.d.	5.81	0.133	34.33	0.24	N.d.	0.002	N.d.	N.d.	98.80	0.91
,,	,,	Cpx*	54.70	0.020	0.59	0.471	N.d.	1.37	0.058	17.98	24.78	0.133	N.d.	N.d.	N.d.	100.10	0.96
,,	,,	Cpx*	55.47	0.008	0.26	0.163	N.d.	1.38	0.071	16.99	24.94	N.d.	N.d.	N.d.	N.d.	99.28	0.96
,,	,,	CrSp*	0.017	0.079	15.36	54.80	N.d.	19.23	0.287	10.61	0.002	0.010	0.004	N.d.	N.d.	100.39	Cr#−0.71
,,	,,	CrSp*	N.d.	0.071	14.97	54.71	N.d.	20.08	0.304	10.28	0.018	N.d.	N.d.	N.d.	N.d.	100.42	Cr#−0.71

Note: Analyses were carried out at the Analytical Center of the Institute of Geology and Mineralogy, Novosibirsk;* – X-ray spectrometry, on a JXA-8100 Camebax-Micro probe (analyst V.N. Korolyuk); the rest – wet chemical technique. Here and in Tables 8.2–8.10: Harz – harzburgites; Cpx Harz – clinopyroxene harzburgites; Lherz – lherzolites; Lherz – plagiolherzolites; Wehrl – wehrlites; Pl Wehrl – plagiowehrlites; Web – websterites; Ol Cpx – olivine clinopyroxenites; Ol Web – olivine websterites; Ol Enst – olivine enstatites; Cpx–Ol Enst – clinopyroxene–olivine enstatitites; Cord – cortlandites; Ol Gb – olivine gabbro; Gbn – gabbronorites; Ol Gbn – olivine gabbronorites; Leuc Ol Gbn – leucocratic olivine gabbronorites; Leuc Gb – leucocratic gabbro; Gb-Pegm – gabbro-pegmatites. Minerals: Ol – olivine, Opx – orthopyroxene, Cpx – clinopyroxene, Amph – amphibole, Pl – plagioclase, Srp – serpentine, CrSp – Cr-spinel, Sulf – sulfide minerals, Mt – magnetite, Ti-Mt – Ti-magnetite, Ilm – ilmenite. The contents of H₂O in amphiboles and serpentines were calculated from stoichiometric formulas. Mg# = Mg/(Mg + Fe); Cr# = Cr/(Cr + Al), An, % = 100Ca/(Ca + Na + K) (f.u.). Hereafter, N.d. – no data, N.f. – not found.

Table 8.2 Chemical composition of olivines from rocks of the Berezovka, Shel'ting, Komsomol'sk, and South Schmidt massifs, wt.%.

Sample 1	Rock 2	SiO_2 3	TiO_2 4	Al_2O_3 5	Cr_2O_3 6	Fe_2O_3 7	FeO 8	MnO 9	MgO 10	CaO 11	Na_2O 12	K_2O 13	NiO 14	H_2O 15	Total 16	$Mg^{\#}$ 17
Berezovka massif																
133*	Cpx Harz	39.55	N.d.	N.d.	N.d.	N.d.	17.46	0.25	42.25	0.02	N.d.	N.d.	0.31	N.d.	99.84	0.81
-"-	-"-	39.61	N.d.	N.d.	N.d.	N.d.	18.02	0.27	42.82	0.01	N.d.	N.d.	0.25	N.d.	100.98	0.81
-"-	-"-	38.78	N.d.	N.d.	N.d.	N.d.	17.68	0.23	43.06	0.01	N.d.	N.d.	0.27	N.d.	100.03	0.81
133-2a	Harz	39.27	N.d.	N.d.	N.d.	N.d.	17.81	0.23	42.64	0.02	N.d.	N.d.	0.26	N.d.	100.23	0.81
1610-2*	Harz	39.85	0.026	0.001	0.001	N.d.	13.45	0.216	46.96	0.005	N.f.	0.002	N.d.	N.d.	100.51	0.86
-"-	-"-	39.88	0.011	0.003	N.f.	N.d.	13.34	0.194	46.92	0.016	0.032	N.f.	N.d.	N.d.	100.39	0.86
-"-	-"-	39.57	N.f.	0.002	0.286	N.d.	13.30	0.203	46.23	0.014	0.003	0.004	N.d.	N.d.	99.61	0.86
-"-	-"-	39.76	0.015	0.003	0.063	N.d.	13.42	0.226	46.46	0.007	0.028	N.f.	N.d.	N.d.	99.97	0.86
1596*	Harz	40.36	N.d.	N.f.	0.01	N.d.	13.49	0.19	46.58	0.01	0.02	N.d.	0.24	N.d.	100.90	0.86
-"-	-"-	40.06	N.d.	N.f.	0.02	N.d.	13.78	0.24	45.75	N.f.	N.f.	N.d.	0.25	N.d.	100.66	0.86
-"-	-"-	40.30	N.d.	N.f.	0.01	N.d.	13.77	0.22	46.81	0.02	0.02	N.d.	0.25	N.d.	100.31	0.86
-"-	-"-	40.67	N.d.	N.f.	0.01	N.f.	13.59	0.23	46.56	0.01	0.05	N.d.	0.26	N.d.	101.37	0.86
-"-	-"-	39.63	N.d.	0.01	N.f.	N.d.	13.51	0.23	47.53	N.f.	N.f.	N.d.	0.26	N.d.	101.18	0.86
-"-	-"-	40.34	N.d.	0.02	0.01	N.d.	13.54	0.23	46.28	0.02	N.f.	N.d.	0.24	N.d.	100.69	0.86
138	Lherz	39.88	N.d.	N.d.	N.d.	N.d.	16.52	0.25	43.96	0.02	N.d.	N.d.	N.d.	0.004	100.53	0.83
142*	Pl Lherz	40.27	0.004	0.045	0.118	N.d.	9.04	0.149	48.97	0.036	0.007	0.018	N.d.	N.d.	98.73	0.91
-"-	-"-	41.03	N.d.	N.d.	N.d.	N.d.	9.17	0.150	49.24	0.010	N.d.	0.005	N.d.	N.d.	99.60	0.91
-"-	-"-	40.67	N.d.	0.012	0.008	N.d.	9.03	0.145	49.42	0.009	0.050	0.014	N.d.	N.d.	99.36	0.91
139	Wehrl	40.02	N.d.	N.d.	N.d.	N.d.	13.15	0.27	46.34	0.03	N.d.	N.d.	N.d.	0.13	99.93	0.86
155*	Ol Web	39.82	N.d.	N.d.	N.d.	N.d.	15.74	0.18	44.26	0.011	N.d.	N.d.	0.26	N.d.	100.27	0.83
-"-	-"-	39.86	N.d.	N.d.	N.d.	N.d.	15.63	0.20	44.46	0.013	N.d.	N.d.	0.28	N.d.	100.47	0.84
140	Ol Web	39.74	N.d.	N.d.	N.d.	N.d.	17.17	0.29	43.70	0.020	N.d.	N.d.	N.d.	N.d.	100.49	0.82
155-3*	Ol Gb	38.38	N.d.	0.007	0.016	N.d.	20.64	0.336	40.61	0.018	0.022	N.d.	N.d.	N.d.	100.02	0.78
-"-	-"-	38.60	0.007	0.013	0.010	N.d.	20.87	0.333	40.31	0.013	0.039	N.d.	N.d.	N.d.	100.19	0.77
155-5*	Ol Gb	38.01	N.f.	0.006	N.f.	N.d.	24.61	0.416	36.88	0.022	0.035	N.f.	N.d.	N.d.	99.98	0.73
-"-	-"-	37.99	N.f.	0.012	N.f.	N.d.	23.85	0.371	37.25	0.023	N.f.	N.f.	N.d.	N.d.	99.50	0.74
-"-	-"-	37.89	N.f.	0.012	N.f.	N.d.	24.36	0.394	36.82	0.023	0.001	N.f.	N.d.	N.d.	99.49	0.73
-"-	-"-	37.77	N.f.	0.008	0.027	N.d.	23.52	0.387	37.61	0.023	0.036	N.f.	N.d.	N.d.	99.37	0.74

(Continued)

Table 8.2 Continued.

Sample 1	Rock 2	SiO$_2$ 3	TiO$_2$ 4	Al$_2$O$_3$ 5	Cr$_2$O$_3$ 6	Fe$_2$O$_3$ 7	FeO 8	MnO 9	MgO 10	CaO 11	Na$_2$O 12	K$_2$O 13	NiO 14	H$_2$O 15	Total 16	Mg# 17
155-1*	Leuc Ol Gbn	38.89	N.d.	N.d.	N.d.	N.d.	20.08	0.32	41.03	0.007	N.d.	N.d.	0.19	N.d.	100.51	0.78
-"-	-"-	39.26	N.d.	N.d.	N.d.	N.d.	19.37	0.29	41.84	0.011	N.d.	N.d.	0.19	N.d.	100.96	0.79
-"-	-"-	39.14	N.d.	N.d.	N.d.	N.d.	18.90	0.30	41.74	0.010	N.d.	N.d.	0.20	N.d.	100.29	0.80
Shel'ting massif																
174*	Ol Enst	40.46	0.002	N.d.	0.007	N.d.	13.54	0.183	46.16	0.024	0.044	N.d.	N.d.	N.d.	100.42	0.86
-"-	-"-	40.41	N.d.	0.006	0.015	N.d.	13.77	0.210	46.21	0.013	N.d.	0.005	N.d.	N.d.	100.63	0.96
176*	Cpx–Ol Enst	38.44	N.d.	N.d.	N.d.	N.d.	23.83	0.34	37.42	0.016	N.d.	N.d.	0.18	N.d.	100.22	0.74
-"-	-"-	38.34	N.d.	N.d.	N.d.	N.d.	24.26	0.31	37.61	0.016	N.d.	N.d.	0.19	N.d.	100.73	0.73
-"-	-"-	38.54	N.d.	N.d.	N.d.	N.d.	24.09	0.35	37.45	0.014	N.d.	N.d.	0.19	N.d.	100.63	0.74
Komsomol'sk massif																
191*	Ol Gbn	39.80	N.d.	N.d.	0.016	N.d.	16.43	0.326	43.59	0.010	0.057	0.005	N.d.	N.d.	100.23	0.83
-"-	-"-	40.25	N.d.	0.015	N.d.	N.d.	16.68	0.285	43.70	0.043	N.d.	N.d.	N.d.	N.d.	100.98	0.82
-"-	-"-	40.09	0.013	N.d.	0.016	N.d.	16.54	0.313	43.51	0.019	0.002	N.d.	N.d.	N.d.	100.51	0.82
South Schmidt massif																
161*	Lherz	40.17	0.004	0.017	N.d.	N.d.	9.33	0.118	49.45	0.036	N.d.	0.005	N.d.	N.d.	99.14	0.90
163*	Lherz	40.61	0.008	0.007	0.012	N.d.	9.77	0.157	49.03	0.014	0.006	N.f.	N.d.	N.d.	99.62	0.90
-"-	-"-	40.48	N.f.	0.019	0.001	N.d.	9.57	0.137	49.22	0.011	0.003	0.012	N.d.	N.d.	99.46	0.90
164*	Ol Web	40.98	N.d.	N.d.	N.d.	N.d.	9.50	0.13	49.23	0.019	N.d.	N.d.	0.34	N.d.	100.19	0.90

Note: Compiled after the data in Table 8.1.

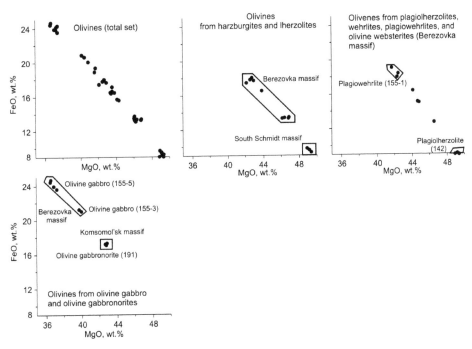

Figure 8.1 Variations in FeO and MgO contents in olivines from ultramafites and mafites of the Berezovka and other massifs (data from Table 8.2).

Study of rocks of the Berezovka massif showed that olivines from restite ultra-mafites and hybrid gabbroids have the same Mg$^{\#}$ values: 0.81–0.83 in the mineral from harzburgites and lherzolites and 0.82–0.83 in that from olivine gabbronorites. This suggests that olivines of olivine gabbronorites are xenogenous, i.e., trapped by the parental melt of these rocks during its percolation through restite ultramafites and its contamination with their substance.

Some olivine grains from different rocks were analyzed for CaO, MnO, and NiO impurities. The CaO content in the mineral varies from 0.007 wt.% (plagiowehrlite) to 0.043 wt.% (olivine gabbronorite) (Fig. 8.2). It shows no dependence on the olivine-hosting rock and on FeO and MgO contents. At the same time, the MgO content in olivines depends on the mineral-hosting rock. For example, it successively increases from 0.118 wt.% in olivine from harzburgites and lherzolites to 0.416 wt.% in the mineral from olivine orthopyroxenites, olivine gabbronorites, and olivine gabbro. The MnO content in olivines is directly related to the FeO content and is inversely related to the MgO content. The NiO content decreases from 0.25–0.31 wt.% in olivine from harzburgites and olivine websterites to 0.19–0.20 wt.% in the mineral from plagiowehrlites and olivine gabbro.

In most of the studied rocks, olivine is partly or, seldom, totally replaced by ser-pentine. Parallel analyses of two olivine grains and two grains of partly replacing serpentine showed that the serpentine is much poorer in FeO (10.22 and 14.31 wt.%)

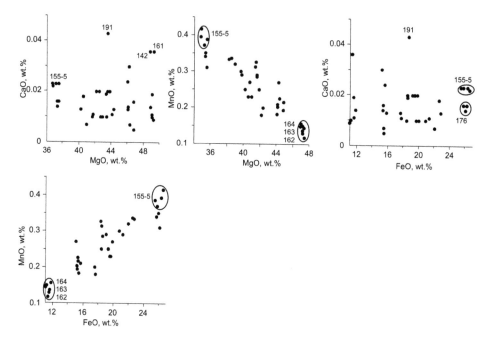

Figure 8.2 Variations in FeO, MgO, CaO, and MnO contents in olivines from ultramafic and gabbroid rocks of the Berezovka and other massifs (data from Table 8.2).

and MgO (36.14 and 34.57 wt.%) than the olivine (FeO = 17.46 and 18.02 wt.%; MgO = 42.25 and 43.06 wt.%). Nevertheless, $Mg^#$ = 0.81 in both minerals.

In general, olivines from rocks of the studied Sakhalin massifs are similar in chemical composition to olivines from analogous rocks of mafic-ultramafic massifs of other ophiolite associations [Lesnov *et al.*, 2005].

8.1.2 Orthopyroxenes

The chemical composition of orthopyroxenes was studied on their grains from harzburgites, lherzolites, plagiolherzolites, olivine websterites, websterites, olivine orthopyroxenites (enstatitites), and olivine and olivine-free gabbronorites of the Berezovka, Shel'ting, South Schmidt, and Komsomol'sk massifs (Table 8.3). The content of FeO in the mineral samples is within 5.5–16 wt.%, and that of MgO is within 26–35 wt.%. The contents of these components are in inverse relationship due to their mutual isomorphous substitution (Fig. 8.3). Orthopyroxenes from restite ultramafites of the South Schmidt massif contain ~6 wt.% FeO, and the MgO content is within 32–35 wt.%. The $Mg^#$ value varies from 0.75 in bronzite from olivine orthopyroxenites of the Shel'ting massif to 0.91 in ferroan enstatite from websterites of the South Schmidt massif. In bronzites from gabbronorites of the Berezovka massif, $Mg^#$ is 0.77–0.82.

The chemical compositions of orthopyroxenes converted to the wollastonite (Wo), enstatite (En), and ferrosilite (Fs) end-members are shown as ternary diagrams. One can

Table 8.3 Chemical composition of orthopyroxenes from rocks of the Berezovka, Shel'ting, Komsomol'sk, and South Schmidt massifs, wt.%.

Sample	Rock	SiO_2	TiO_2	Al_2O_3	Cr_2O_3	Fe_2O_3	FeO	MnO	MgO	CaO	Na_2O	K_2O	NiO	H_2O	Total	$Mg^\#$
Berezovka massif																
133*	Cpx Harz	55.65	0.02	0.86	0.16	N.d.	11.84	0.30	30.68	0.41	N.d.	N.d.	N.d.	N.d.	99.92	0.82
1610-2*	Harz	55.62	0.030	1.69	0.512	N.d.	8.71	0.208	31.55	2.06	0.003	N.f.	N.d.	N.d.	100.38	0.87
-"-	-"-	55.25	0.030	1.50	0.634	N.d.	8.67	0.239	31.16	1.77	N.f.	N.f.	N.d.	N.d.	99.25	0.87
-"-	-"-	55.97	0.021	1.53	0.437	N.d.	8.96	0.220	32.42	0.377	0.025	N.f.	N.d.	N.d.	99.96	0.87
-"-	-"-	55.56	0.037	1.65	0.493	N.d.	9.04	0.223	32.14	0.548	N.f.	0.006	N.d.	N.d.	99.68	0.86
138*	Lherz	50.00	0.08	3.03	0.38	N.d.	12.23	0.14	30.88	1.43	N.d.	N.d.	0.06	N.d.	98.23	0.91
-"-	-"-	55.74	0.05	1.91	0.19	N.d.	9.66	0.25	31.65	0.59	0.02	N.d.	N.d.	N.d.	100.21	0.85
154*	Lherz	55.80	N.d.	2.45	N.d.	N.d.	7.53	N.d.	33.93	0.58	N.d.	N.d.	N.d.	N.d.	100.29	0.89
142	Pl Lherz	51.80	0.12	0.86	N.d.	0.96	14.42	0.43	27.62	1.67	N.d.	N.d.	0.06	N.d.	97.94	0.76
-"-	Pl Lherz	55.01	0.07	0.79	N.d.	N.d.	14.23	N.d.	29.12	0.46	N.d.	N.d.	0.06	N.d.	99.68	0.78
145	Ol Web	55.30	0.15	1.10	0.19	4.02	7.54	0.43	28.57	1.38	N.d.	N.d.	0.03	N.d.	98.71	0.82
-"-	-"-	55.69	0.064	1.57	0.266	N.d.	11.13	0.278	30.64	0.646	N.f.	0.011	N.d.	N.d.	100.30	0.89
-"-	-"-	55.09	0.047	1.42	0.207	N.d.	11.00	0.283	30.45	0.827	0.029	0.011	N.d.	N.d.	99.36	0.88
155	Ol Web	53.50	0.13	0.80	0.30	1.15	8.43	0.50	30.23	0.92	N.d.	N.d.	0.04	N.d.	98.00	0.85
134	Ol Web	55.00	N.d.	1.35	0.07	0.42	9.98	0.18	30.56	1.38	N.d.	N.d.	0.04	N.d.	98.98	0.84
137	Ol Web	55.00	N.d.	1.75	0.17	0.27	10.05	0.20	30.89	0.92	N.d.	N.d.	0.03	N.d.	99.28	0.84
140	Ol Web	52.20	0.07	1.51	0.43	1.71	9.77	0.21	29.89	2.86	N.d.	N.d.	0.02	N.d.	98.67	0.86
-"-	-"-	54.42	0.10	2.13	N.d.	N.d.	11.54	0.31	31.30	0.53	N.d.	N.d.	N.d.	N.d.	100.33	0.83
-"-	-"-	54.71	0.02	2.09	0.33	N.d.	11.57	0.28	30.82	0.49	0.03	0.01	N.d.	N.d.	100.32	0.83
1603-1*	Ol Web	55.46	0.039	1.74	0.264	N.d.	10.30	0.260	31.60	0.506	0.010	N.f.	N.d.	N.d.	100.17	0.85
-"-	-"-	55.98	0.039	1.63	0.249	N.d.	10.37	0.271	31.35	0.567	0.012	0.006	N.d.	N.d.	100.47	0.84
-"-	-"-	55.70	0.028	1.53	0.242	N.d.	10.53	0.272	31.20	0.591	0.009	N.f.	N.d.	N.d.	100.11	0.84
-"-	-"-	55.94	0.069	1.65	0.220	N.d.	10.65	0.263	31.14	0.425	0.010	0.006	N.d.	N.d.	100.37	0.84
-"-	-"-	55.24	0.017	1.62	0.290	N.d.	11.13	0.295	30.94	0.388	N.d.	0.004	N.d.	N.d.	99.93	0.83
-"-	-"-	55.25	0.047	1.49	0.280	N.d.	10.78	0.275	31.25	0.528	0.007	0.006	N.d.	N.d.	99.92	0.84
143*	Web	54.86	0.065	0.876	0.089	N.d.	15.06	0.413	27.96	0.821	0.036	0.011	N.d.	N.d.	100.18	0.77
-"-	-"-	53.78	0.078	0.812	0.104	N.d.	15.36	0.426	28.33	0.583	N.d.	N.d.	N.d.	N.d.	99.47	0.77
-"-	-"-	54.11	0.077	0.719	0.083	N.d.	15.27	0.454	28.21	0.572	0.014	0.003	N.d.	N.d.	99.52	0.77

(Continued)

Table 8.3 Continued.

Sample	Rock	SiO$_2$	TiO$_2$	Al$_2$O$_3$	Cr$_2$O$_3$	Fe$_2$O$_3$	FeO	MnO	MgO	CaO	Na$_2$O	K$_2$O	NiO	H$_2$O	Total	Mg$^\#$
152*	Web	56.58	0.01	1.09	0.33	N.d.	6.63	0.17	34.19	0.22	0.03	0.01	N.d.	N.d.	99.22	0.90
148-1*	Ol Gb	54.23	0.052	1.63	0.221	N.d.	15.23	0.318	27.88	0.633	0.051	N.f.	N.d.	N.d.	100.24	0.77
-"-	-"-	54.16	0.036	1.69	0.235	N.d.	16.20	0.377	26.90	0.685	0.050	0.001	N.d.	N.d.	100.34	0.75
130*	Gbn	55.07	0.03	1.35	0.07	N.d.	14.24	0.33	27.42	0.84	0.01	0.01	N.d.	N.d.	99.34	0.77
-"-	-"-	55.04	0.03	1.39	0.06	N.d.	14.27	0.33	27.25	0.86	N.f.	0.02	N.d.	N.d.	99.24	0.77
-"-	-"-	55.19	0.03	1.23	0.08	N.d.	14.22	0.31	27.64	0.60	N.f.	N.f.	N.d.	N.d.	99.24	0.77
-"-	-"-	54.65	0.03	1.47	0.07	N.d.	13.82	0.33	27.19	1.27	0.01	N.f.	N.d.	N.d.	99.30	0.78
-"-	-"-	55.39	0.03	1.31	0.08	N.d.	14.63	0.34	27.58	0.63	0.01	0.01	N.d.	N.d.	98.85	0.77
138a	Gbn	54.96	0.06	2.77	N.d.	N.d.	11.20	0.28	30.90	N.d.	0.03	0.01	N.d.	N.d.	100.02	0.85
141*	Gbn	55.49	N.d.	2.17	N.d.	N.d.	10.35	N.d.	31.89	0.61	0.02	N.d.	N.d.	N.d.	101.01	0.85
144	Gbn	51.20	0.40	2.59	0.05	1.31	12.58	0.28	28.82	1.43	N.d.	N.d.	0.04	N.d.	98.70	0.79
146	Gbn	54.60	N.d.	1.20	N.d.	5.22	9.25	0.43	25.90	0.92	N.d.	N.d.	0.03	N.d.	97.55	0.77
147*	Gbn	56.04	0.03	0.96	N.d.	N.d.	12.97	N.d.	30.30	0.49	N.d.	N.d.	N.d.	N.d.	100.79	0.81
149	Gbn	53.20	N.d.	2.80	0.20	1.79	9.10	0.50	27.90	1.84	N.d.	N.d.	0.03	N.d.	97.36	0.82
1605*	Gbn	55.22	N.f.	1.81	0.078	N.d.	12.25	0.336	30.37	0.441	0.056	0.009	N.d.	N.d.	100.58	0.82
-"-	-"-	55.33	0.032	1.80	0.078	N.d.	12.27	0.321	30.19	0.406	0.017	0.006	N.d.	N.d.	100.45	0.81
-"-	-"-	55.17	0.011	1.98	0.109	N.d.	11.86	0.287	30.30	0.518	N.f.	0.001	N.d.	N.d.	100.22	0.83
-"-	-"-	55.14	0.002	1.76	0.103	N.d.	12.03	0.309	29.83	0.682	N.f.	0.002	N.d.	N.d.	99.86	0.82
155-1*	Leuc Ol Gbn	53.85	0.143	2.58	0.335	N.d.	13.45	0.312	28.83	0.691	N.f.	0.001	N.d.	N.d.	100.19	0.79
Shel'ting massif																
174*	Ol Enst	56.07	0.008	0.45	0.29	N.d.	8.92	0.210	32.53	0.887	0.015	0.003	N.d.	N.d.	99.38	0.87
-"-	-"-	57.02	0.018	0.44	0.29	N.d.	8.93	0.201	32.32	0.982	0.011	0.005	N.d.	N.d.	100.22	0.87
-"-	-"-	56.65	0.016	0.50	0.24	N.d.	8.99	0.218	32.10	0.929	0.028	N.d.	N.d.	N.d.	99.68	0.86
-"-	-"-	56.76	0.032	0.47	0.28	N.d.	8.91	0.236	32.08	1.031	N.d.	N.d.	N.d.	N.d.	99.81	0.87

176*	Cpx–Ol Enst	54.21	0.029	1.22	0.116	N.d.	16.43	0.36	27.02	0.92	N.d.	N.d.	N.d.	100.30	0.75
-"-		54.88	0.050	1.28	0.161	N.d.	16.25	0.34	27.01	1.07	0.04	N.d.	N.d.	101.08	0.75
-"-		54.16	0.032	1.36	0.181	N.d.	16.10	0.34	27.03	1.00	N.d.	N.d.	N.d.	100.22	0.75
-"-		54.68	0.053	1.32	0.174	N.d.	16.12	0.37	26.97	1.14	0.04	N.d.	N.d.	100.82	0.75
-"-		54.61	0.047	1.28	0.152	N.d.	16.18	0.32	27.01	1.25	0.04	N.d.	N.d.	100.88	0.75
-"-		54.69	0.057	1.26	1.160	N.d.	16.30	0.36	26.97	1.15	0.03	N.d.	N.d.	100.98	0.75
-"-		55.13	0.037	1.21	0.188	N.d.	16.15	0.34	28.80	1.01	0.03	N.d.	N.d.	102.90	0.76
182*	Gbn	53.76	0.069	1.39	0.264	N.d.	11.98	0.233	31.22	0.775	0.005	0.002	N.d.	99.71	0.82
-"-		53.91	0.035	1.49	0.248	N.d.	12.04	0.237	31.16	0.810	0.038	N.d.	N.d.	99.96	0.82
183*	Gbn	54.54	0.06	1.34	0.08	N.d.	13.23	0.26	28.85	1.02	N.d.	N.d.	N.d.	99.39	0.80
-"-		54.30	0.07	1.13	0.11	N.d.	13.24	0.27	29.32	0.70	0.02	0.004	N.d.	99.16	0.80
-"-		54.27	0.07	1.40	0.11	N.d.	13.24	0.28	28.90	0.02	0.03	0.003	N.d.	99.31	0.80
-"-		55.14	0.06	1.28	0.07	N.d.	12.68	0.30	28.82	0.73	0.01	0.004	N.d.	99.10	0.80

Komsomol'sk massif

189*	Gbn	53.09	0.039	2.30	0.073	N.d.	15.71	N.d.	28.53	0.66	N.d.	N.d.	N.d.	100.74	0.80

South Schmidt massif

161*	Lherz	55.18	0.008	1.79	0.629	N.d.	5.93	0.151	34.88	0.775	N.d.	0.017	N.d.	99.35	0.91
-"-		55.34	0.029	1.73	0.614	N.d.	5.83	0.136	34.16	1.567	0.046	N.d.	N.d.	99.46	0.91
163*	Lherz	55.64	0.018	3.05	0.658	N.d.	6.38	0.154	32.71	0.648	0.032	0.003	N.d.	99.30	0.90
-"-		54.92	0.036	3.36	0.855	N.d.	6.29	0.126	32.05	0.881	0.002	0.011	N.d.	100.55	0.90
-"-		57.15	0.018	2.39	0.575	N.d.	6.18	0.158	33.73	0.200	N.f.	N.f.	N.d.	100.39	0.91
-"-		54.83	0.047	3.35	1.17	N.d.	6.36	0.145	32.50	0.564	0.075	0.047	N.d.	99.07	0.90
-"-		54.93	0.035	3.27	1.17	N.d.	6.44	0.161	32.46	0.560	0.070	0.035	N.d.	99.12	0.90
164*	OlWeb	56.98	0.012	0.927	0.480	N.d.	6.03	0.151	34.00	1.03	0.069	0.007	N.d.	99.69	0.91
-"-		56.59	0.017	0.812	0.534	N.d.	6.07	0.127	34.30	0.901	0.011	0.010	N.d.	99.47	0.91
-"-		56.92	0.006	0.830	0.501	N.d.	6.11	0.155	34.26	0.681	0.020	0.001	N.d.	99.48	0.91
162*	Web	57.04	0.027	0.93	0.45	N.d.	5.64	0.149	33.30	1.14	0.221	0.021	N.d.	98.91	0.91
-"-		57.71	N.d.	0.46	0.12	N.d.	5.81	0.133	34.33	0.24	N.d.	0.002	N.d.	98.80	0.91

Note: Compiled after the data in Table 8.1.

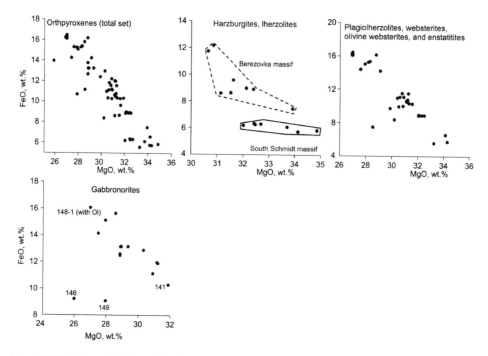

Figure 8.3 Variations in FeO and MgO contents in orthopyroxenes from ultramafites and gabbroids of the Berezovka, Shel'ting, Komsomol'sk, and South Schmidt massifs (data from Table 8.3).

Table 8.4 Limiting contents of wollastonite (Wo), enstatite (En), and ferrosilite (Fs) end-members in orthopyroxenes from ultramafic rocks and gabbroids of the Berezovka and other massifs.

Rock	Number of analyses	Wo, %	En, %	Fs, %
Harzburgites, lherzolites	10	0.4–3.5	81.3–90.4	8.5–17.7
Plagiolherzolites, olivine and olivine-free websterites, enstatites, olivine gabbro	32	0.5–2.4	72.9–90.9	8.5–25.1
Gabbronorites, gabbro	15	0.8–2.6	60.3–81.2	17.5–38.2

Note: The contents of end-members were calculated from the data in Table 8.3.

see a slight difference in composition between orthopyroxenes from restite harzburgites and lherzolites and from hybrid ultramafites and gabbroids (Table 8.4, Fig. 8.4). The contents of the ferrosilite and enstatite end-members in these orthopyroxenes differ more noticeably than the contents of the wollastonite end-member.

The difference in the composition of orthopyroxenes from different types of rocks is also depicted on covariation diagrams of CaO, Al$_2$O$_3$, Cr$_2$O$_3$, TiO$_2$, and MnO contents (Fig. 8.5). The CaO contents in the studied orthopyroxene samples vary from 0.22 to 2.06 wt.%. Nonuniform distribution of this component is also

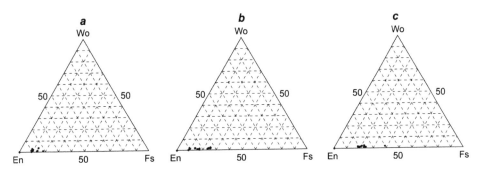

Figure 8.4 En–Wo–Fs composition (%) diagrams for orthopyroxenes from ultramafites and gabbroids of the Berezovka, Shel'ting, Komsomol'sk, and South Schmidt massifs (data from Tables 8.3 and 8.4).
a – from harzburgites and lherzolites; *b* – from olivine and olivine-free websterites; *c* – from gabbronorites.

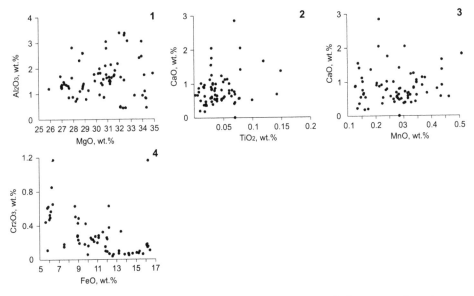

Figure 8.5 Variations in CaO, Al_2O_3, TiO_2, MnO, Cr_2O_3, MgO, and FeO contents in orthopyroxenes from ultramafites and gabbroids of the Berezovka, Shel'ting, Komsomol'sk, and South Schmidt massifs (data from Table 8.3).

observed in different orthopyroxene grains from the same rock sample. For example, four orthopyroxene grains from harzburgite show CaO = 0.337–2.060 wt.%; five grains from lherzolite, CaO = 0.200–0.881 wt.%; and four grains from gabbronorite, CaO = 0.02–1.02 wt.%. In general, orthopyroxenes from harzburgites, lherzolites, websterites, and orthopyroxenites are characterized by higher contents of CaO as compared with those from olivine and olivine-free gabbronorites. The nonuniform distribution of CaO in the orthopyroxenes might be explained as follows. As mentioned

Table 8.5 Chemical composition of orthopyroxene crystal from olivine websterite (sample 140, Berezovka massif), clinopyroxene lamellae in it, and clinopyroxene-substituting amphibole, wt.%.

Analyses No.	SiO_2	TiO_2	Al_2O_3	FeO	MnO	MgO	CaO	K_2O	H_2O	Total	$Mg^{\#}$
Orthopyroxene											
1	55.61	0.02	1.93	11.14	0.30	30.98	0.51	N.f.	N.d.	100.83	83.2
2	55.36	0.03	1.89	10.98	0.28	30.79	0.53	N.f.	N.d.	100.17	83.2
3	54.95	0.03	1.86	11.13	0.28	30.82	0.53	N.f.	N.d.	99.87	83.2
Clinopyroxene											
4	52.78	0.08	2.47	4.01	0.14	16.67	23.79	0.01	N.d.	100.54	88.1
5	52.86	0.07	2.53	4.03	0.13	16.62	23.87	N.f.	N.d.	100.71	88.0
6	52.73	0.09	2.57	4.08	0.15	16.59	23.82	N.f.	N.d.	100.68	87.9
7	52.72	0.07	2.45	4.05	0.14	16.62	23.82	0.01	N.d.	100.44	87.9
Amphibole											
8	54.29	0.07	3.96	4.23	0.09	21.07	12.92	0.01	2.00	99.00	89.9
9	52.77	0.07	5.63	5.04	0.09	20.34	13.02	N.f.	2.00	99.54	87.8

Note: Microphotograph of orthopyroxene crystal with clinopyroxene lamellae is presented in Fig. 4.5. Analyses were carried out by X-ray spectrometry on a JXA-8100 Camebax-Micro probe at the Analytical Center of the Institute of Geology and Mineralogy, Novosibirsk (analyst V.N. Korolyuk). $Mg^{\#} = 100Mg/(Mg + Fe)$ (formula coefficients). The content of H_2O in amphiboles (analyses 8 and 9) was calculated from stoichiometric formulas.

in Chapter 4, orthopyroxenes often contain nonuniformly distributed clinopyroxene lamellae of different widths produced during exsolution (Fig. 4.5). The content of Ca in the lamellae is higher than the content of Ca present as an isomorphous substituent in the orthopyroxene crystal structure (Table 8.5). Therefore, during a probe microanalysis of orthopyroxenes, an electron beam can excite not only the Ca-depleted mineral matrix but also a particular quantity of Ca-enriched clinopyroxene lamellae in it, which results in more significant fluctuations in estimated CaO contents.

The Al_2O_3 impurity in the analyzed orthopyroxenes is also nonuniformly distributed. The highest content of this component (2.39–3.36 wt.%) was found in the mineral grains from lherzolites of the South Schmidt massif. Most likely, these orthopyroxenes crystallized at ultrahigh pressure. Lower Al_2O_3 contents were detected in orthopyroxenes from harzburgites (0.86–3.03 wt.%), gabbronorites (0.96–2.98 wt.%), and websterites (0.72–2.13 wt.%) of the Berezovka massif and from orthopyroxenites of the Shel'ting massif (0.44–2.58 wt.%), which probably crystallized at lower pressures than the above-mentioned orthopyroxenes from the South Schmidt massif lherzolites.

Analysis of few orthopyroxene grains from rocks of the Berezovka, Shel'ting, and South Schmidt massifs showed variations in Cr_2O_3 content from 0.05 to 1.17 wt.%. Orthopyroxenes from harzburgites of the Berezovka massif are somewhat richer in this component (0.35–0.63 wt.%) than those from olivine websterites (0.19–0.43 wt.%) and gabbronorites (0.07–0.24 wt.%). The arrangement of figurative points on the composition diagram (Fig. 8.5,4) suggests an inverse relationship between Cr_2O_3 and FeO contents. The TiO_2 contents in most of the orthopyroxene grains vary from 0.002 to 0.15 wt.%, reaching 0.4 wt.% in occasional grains. Orthopyroxenes from harzburgites and lherzolites are often poorer in TiO_2 than those from websterites, orthopyroxenites, and gabbroids. The MnO contents in orthopyroxenes from different rocks vary from

0.13 to 0.50 wt.%, being maximum in the mineral grains from websterites and gab-bronorites. Note that orthopyroxenes from harzburgites of the Berezovka massif are richer in MnO than those from lherzolites of the South Schmidt massif and that orthopyroxenes from two-pyroxene rocks contain nearly twice more MnO than the coexisting clinopyroxenes.

8.1.3 Clinopyroxenes

The chemical composition of clinopyroxenes was studied on a more representative sample of grains as compared with other minerals from the Sakhalin mafic-ultramafic massifs (Table 8.6). Clinopyroxene grains from restite ultramafites (harzburgites and lherzolites) amount to ~19% of the analyzed grains; those from hybrid ultramafites and hybrid gabbroids (plagiolherzolites, olivine and olivine-free websterites, clinopy-roxenites, and olivine gabbroids), ~50%; and grains from orthomagmatic gabbroids (gabbronorites and gabbro) amounted to ~30%.

In the total sample of clinopyroxenes the MgO contents vary from 14.7 to 17.2 wt.%, and the FeO contents, from 0.9 to 8.6 wt.% (Fig. 8.6). The lowest FeO contents were detected in clinopyroxenes from lherzolites and websterites of the South Schmidt massif, and the highest ones, in certain mineral grains from gab-broids of the Berezovka and other massifs. The location of figurative points on the composition diagrams for the total set of clinopyroxenes indicates a weak inverse relationship between MgO and FeO contents. The Al_2O_3 contents in the total set of clinopyroxenes are within 0.59–3.59 wt.% (Fig. 8.7); the grains from clinopyroxene-containing harzburgites and lherzolites of the Berezovka massif are characterized by $Al_2O_3 = 0.91–3.46$ wt.%, and those from lherzolites of the South Schmidt massif, by $Al_2O_3 = 1.75–3.59$ wt.%. For comparison, the Al_2O_3 contents in clinopyroxenes from harzburgites and lherzolites of the Troodos massif, Cyprus, are nearly the same or somewhat higher (1.44–5.22 wt.%); most of the mineral grains contain <3 wt.% Al_2O_3 [Sobolev and Batanova, 1995]. The clinopyroxenes from rocks of the Sakhalin massifs, like the Troodos ones, show a direct relationship between Al_2O_3 and TiO_2 contents (Fig. 8.7, 1) but no relationship of Al_2O_3 contents with MgO, FeO, CaO, and Na_2O ones (Fig. 8.7, 2–5).

The $Mg^\#$ value in clinopyroxenes from the studied rocks of the Sakhalin massifs varies from 0.81 to 0.96. High values are specific for the mineral grains from lherzolites and websterites of the South Schmidt massif, and low ones, for the grains from olivine and olivine-free websterites, olivine gabbro, olivine-free gabbronorites, and gabbro of the same massif.

The calculated contents of the Wo, En, and Fs end-members in the total set of clinopyroxenes vary over a relatively narrow range (Table 8.7, Fig. 8.8). Clinopyrox-enes from the Sakhalin restite ultramafites, like those from most of restite ultramafites of other ophiolite associations, correspond in composition to diopsides (Fig. 8.8, a), and clinopyroxenes from hybrid ultramafites and gabbroids (Fig. 8.8, b) and from orthomagmatic gabbroids (Fig. 8.8, c) correspond to diopsides and augites.

Clinopyroxenes from some rock varieties differ in contents of Cr_2O_3, TiO_2, Na_2O, and MnO impurities. For example, clinopyroxenes from ultramafites and gabbroids of the Berezovka massif contain 0.080–0.966 wt.% Cr_2O_3; some mineral grains from lherzolites, plagiolherzolites, plagiowehrlites, and olivine and olivine-free websterites

Table 8.6 Chemical composition of clinopyroxenes from rocks of the Berezovka, Shel'ting, Komsomol'sk, and South Schmidt massifs, wt.%.

Sample 1	Rock 2	SiO₂ 3	TiO₂ 4	Al₂O₃ 5	Cr₂O₃ 6	Fe₂O₃ 7	FeO 8	MnO 9	MgO 10	CaO 11	Na₂O 12	K₂O 13	NiO 14	H₂O 15	Total 16	Mg# 17
Berezovka massif																
133*	Cpx Harz	54.29	0.01	0.91	0.25	N.d.	3.52	0.07	17.50	24.04	0.07	N.d.	N.d.	N.d.	100.66	0.89
-"-		53.69	0.04	1.03	0.15	N.d.	4.19	0.12	16.23	23.44	0.13	N.d.	N.d.	N.d.	99.02	0.87
-"-		54.00	0.02	1.03	0.13	N.d.	4.25	0.11	17.47	23.56	0.12	N.d.	N.d.	N.d.	100.71	0.88
1596*	Harz	54.06	0.01	0.89	0.14	N.d.	2.76	0.13	16.76	24.37	0.02	N.f.	N.d.	N.d.	99.14	0.92
-"-		52.23	0.08	2.43	0.88	N.d.	3.40	0.11	16.37	23.60	0.10	0.01	N.d.	N.d.	99.20	0.90
-"-		53.88	0.08	1.21	0.26	N.d.	2.87	0.11	17.13	24.27	0.05	N.f.	N.d.	N.d.	99.87	0.92
138*	Lherz	52.64	0.19	2.53	0.29	N.d.	4.22	0.15	16.40	23.46	0.07	N.d.	N.d.	N.d.	99.95	0.87
-"-		48.90	0.21	3.46	0.46	N.d.	1.94	0.11	16.47	23.86	0.29	N.d.	0.04	N.d.	98.88	0.86
154*	Lherz	53.57	0.04	2.79	N.d.	N.d.	2.79	0.11	17.95	24.58	0.10	N.d.	N.d.	N.d.	101.82	0.92
-"-		50.09	N.d.	3.31	N.d.	N.d.	4.09	0.09	15.43	25.66	0.14	N.d.	N.d.	N.d.	98.81	0.87
142	Pl Lherz	53.83	0.01	2.40	N.d.	2.53	N.d.	N.d.	17.22	23.64	0.17	N.d.	N.d.	N.d.	99.61	0.93
-"-		51.85	0.059	2.58	0.930	N.d.	2.48	0.091	16.44	22.41	0.098	0.004	N.d.	N.d.	96.95	0.92
-"-		52.10	0.049	2.63	0.966	N.d.	2.47	0.090	16.43	22.72	0.144	N.d.	N.d.	N.d.	97.60	0.92
-"-		52.48	0.039	2.30	0.378	N.d.	2.48	0.106	17.02	21.65	0.139	0.022	N.d.	N.d.	96.62	0.92
-"-		52.96	0.035	2.18	0.384	N.d.	2.47	0.115	16.59	22.37	0.230	0.008	N.d.	N.d.	97.33	0.92
139*	Wehrl	54.41	0.03	1.68	0.30	N.d.	2.92	0.10	17.92	22.51	0.04	N.d.	N.d.	N.d.	99.90	0.92
-"-		54.74	0.05	1.94	0.027	N.d.	2.73	0.07	18.79	20.75	0.06	N.d.	N.d.	N.d.	99.40	0.92
-"-		54.41	0.03	1.70	0.26	N.d.	2.79	0.08	18.34	21.63	0.01	N.d.	N.d.	N.d.	99.26	0.93
-"-		54.03	0.04	1.64	0.40	N.d.	3.21	0.12	17.26	23.67	0.05	N.d.	N.d.	N.d.	100.43	0.91
-"-		54.45	N.d.	1.42	0.26	N.d.	2.93	0.13	17.29	23.99	0.04	N.d.	N.d.	N.d.	100.51	0.91
145	Ol Web	52.90	0.05	1.30	0.28	1.53	3.36	0.14	17.60	21.71	0.23	N.d.	0.01	N.d.	99.11	0.87
-"-		53.60	0.05	1.63	N.d.	N.d.	4.10	N.d.	16.85	23.65	0.10	N.d.	N.d.	N.d.	99.98	0.88
-"-		53.09	0.08	1.47	0.29	N.d.	3.70	0.122	16.40	24.25	0.068	0.001	N.d.	N.d.	99.48	0.83
-"-		52.76	0.072	1.82	0.432	N.d.	4.00	0.15	16.65	23.40	0.080	0.010	N.d.	N.d.	99.42	0.83
155	Ol Web	52.90	N.d.	1.20	0.45	0.11	2.96	0.14	18.27	22.64	0.27	N.d.	0.01	N.d.	98.95	0.91
-"-		53.40	0.08	1.50	0.45	N.d.	3.67	0.12	16.30	23.33	0.13	N.d.	N.d.	N.d.	98.97	0.89
134	Ol Web	53.83	0.06	1.50	0.19	0.09	4.07	0.13	17.27	21.72	0.14	N.d.	0.02	N.d.	99.02	0.88
134*	Ol Web	49.68	N.d.	1.79	0.89	N.d.	2.44	0.04	17.36	24.12	0.24	N.d.	N.d.	N.d.	96.56	0.93

Sample	Type														
137	Ol Web	53.00	0.06	2.10	0.33	0.09	4.07	0.07	17.94	21.25	0.16	N.d.	0.03	N.d.	99.10	0.89
140	Ol Web	50.40	0.10	2.37	0.38	1.22	3.48	0.14	16.98	23.38	0.25	N.d.	0.01	N.d.	98.71	0.87
"	"	52.28	0.08	2.41	0.49	N.d.	4.30	0.14	16.41	23.58	0.11	N.d.	N.d.	N.d.	99.79	0.87
1603-1*	Ol Web	52.93	0.107	2.00	0.459	N.d.	4.04	0.123	16.25	23.27	0.089	N.d.	N.d.	N.d.	99.47	0.88
"	"	52.97	0.100	2.14	0.495	N.d.	3.73	0.154	15.94	24.05	0.042	N.d.	N.d.	N.d.	99.61	0.88
1593*	Ol Web	52.90	0.07	2.63	0.95	N.d.	3.62	0.12	15.65	23.72	0.13	N.d.	N.d.	N.d.	99.78	0.87
"	"	52.68	0.06	2.71	0.82	N.d.	3.73	0.11	15.71	23.49	0.12	N.d.	N.d.	N.d.	99.41	0.88
"	"	52.95	0.07	2.42	0.65	N.d.	3.80	0.12	16.17	23.11	0.13	N.d.	N.d.	N.d.	99.41	0.88
"	"	52.97	0.07	2.45	0.64	N.d.	3.42	0.11	15.77	23.84	0.10	N.d.	N.d.	N.d.	99.37	0.89
"	"	52.57	0.09	2.51	0.70	N.d.	3.31	0.10	15.81	23.67	0.06	N.d.	N.d.	N.d.	98.80	0.90
"	"	52.59	0.07	2.63	0.72	N.d.	3.58	0.12	15.79	23.31	0.12	N.d.	N.d.	N.d.	98.93	0.89
"	"	53.05	0.10	2.64	0.70	N.d.	3.74	0.12	15.66	23.32	0.13	0.01	N.d.	N.d.	99.47	0.88
143*	Web	51.85	0.09	1.29	N.d.	N.d.	5.40	N.d.	15.00	23.03	0.18	N.d.	N.d.	N.d.	96.84	0.83
"	"	53.27	0.117	1.22	0.216	N.d.	5.57	0.179	15.57	22.97	0.225	0.010	N.d.	N.d.	99.36	0.83
"	"	53.53	0.112	1.14	0.214	N.d.	5.60	0.208	15.46	23.03	0.205	0.002	N.d.	N.d.	99.60	0.83
"	"	53.37	0.094	1.169	0.121	N.d.	5.43	0.181	15.33	23.16	0.269	N.d.	N.d.	N.d.	99.12	0.83
"	"	53.20	0.143	1.429	0.198	N.d.	7.13	0.220	16.75	20.30	0.239	N.d.	N.d.	N.d.	99.61	0.81
150	Web	51.50	N.d.	2.40	0.17	1.72	4.66	0.21	14.95	23.10	0.24	N.d.	0.01	N.d.	98.96	0.81
152*	Web	53.87	0.04	1.86	N.d.	N.d.	2.03	N.d.	17.55	24.30	0.20	N.d.	N.d.	N.d.	99.85	0.94
"	"	53.98	0.02	1.41	0.76	N.d.	2.12	0.08	17.48	24.13	0.14	0.01	N.d.	N.d.	100.10	0.94
154-2*	Web	51.52	N.d.	1.80	N.d.	N.d.	4.33	0.09	16.20	26.08	0.16	N.d.	N.d.	N.d.	100.16	0.87
Cx-2*	Ol Cpx	52.90	0.040	2.54	1.05	N.d.	3.01	0.094	17.02	23.25	0.137	0.011	N.d.	N.d.	100.06	0.91
"	"	52.56	0.031	2.45	0.938	N.d.	3.05	0.079	17.60	23.45	0.099	N.f.	N.d.	N.d.	99.26	0.91
"	"	53.05	0.033	2.49	0.910	N.d.	3.12	0.087	17.17	23.54	0.178	N.f.	N.d.	N.d.	100.58	0.91
"	"	53.65	0.045	1.77	1.45	N.d.	2.66	0.92	16.76	23.68	0.065	N.d.	N.d.	N.d.	100.18	0.92
"	"	51.47	0.037	2.54	3.12	N.d.	3.09	0.066	16.16	22.72	0.124	N.d.	N.d.	N.d.	99.33	0.90
1606-1*	Ol Cpx	52.48	0.039	2.28	0.982	N.d.	2.94	0.093	16.74	23.46	0.106	0.005	N.d.	N.d.	99.23	0.91
155-3*	Ol Gb	51.79	0.047	1.63	0.057	N.d.	5.65	0.156	15.79	21.40	0.179	0.005	N.d.	N.d.	96.69	0.83
155-5*	Ol Gb	51.22	0.295	2.69	0.179	N.d.	6.88	0.200	15.51	22.11	0.232	N.f.	N.d.	N.d.	99.31	0.80
"	"	51.53	0.320	2.62	0.226	N.d.	8.64	0.232	16.57	19.42	0.172	0.005	N.d.	N.d.	99.74	0.77
"	"	52.34	0.228	2.23	0.108	N.d.	8.42	0.63	16.94	18.43	0.154	0.005	N.d.	N.d.	99.13	0.78

(Continued)

Table 8.6 Continued.

Sample 1	Rock 2	SiO₂ 3	TiO₂ 4	Al₂O₃ 5	Cr₂O₃ 6	Fe₂O₃ 7	FeO 8	MnO 9	MgO 10	CaO 11	Na₂O 12	K₂O 13	NiO 14	H₂O 15	Total 16	Mg# 17
148-1*	Ol Gb	52.43	0.081	1.97	0.353	N.d.	5.34	0.131	15.87	23.06	0.152	0.010	N.d.	N.d.	99.40	0.84
-"-		52.67	0.024	1.74	0.373	N.d.	5.29	0.142	15.99	23.25	0.207	0.002	N.d.	N.d.	99.68	0.84
128	Gbn	53.00	0.07	2.62	0.16	0.01	5.32	N.d.	15.45	22.18	0.30	N.d.	0.03	N.d.	99.30	0.84
130*	Gbn	53.14	0.03	1.56	0.08	N.d.	5.19	0.14	15.99	22.99	0.13	N.d.	N.d.	N.d.	99.25	0.85
-"-		54.03	0.01	2.02	N.d.	0.20	4.81	0.14	15.95	21.72	0.30	N.d.	0.03	N.d.	99.21	0.85
131	Gbn	54.03	0.05	2.16	0.03	0.21	4.66	0.14	15.45	22.18	0.30	N.d.	0.03	N.d.	99.24	0.85
131a*	Gbn	54.50	0.06	2.12	N.d.	N.d.	4.53	0.20	16.91	20.26	0.17	N.d.	N.d.	N.d.	98.55	0.87
132a*	Gbn	52.46	0.14	2.20	N.d.	N.d.	4.60	0.17	16.81	23.33	0.19	N.d.	N.d.	N.d.	99.93	0.87
138a*	Gbn	50.97	0.12	3.02	0.62	N.d.	5.57	0.11	16.51	23.04	0.23	N.d.	N.d.	N.d.	98.81	0.84
141	Gbn	49.90	0.12	3.02	0.62	2.96	1.80	N.d.	16.81	23.38	0.43	N.d.	0.01	N.d.	99.16	0.87
144	Gbn	53.82	0.12	2.46	N.d.	N.d.	4.76	0.11	16.26	22.17	0.21	N.d.	N.d.	N.d.	99.80	0.86
146	Gbn	53.30	N.d.	2.33	0.15	0.61	4.58	0.21	15.61	22.18	0.35	N.d.	0.01	N.d.	99.33	0.84
147	Gbn	50.40	0.12	2.38	0.13	1.51	4.14	0.11	15.76	23.86	0.39	N.d.	0.01	N.d.	98.81	0.84
-"-		54.04	0.08	1.30	N.d.	N.d.	4.69	N.d.	16.13	23.33	0.23	N.d.	0.01	N.d.	100.00	0.86
149	Gbn	52.00	0.26	2.40	0.30	2.47	3.11	N.d.	14.95	23.10	0.32	N.d.	0.01	N.d.	98.92	0.83
151	Gbn	53.40	N.d.	1.80	0.18	0.56	3.55	0.21	15.61	22.64	0.34	N.d.	0.01	N.d.	98.30	0.87
1607*	Gbn	54.10	0.024	0.75	0.04	N.d.	2.52	0.164	16.65	24.93	N.f.	N.f.	N.d.	N.d.	99.18	0.92
-"-		53.93	N.f.	1.84	0.183	N.d.	3.57	0.125	17.02	22.39	0.124	0.006	N.d.	N.d.	99.18	0.89
-"-		53.76	0.021	0.73	0.169	N.d.	2.89	0.146	15.75	25.23	0.041	N.f.	N.d.	N.d.	98.75	0.91
-"-		54.22	0.007	1.41	0.248	N.d.	3.18	0.150	16.23	24.77	0.075	N.f.	N.d.	N.d.	100.29	0.90
-"-		53.48	0.001	1.64	0.158	N.d.	3.37	0.141	16.85	24.60	0.027	N.f.	N.d.	N.d.	100.25	0.90
-"-		53.88	0.017	0.99	0.137	N.d.	3.09	0.137	15.85	25.34	0.027	0.009	N.d.	N.d.	99.49	0.90
1605*	Gbn	52.97	0.072	2.24	0.171	N.d.	4.37	0.121	16.53	23.62	N.f.	0.104	N.d.	N.d.	100.19	0.87
-"-		52.80	0.068	2.18	0.360	N.d.	4.42	0.152	16.32	23.13	N.f.	0.163	N.d.	N.d.	99.60	0.87
-"-		52.58	0.051	2.16	0.209	N.d.	5.11	0.129	15.40	23.59	0.215	0.003	N.d.	N.d.	99.46	0.84
129*	Gb	52.96	0.03	2.19	0.17	N.d.	4.52	0.17	16.53	23.59	0.13	0.01	N.d.	N.d.	100.30	0.87
-"-		53.69	0.10	2.46	0.18	N.d.	4.62	0.14	15.52	22.18	0.14	N.d.	N.d.	N.d.	99.03	0.86
-"-		52.96	0.03	2.19	0.17	N.d.	4.52	0.17	16.53	23.59	0.13	0.01	N.d.	N.d.	100.30	0.87
1612*	Gb	52.60	0.066	2.01	0.44	N.d.	4.92	0.136	18.60	20.85	0.135	N.d.	N.d.	N.d.	99.74	0.87
-"-		53.51	0.070	2.16	0.46	N.d.	4.25	0.147	16.45	23.39	0.154	N.d.	N.d.	N.d.	100.61	0.87
155-4*	Leuc Gb	52.96	0.041	1.866	0.338	N.d.	4.17	0.126	15.30	24.02	0.204	0.017	N.d.	N.d.	99.03	0.87
155-1*	Leuc Ol Gnb	52.30	0.474	3.07	0.33	N.d.	4.93	0.14	16.26	21.89	0.243	0.007	N.d.	N.d.	99.64	0.85

Shel'ting massif

176*	Cpx–Ol Enst	51.84	0.08	1.91	0.32	N.d.	6.75	0.17	15.34	21.99	0.15	N.d.	N.d.	N.d.	98.55	0.84
182*	Gbn	52.59	0.099	1.95	0.487	N.d.	4.77	0.134	16.65	19.91	0.227	0.007	N.d.	N.d.	96.83	0.86
„	„	52.18	0.120	2.40	0.461	N.d.	4.86	0.125	16.22	20.16	0.351	0.010	N.d.	N.d.	96.88	0.86
„	„	52.35	0.092	2.11	0.499	N.d.	4.66	0.112	16.56	20.56	0.316	N.d.	N.d.	N.d.	97.29	0.86
„	„	52.24	0.125	2.17	0.471	N.d.	5.11	0.116	16.25	20.07	0.264	0.019	N.d.	N.d.	96.83	0.85
183*	Gbn	52.63	0.11	2.11	0.24	N.d.	5.65	0.16	16.02	22.78	0.24	0.004	N.d.	N.d.	99.94	0.83
„	„	52.71	0.10	1.97	0.20	N.d.	5.38	0.16	16.02	23.03	0.24	0.009	N.d.	N.d.	99.82	0.83
„	„	52.28	0.12	2.17	0.21	N.d.	5.46	0.16	16.40	22.98	0.22	N.d.	N.d.	N.d.	100.00	0.84

Komsomol'sk massif

191*	Ol Gbn	51.91	0.087	3.57	0.323	N.d.	4.46	0.150	15.16	23.49	0.139	0.003	N.d.	N.d.	99.29	0.86
„	„	53.14	0.058	2.63	0.182	N.d.	4.10	0.141	15.47	23.75	0.100	0.011	N.d.	N.d.	99.59	0.87
189*	Gbn	52.33	0.11	2.46	0.15	N.d.	5.80	0.20	15.73	22.43	0.15	N.d.	N.d.	N.d.	99.35	0.83
„	„	52.07	0.11	2.88	0.16	N.d.	5.91	0.17	15.17	22.37	0.21	N.d.	N.d.	N.d.	99.05	0.82
„	„	52.45	0.35	2.49	0.50	N.d.	6.25	0.16	15.91	20.64	0.29	N.d.	N.d.	N.d.	99.04	0.82
„	„	52.15	0.13	2.93	0.18	N.d.	6.15	0.19	14.86	22.63	0.28	0.004	N.d.	N.d.	99.51	0.81
„	„	52.03	0.11	2.93	0.14	N.d.	6.07	0.16	14.99	23.30	0.23	0.01	N.d.	N.d.	99.98	0.81

South Schmidt massif

161*	Lherz	54.24	0.004	1.39	0.501	N.d.	2.03	0.092	17.07	23.64	0.200	0.006	N.d.	N.d.	99.18	0.94
„	„	53.68	0.024	1.90	0.948	N.d.	1.98	0.088	16.52	23.77	0.181	N.d.	N.d.	N.d.	99.09	0.94
„	„	53.36	0.022	1.75	0.889	N.d.	2.55	0.089	16.97	23.27	0.196	0.006	N.d.	N.d.	99.09	0.92
163*	Lherz	51.99	0.092	3.44	1.05	N.d.	2.31	0.057	15.86	23.95	0.081	N.f.	N.d.	N.d.	98.83	0.92
„	„	53.09	0.079	2.59	0.724	N.d.	2.11	0.081	16.17	24.07	0.102	0.003	N.d.	N.d.	99.00	0.93
„	„	52.51	0.062	3.04	0.950	N.d.	2.13	0.082	16.27	24.13	0.101	N.f.	N.d.	N.d.	99.28	0.93
„	„	52.04	0.092	3.59	1.166	N.d.	2.16	0.070	15.81	24.35	0.102	0.009	N.d.	N.d.	99.38	0.93
„	„	52.36	0.075	3.44	1.007	N.d.	2.28	0.082	15.74	23.98	0.096	N.f.	N.d.	N.d.	99.06	0.92
162*	Web	54.70	0.020	0.59	0.471	N.d.	1.37	0.058	17.98	24.78	0.133	N.d.	N.d.	N.d.	100.10	0.96
„	„	55.47	0.008	0.26	0.163	N.d.	1.38	0.071	16.99	24.94	N.d.	N.d.	N.d.	N.d.	99.28	0.96

Note: Compiled after the data in Table 8.1.

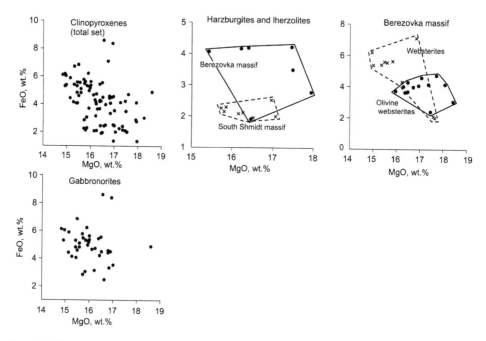

Figure 8.6 Variations in MgO and FeO contents in clinopyroxenes from ultramafites and gabbroids of the Berezovka, Shel'ting, Komsomol'sk, and South Schmidt massifs (data from Table 8.6).

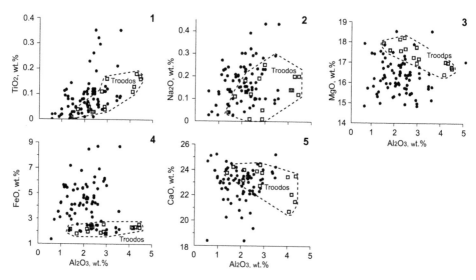

Figure 8.7 Variations in MgO, FeO, Al$_2$O$_3$, CaO, TiO$_2$, and Na$_2$O contents in clinopyroxenes from ultramafites and gabbroids of the Berezovka, Shel'ting, Komsomol'sk, and South Schmidt massifs (data from Table 8.6) and of the Troodos massif, Cyprus (after [Sobolev and Batanova, 1995]).

Table 8.7 Limited contents of wollastonite (Wo), enstatite (En), and ferrosilite (Fs) end-members in clinopyroxenes from ultramafic rocks and gabbroids of the Berezovka and other massifs, %.

Rock	Number of analyses	Wo	En	Fs
Harzburgite, lherzolites	17	50.69–46.00	48.49–45.64	6.93–3.22
Wehrlites, pyroxenites, olivine gabbro	45	50.23–39.48	53.34–43.78	13.69–2.08
Gabbronorites, gabbro	28	51.07–41.21	51.18–42.64	10.25–3.96

Note: The contents of end-members were calculated from the data in Table 8.6.

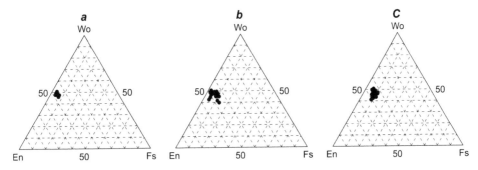

Figure 8.8 En–Wo–Fs composition (%) diagrams for clinopyroxenes from ultramafic rocks and gabbroids of the Berezovka, Shel'ting, Komsomol'sk, and South Schmidt massifs (data from Tables 8.6 and 8.7).
a – from harzburgites and lherzolites; *b* – from plagiolherzolites, wehrlites, olivine and olivine-free websterites, clinopyroxenites, olivine gabbronorites, and olivine gabbro; *c* – from gabbronorites and gabbro.

have elevated contents of this impurity. The Cr_2O_3 contents in clinopyroxenes from two-pyroxene rocks are nearly twice as high as those in the coexisting orthopyroxenes. The TiO_2 contents in the total set of clinopyroxenes vary from 0.001 to 0.350 wt.% (Fig. 8.9) but do not exceed 0.10 wt.% in most of the grains. The Na_2O contents in clinopyroxenes are within 0.027–0.430 wt.%, not exceeding 0.25 wt.% in most of the grains (Fig. 8.9). High contents of this impurity were detected in some mineral grains from gabbroids.

The location of figurative points on the composition diagrams suggests a direct relationship between TiO_2 and Na_2O contents in clinopyroxenes from gabbroids, wehrlites, and websterites (Fig. 8.9, 3,4). The contents of MnO impurity vary from the lowest in the mineral grains from lherzolites of the South Schmidt massif (0.057–0.092 wt.%) to high in the grains from olivine and olivine-free gabbroids of the Berezovka massif (0.14–0.63 wt.%).

8.1.4 Plagioclases

The chemical composition of plagioclases was studied on a sample of grains from gabbronorites, gabbro, olivine gabbro, olivine gabbronorites, and gabbro-pegmatites

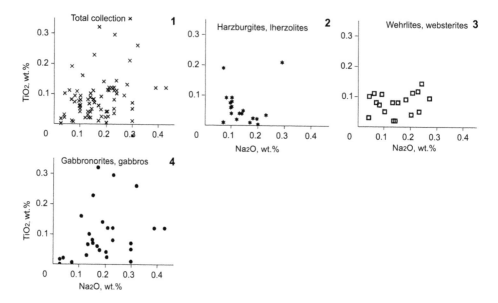

Figure 8.9 Variations in Na$_2$O and TiO$_2$ contents in clinopyroxenes from ultramafites and gabbroids of the Berezovka, Shel'ting, Komsomol'sk, and South Schmidt massifs (data from Table 8.7). 1 – total set of clinopyroxenes; 2 – clinopyroxenes from harzburgites and lherzolites; 3 – clinopyroxenes from wehrlites and websterites; 4 – clinopyroxenes from gabbronorites and gabbro.

of the Berezovka, Shel'ting, and Komsomol'sk massifs (Table 8.8). The CaO contents in the total sample of plagioclases are within 14.67–20.13 wt.%, and the Al$_2$O$_3$ contents are within 32.87–36.27 wt.%. Plagioclases from gabbroids of the Berezovka and Komsomol'sk massifs are richer in these components than plagioclases from gabbronorites of the Shel'ting massif (Fig. 8.10). The position of figurative points on the CaO–Al$_2$O$_3$ composition diagram suggests a direct relationship between the contents of these components. The contents of the anorthite end-member (An) in plagioclases from rocks of the Berezovka massif vary from 75.7% (bytownite from gabbronorites) to 95.8–97.1% (anorthite from gabbro and leucocratic gabbro).

Plagioclases from the studied collection of rocks differ slightly in the contents of structural FeO impurity. For example, plagioclase from gabbronorites of the Berezovka massif contains 0.007–0.550 wt.% FeO, and that from similar rocks of the Shel'ting massif, 0.293–0.495 wt.%. Earlier it was established that the contents of this isomorphous structural impurity in the high-Ca plagioclases decrease from ~1 wt.% in the mineral microlites from bulk rapidly crystallized basalts to 0.20–0.40 wt.%. in the grains from more slowly crystallized mesoabyssal gabbroids and to few hundredths of wt.% in plagioclases from gabbroids crystallized during the very slow cooling of melt at great depths [Lesnov and Korolyuk, 1977; Lesnov, 1991]. The above observations and the obtained data on FeO contents in plagioclases from gabbroids of the Sakhalin massifs suggest that a considerable portion of these gabbroids crystallized during the moderate cooling of basaltoid melt, i.e., in mesoabyssal conditions. In particular, plagioclases from gabbronorite (sample 1607), containing 0.007–0.160 wt.% FeO, probably crystallized more slowly than those from gabbroids, richer in this

Table 8.8 Chemical composition of plagioclases from rocks of the Berezovka, Shel'ting, and Komsomol'sk massifs, wt.%.

Sample 1	Rock 2	SiO$_2$ 3	TiO$_2$ 4	Al$_2$O$_3$ 5	Cr$_2$O$_3$ 6	Fe$_2$O$_3$ 7	FeO 8	MnO 9	MgO 10	CaO 11	Na$_2$O 12	K$_2$O 13	NiO 14	H$_2$O 15	Total 16	An, % 17
Berezovka massif																
155-3*	Ol Gb	45.46	N.d.	34.98	N.d.	N.d.	0.342	0.003	N.d.	18.23	0.647	0.020	N.d.	N.d.	99.68	93.9
-"-	-"-	44.98	0.010	34.63	0.003	N.d.	0.411	N.d.	N.d.	18.29	0.732	0.012	N.d.	N.d.	99.07	93.2
148-1*	Ol Gb	44.72	N.f.	35.00	0.038	N.d.	0.218	0.002	N.f.	18.97	0.792	0.013	N.d.	N.d.	99.75	92.9
-"-	-"-	43.96	N.f.	35.41	0.021	N.d.	0.109	0.007	N.f.	19.40	0.529	N.f.	N.d.	N.d.	99.44	95.3
155-5*	Ol Gb	45.38	0.019	34.07	0.005	N.d.	0.315	0.003	N.f.	17.93	1.17	0.012	N.d.	N.d.	98.90	89.4
-"-	-"-	45.39	0.016	34.68	0.022	N.d.	0.381	N.f.	N.f.	18.53	1.00	0.018	N.d.	N.d.	100.03	91.0
-"-	-"-	43.63	0.011	35.28	0.019	N.d.	0.218	0.007	N.f.	19.32	0.42	0.016	N.d.	N.d.	98.92	96.1
-"-	-"-	45.64	0.006	34.16	0.015	N.d.	0.353	N.f.	N.d.	18.17	1.20	0.014	N.d.	N.d.	99.56	89.3
128	Gbn	44.10	N.d.	34.00	N.d.	0.07	0.55	N.d.	0.55	18.53	1.20	0.30	N.d.	1.28	100.03	88.2
130	Gbn	44.26	N.d.	34.96	N.d.	0.46	0.11	N.d.	0.74	16.21	1.05	0.64	N.d.	1.80	100.04	85.7
131	Gbn	43.30	N.d.	33.23	N.d.	N.d.	0.55	N.d.	0.37	14.67	1.56	1.45	N.d.	4.00	99.50	75.7
131a	Gbn	42.78	N.d.	35.17	N.d.	0.65	0.21	N.d.	0.37	17.75	0.76	0.24	N.d.	2.16	100.09	91.4
132a	Gbn	42.66	N.d.	35.39	N.d.	0.71	0.15	N.d.	0.37	17.75	0.76	0.42	N.d.	2.32	100.53	90.3
141	Gbn	41.51	N.d.	33.66	N.d.	0.34	0.09	N.d.	0.37	18.57	1.30	0.20	N.d.	3.76	99.80	87.8
144	Gbn	43.40	N.d.	34.53	N.d.	0.31	0.24	N.d.	0.74	16.98	1.54	0.64	N.d.	2.12	100.50	83.5
146	Gbn	43.50	N.d.	33.90	N.d.	0.19	0.55	N.d.	0.55	16.47	1.55	0.20	N.d.	1.90	98.81	84.0
147	Gbn	43.38	N.d.	34.53	N.d.	0.65	0.21	N.d.	0.37	16.97	1.34	0.80	N.d.	1.68	99.94	83.9
151	Gbn	43.22	N.d.	34.52	N.d.	0.27	0.15	N.d.	0.18	17.75	0.84	0.64	N.d.	N.d.	99.55	89.3
1607*	Gbn	43.15	0.026	34.16	0.057	N.d.	0.007	N.f.	N.f.	18.99	0.703	0.011	N.d.	N.d.	97.10	93.7
-"-	-"-	43.32	N.f.	34.85	0.042	N.d.	0.011	N.f.	N.f.	19.70	0.269	N.f.	N.d.	N.d.	98.20	97.6
-"-	-"-	43.05	N.f.	34.92	0.021	N.d.	0.162	0.013	N.f.	20.04	0.192	N.f.	N.d.	N.d.	98.39	98.3
-"-	-"-	43.60	0.012	35.89	0.073	N.d.	0.107	N.f.	N.f.	19.85	0.251	N.f.	N.d.	N.d.	99.78	97.8
-"-	-"-	43.47	N.f.	36.27	0.019	N.d.	0.054	N.f.	N.f.	20.04	0.129	N.f.	N.d.	N.d.	99.98	98.8
-"-	-"-	43.16	N.f.	36.06	0.056	N.d.	0.033	N.f.	N.f.	20.13	0.056	N.f.	N.d.	N.d.	99.49	99.5

(Continued)

Table 8.8 Continued.

Sample 1	Rock 2	SiO₂ 3	TiO₂ 4	Al₂O₃ 5	Cr₂O₃ 6	Fe₂O₃ 7	FeO 8	MnO 9	MgO 10	CaO 11	Na₂O 12	K₂O 13	NiO 14	H₂O 15	Total 16	An, % 17
1605*	Gbn	43.12	0.020	36.01	0.015	N.d.	0.186	N.f.	N.f.	19.58	0.216	N.f.	N.d.	N.d.	99.16	98.0
1612*	Gbn	44.16	N.f.	35.73	N.f.	N.d.	0.133	0.005	N.f.	19.57	0.329	N.f.	N.d.	N.d.	99.92	97.1
-"-	-"-	43.63	N.f.	35.52	0.015	N.d.	0.135	0.013	N.f.	19.49	0.495	N.f.	N.d.	N.d.	99.30	95.6
155-4*	Leuc Gb	45.05	0.014	35.12	N.d.	N.d.	0.234	N.d.	N.d.	19.07	0.672	0.014	N.d.	N.d.	100.17	93.9
-"-	-"-	44.80	N.d.	35.42	0.006	N.d.	0.159	N.d.	N.d.	19.11	0.484	0.009	N.d.	N.d.	99.99	95.8
155-1*	Leuc Ol Gbn	47.56	0.031	33.01	0.002	N.d.	0.158	0.037	N.d.	16.54	1.96	N.d.	N.d.	N.d.	99.30	82.3
-"-	-"-	45.19	0.003	34.72	N.d.	N.d.	0.030	0.003	N.d.	18.29	0.798	0.005	N.d.	N.d.	99.05	92.7
Shel'ting massif																
182*	Gbn	47.25	0.011	32.87	0.014	N.d.	0.293	N.d.	N.d.	16.95	1.74	0.021	N.d.	N.d.	99.15	84.2
-"-	-"-	47.40	0.015	33.17	N.d.	N.d.	0.312	0.011	N.d.	16.88	1.91	0.025	N.d.	N.d.	99.73	82.9
183*	Gbn	45.28	N.d.	33.27	0.013	N.d.	0.495	0.009	N.d.	17.99	1.29	0.011	N.d.	N.d.	98.35	88.5
-"-	-"-	45.87	0.032	34.12	N.d.	N.d.	0.487	N.d.	N.d.	17.94	1.22	0.009	N.d.	N.d.	99.97	89.0
-"-	-"-	45.83	0.028	34.18	0.004	N.d.	0.451	0.004	N.d.	18.21	1.29	0.019	N.d.	N.d.	100.01	88.6
-"-	-"-	45.49	N.d.	33.98	0.018	N.d.	0.370	0.013	N.d.	17.77	1.34	0.005	N.d.	N.d.	98.99	88.0
-"-	-"-	46.16	N.d.	33.74	0.007	N.d.	0.328	0.006	N.d.	17.62	1.63	0.003	N.d.	N.d.	99.40	86.4
-"-	-"-	45.75	0.006	34.05	N.d.	N.d.	0.376	N.d.	N.d.	18.16	1.181	0.020	N.d.	N.d.	99.54	89.4
-"-	-"-	45.29	0.031	34.17	0.012	N.d.	0.325	0.018	N.d.	18.09	1.038	0.011	N.d.	N.d.	98.98	90.6
Komsomol'sk massif																
189*	Gbn	44.61	0.019	35.39	N.d.	N.d.	0.322	N.d.	N.d.	19.93	0.62	0.003	N.d.	N.d.	99.90	94.4
-"-	-"-	44.13	0.018	35.97	N.d.	N.d.	0.204	N.d.	N.d.	19.25	0.39	0.005	N.d.	N.d.	99.96	96.4
192*	Pegm Gb	45.51	N.d.	35.23	0.016	N.d.	0.285	N.d.	N.d.	17.21	0.653	0.016	N.d.	N.d.	98.92	93.5

Note: Compiled after the data in Table 8.1.

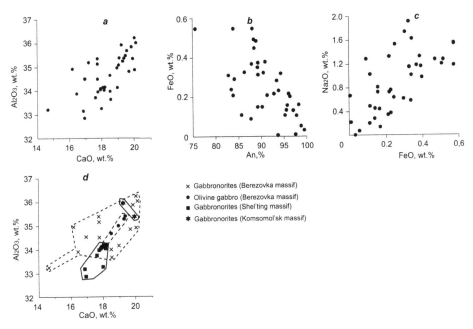

Figure 8.10 Variations in CaO,Al₂O₃, Na₂O,and FeO contents and anorthite end-member (An) content in plagioclases from gabbroids of the Berezovka, Shel'ting, and Komsomol'sk massifs. *a, c, d* – total set of plagioclases; *b* – plagioclases from gabbroids of different massifs (data from Table 8.8).

impurity (wt.%): sample 182 (0.29–0.31), sample 183 (0.33–0.50), and samples 128, 131, and 146 (0.55). The location of figurative points on the composition diagram (Fig. 8.10) shows a tendency for a direct relationship between FeO and Na₂O contents and an inverse relationship between FeO and An contents.

Let us consider specific variations in the chemical composition of plagioclases in parallel-banded rocks usually present in contact reaction zones of polygenic mafic-ultramafic massifs. As mentioned above, the massifs of the East Sakhalin ophiolite association (particularly, the Berezovka massif) are formed by hybrid gabbroids and plagioclase-containing ultramafites of different quantitative mineral compositions and of taxitic, including parallel-banded, texture. The genesis of such banded rocks was treated by many petrologists based on the model of gravitative crystallization differentiation of mafic melts. According to this model, such rocks are regarded as "cumulates" [Wager and Brown, 1970]. The model rests upon the concept that the cooling and crystallization of mafic melts at different levels of magma chamber is accompanied by a rhythmic change in the quantitative mineral composition of rocks and the chemical composition of the hosted minerals, in particular, plagioclases, in the series from early high-temperature to late low-temperature ones. These specific variations in the composition of minerals in mafic-ultramafic rocks are the major evidence for their "latent foliation".

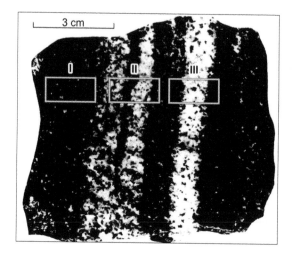

Figure 8.11 Cross section of ultramafic-mafic rock with a parallel-banded structure (sample 115-3, Berezovka massif).
The quantitative mineral composition of alternating rock "bands" varies from plagiowehrlite (dark gray sites) to leucocratic gabbro and anorthosite (light sites). I, II, and III – fragments where CaO, Na₂O, and Al₂O₃ contents in plagioclase grains were determined by X-ray spectral analysis (see Fig. 8.12).

Since the above problem is important for studying the genesis of complex mafic-ultramafic massifs and the "cumulative" mechanism of crystallization and differentiation of mafic melts should be verified, a mineralogical/geochemical experiment was carried out. Chemical analyses of plagioclase grains from a lump of parallel-banded ultramafic-mafic rock (Fig. 8.11, Table 8.9) showed nearly constant CaO and Na₂O contents independently of the quantitative mineral composition of the alternating rock "bands" (Fig. 8.12). This indicates that the parallel-banded rock lacks "latent foliation" caused by gravitative crystallization differentiation of mafic melt. The nearly constant composition of plagioclase grains in the alternating rock "bands" differing in quantitative mineral compositions can be explained based on the mechanism of percolation of mafic melt into ultramafic rock along a system of subparallel fractures in platy jointing [Lesnov, 1981a]. The process was, most likely, accompanied by the more or less active magmatic and metasomatic melt–ultramafite interaction and contamination of the melt with the rock substance. This mechanism of formation of parallel-banded rocks in contact reaction zones of mafic-ultramafic massifs agrees with the concept of the polygenic formation of the Berezovka and similar massifs [Lesnov, 1984, 1986a,b, 1988].

8.1.5 Amphiboles

The chemical composition of amphiboles was studied on a small collection of their grains from olivine gabbro, gabbronorites, gabbro, leucocratic gabbro, gabbro-pegmatites, and schriesheimites (cortlandites) of the Berezovka and Komsomol'sk massifs (Table 8.10). Microscopic examination showed that amphiboles in these

Table 8.9 Incomplete chemical analyses of plagioclase grains from parallel-banded wehrlite–gabbroid (sample 155-3, Berezovka massif), wt.%.

Grain	SiO_2	Al_2O_3	CaO	FeO	Na_2O
1	N.d.	35.31	18.41	N.d.	0.74
2	43.38	36.74	19.23	0.33	0.64
3	N.d.	36.04	18.47	N.d.	0.73
4	N.d.	35.17	18.98	N.d.	0.63
5	N.d.	34.86	19.37	N.d.	0.61
6	N.d.	36.05	19.04	N.d.	0.63
7	N.d.	35.76	19.24	N.d.	0.61
8	N.d.	35.36	19.32	N.d.	0.63
9	N.d.	35.74	19.45	N.d.	0.58
10	N.d.	35.07	19.49	N.d.	0.69
11	43.13	35.35	18.80	0.50	0.61
12	N.d.	34.87	18.62	N.d.	0.56
13	N.d.	35.53	18.54	N.d.	0.49
14	N.d.	35.21	18.48	N.d.	0.59
15	N.d.	35.36	18.64	N.d.	0.56
16	N.d.	36.21	18.51	N.d.	0.54
17	N.d.	35.69	18.79	N.d.	0.38
18	N.d.	35.13	18.52	N.d.	0.55
19	N.d.	35.54	18.70	N.d.	0.45
20	N.d.	35.33	18.70	N.d.	0.53
21	N.d.	35.05	18.44	N.d.	0.58
21a	N.d.	35.05	18.46	N.d.	0.53
22	N.d.	35.42	18.96	N.d.	0.45
23	N.d.	34.96	18.53	N.d.	0.68
24	N.d.	35.08	18.65	N.d.	0.57
25	N.d.	35.41	18.84	N.d.	0.49
26	N.d.	35.44	18.84	N.d.	0.59
27	N.d.	35.58	18.86	N.d.	0.43
Average	–	35.44 ± 0.43	18.82 ± 0.33	–	0.57 ± 0.09

Note: Analyses were carried out by X-ray spectrometry on a Cameca microprobe (along the profile shown in Figs. 8.11 and 8.12) at the Institute of Geology and Geophysics, Novosibirsk (analyst V.N. Korolyuk).

rocks (except for gabbro-pegmatites) resulted from the pseudomorphous replacement of clinopyroxenes at the late magmatic or postmagmatic stages of rock crystallization. Amphiboles from gabbro-pegmatites, whose veins occur in gabbroids of the Komsomol'sk massif, are, on the contrary, a phase crystallized from fluid-enriched residual mafic melt.

The FeO contents in the studied amphibole grains vary from 4.07 wt.% (gabbro, Berezovka massif) to 11.93 wt.% (gabbro-pegmatite, Komsomol'sk massif). The latter grains have the lowest MgO content (15.11 wt.%), and the highest one (22.70 wt.%) was found in the amphibole grain from gabbro of the Berezovka massif. As seen from the location of figurative points on the composition diagram, the FeO and MgO contents are in inverse relationship due to the mutual isomorphous substitution of these oxides (Fig. 8.13). The $Mg^{\#}$ values in the analyzed amphiboles vary from 0.75 to 0.91. The lowest Al_2O_3 content (1.29 wt.%) was detected in the amphibole grain from gabbro, and the highest ones, in the grains from schriesheimite (12.72 wt.%) and

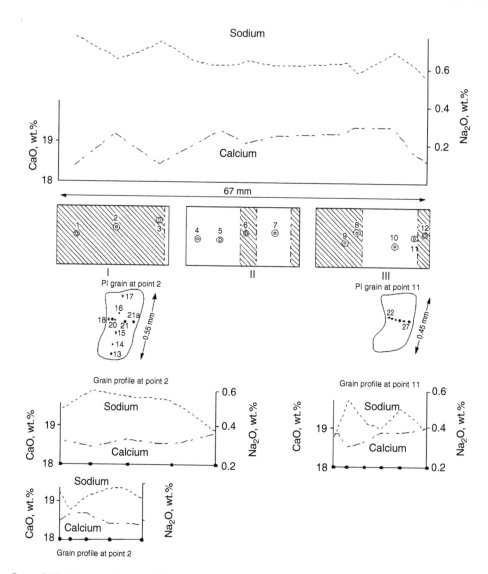

Figure 8.12 Scheme of study of the composition of plagioclase grains in parallel-banded ultramafic-mafic rock (sample 115-3, Berezovka massif) (data from Table 8.9).

1–12 – Points of determination of CaO, Na$_2$O, and Al$_2$O$_3$ contents along the profile in plagioclase grains (fragments I, II, and III). Oblique hatching marks the rock sites composed of plagiowehrlite, and white field marks the rock sites composed of leucocratic gabbro and anorthosite. The curves of CaO and Na$_2$O distribution in plagioclase grains are constructed from data in Table 8.9. The lower part of the figure shows the points of analyses in plagioclase grains (point 2, fragment I, and point 11, fragment III) and the curves of CaO and Na$_2$O distribution in them.

Table 8.10 Chemical composition of amphiboles from rocks of the Berezovka and Komsomol'sk massifs, wt.%.

Sample	Rock	SiO_2	TiO_2	Al_2O_3	Cr_2O_3	Fe_2O_3	FeO	MnO	MgO	CaO	Na_2O	K_2O	NiO	H_2O	Total	$Mg^{\#}$
Berezovka massif																
1596*	Harz	46.28	0.46	10.69	1.37	N.d.	5.85	0.08	17.65	12.43	1.28	0.30	N.d.	2.00	98.37	0.84
-"-	-"-	48.29	0.34	9.28	1.12	N.d.	5.64	0.09	18.44	12.61	1.08	0.19	N.d.	2.00	99.08	0.85
-"-	-"-	46.71	0.34	8.93	1.06	N.d.	6.22	0.07	20.20	11.37	1.00	0.21	N.d.	2.00	98.11	0.85
135*	Cortl	44.66	0.52	12.72	0.64	N.d.	7.14	0.16	17.96	11.32	2.31	0.21	N.d.	2.00	99.64	0.82
130*	Gbn	50.55	0.18	7.48	0.10	N.d.	7.63	0.11	17.40	12.15	0.96	0.09	N.d.	2.00	98.65	0.80
-"-	-"-	50.51	0.18	7.42	0.18	N.d.	7.70	0.11	16.96	11.99	0.91	0.06	N.d.	2.00	98.02	0.80
129*	Gb	57.18	0.01	1.29	0.03	N.d.	4.07	0.20	22.70	12.94	0.15	N.d.	N.d.	2.00	100.57	0.91
-"-	-"-	54.03	0.01	4.84	0.09	N.d.	5.32	0.18	21.54	12.24	0.74	0.02	N.d.	2.00	101.01	0.88
-"-	-"-	54.79	0.05	2.86	0.14	N.d.	3.94	0.09	20.60	12.92	0.29	0.01	N.d.	2.00	97.69	0.90
-"-	-"-	46.01	0.14	11.86	0.11	N.d.	7.13	0.11	16.74	12.47	1.55	0.06	N.d.	2.00	98.18	0.81
148-1*	Ol Gb	48.94	0.080	8.33	0.468	N.d.	8.26	0.131	17.49	12.94	1.26	0.102	N.d.	2.00	99.39	0.84
-"-	-"-	50.66	0.052	6.83	0.329	N.d.	7.40	0.127	18.45	12.25	1.16	0.083	N.d.	2.00	99.34	0.85
155-5*	Ol Gb	53.50	0.16	1.87	0.119	N.d.	12.04	0.304	19.79	12.25	0.110	N.f.	N.d.	N.d.	100.15	0.75
155-4*	Leuc Gb	46.26	0.135	11.83	0.248	N.d.	8.20	0.106	16.88	12.36	1.869	0.122	N.d.	2.00	100.00	0.78
-"-	-"-	45.58	0.176	12.48	0.500	N.d.	7.72	0.108	16.91	12.32	1.886	0.125	N.d.	2.00	99.81	0.78
-"-	-"-	45.29	0.136	13.50	0.411	N.d.	6.66	0.102	17.37	12.06	1.909	0.142	N.d.	2.00	99.58	0.77
Komsomol'sk massif																
189*	Gbn	44.48	0.37	12.14	0.27	N.d.	9.37	0.08	15.58	11.77	1.95	0.06	N.d.	2.00	98.07	0.75
-"-	-"-	44.35	0.403	12.40	0.276	N.d.	9.91	0.096	15.24	12.15	2.05	0.075	N.d.	2.00	98.94	0.73
191*	Ol Gbn	45.12	0.325	13.04	0.838	N.d.	6.91	0.108	16.23	12.58	1.54	0.35	N.d.	2.00	99.04	0.81
192*	Pegm Gb	45.84	0.663	11.81	0.021	N.d.	11.88	0.153	15.30	9.99	1.73	0.052	N.d.	2.00	99.44	0.76
-"-	-"-	45.61	0.682	11.77	0.100	N.d.	11.74	0.156	15.44	10.22	1.75	0.057	N.d.	2.00	99.52	0.76
-"-	-"-	45.91	0.722	12.23	N.d.	N.d.	11.93	0.159	15.11	10.05	1.70	0.055	N.d.	2.00	99.86	0.76

Note: Compiled after the data in Table 8.1.

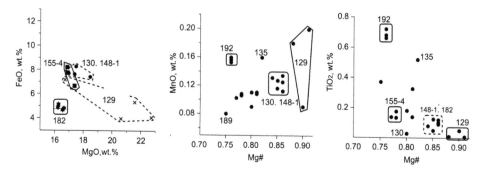

Figure 8.13 Variations in MgO, FeO, MnO, and TiO₂ contents and Mg# = Mg/(Mg + Fe) (f.u.) values in amphiboles from rocks of the Berezovka and Komsomol'sk massifs (data from Table 8.10).

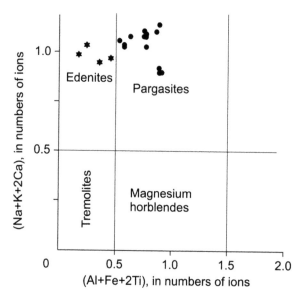

Figure 8.14 Compositions of amphiboles from rocks of the Berezovka and Komsomol'sk massifs (data from Table 8.10) on the classification diagram of Ca-amphiboles (after [Hawthorne et al., 2012]).

olivine gabbronorite (13.04 wt.%). The CaO contents in the studied amphiboles vary from 9.99 wt.% (gabbro-pegmatite) to 12.94 wt.% (gabbro).

The analyzed amphiboles differ in contents of TiO₂ and MnO impurities. The TiO₂ contents increase in the series from one amphibole grain from gabbro (0.01 wt.%) to grains from schriesheimite (0.52 wt.%) and then to grains from gabbro-pegmatites (0.66–0.72 wt.%), being in inverse relationship with Mg#. The MnO contents increase in the series from amphibole grains from gabbronorite (0.084 wt.%) to grains from gabbro-pegmatite (0.153–0.159 wt.%) and schriesheimite (0.16 wt.%). The location of figurative points on the composition diagram for the studied amphiboles shows a direct relationship between MnO contents and Mg# (Fig. 8.13).

All studied amphiboles have high contents of CaO (9.99–12.94 wt.%), which permits them to be assigned to the subgroup of Ca-amphiboles. The location of figurative points on the $(Na + K + 2Ca) - (Al + Fe + 2Ti)$ discrimination diagram indicates that some of the analyzed amphiboles are pargasite and the other are edenite (Fig. 8.14).

8.2   DISTRIBUTION COEFFICIENTS OF CHEMICAL COMPONENTS BETWEEN COEXISTING ROCK-FORMING MINERALS

One of the crucial tasks in geochemical studies of igneous rocks is elucidation of the regularities of distribution of chemical elements between the rock-forming minerals. As reported above, during the X-ray spectral analyses of minerals from rocks of the Sakhalin massifs, special attention was focused on the composition of coexisting phases in the rock samples [Lesnov, 2010a]. For this purpose, two or more grains of each mineral were analyzed, and the average contents of components were estimated in olivines, orthopyroxenes, clinopyroxenes, plagioclases, and amphiboles from representative samples of clinopyroxene-containing harzburgites, lherzolites, wehrlites, olivine and olivine-free websterites, and gabbronorites (Table 8.11). The estimates show that the average contents of FeO in olivines decrease and the average contents of MgO increase in the series from clinopyroxene-containing harzburgites to lherzolites and wehrlites. Orthopyroxenes from clinopyroxene-containing harzburgites are richer in FeO than orthopyroxenes from lherzolites, whereas the MgO contents in these minerals are nearly the same. Clinopyroxenes from clinopyroxene-containing harzburgites, lherzolites, wehrlites, and olivine websterites are characterized by slightly different average contents of MgO but close contents of FeO. The average contents of Al_2O_3 increase in the series from orthopyroxenes to clinopyroxenes and amphiboles. Clinopyroxenes from gabbronorites and gabbro are slightly richer in Na_2O than those from websterites, wehrlites, lherzolites, and clinopyroxene-containing harzburgites. The average $Mg^{\#}$ values increase from olivines (0.81–0.90) to orthopyroxenes (0.75–0.91) and clinopyroxenes (0.79–0.94), and in amphiboles they are much lower (0.10–0.20).

Using the average compositions of coexisting minerals (Table 8.11), we calculated the average distribution coefficients of chemical components (K_d) between olivines and orthopyroxenes, olivines and clinopyroxenes, orthopyroxenes and clinopyroxenes, and clinopyroxenes and plagioclases (Table 8.12). The calculation results were published earlier [Lesnov *et al.*, 2010c]. Let us consider them in more detail.

$K_d(olivine–orthopyroxene)$. The values of this coefficient were calculated for clinopyroxene-containing harzburgites, lherzolites, plagiolherzolites, and olivine websterites. In this series of rocks, the K_d values of components are in the following ranges: SiO_2 (0.7–0.8), FeO (0.6–1.7), MnO (0.3–1.2), MgO (1.2–1.7), and CaO (0.01–0.09). As evidenced from these values, MgO is present mainly in olivines, and SiO_2 and CaO are localized in orthopyroxenes. The distribution of FeO and MnO between olivine and orthopyroxene shows no preference. For example, in lherzolites FeO, like MnO, is concentrated mostly in olivines, whereas in plagiolherzolite both components are present predominantly in orthopyroxenes. The position of figurative points on the K_d covariation diagrams for MgO, FeO, CaO, and MnO testifies to an inverse

Table 8.11 Average chemical compositions of coexisting olivines, orthopyroxenes, clinopyroxenes, plagioclases, and amphiboles from rocks of the Berezovka, Shel'ting, Komsomol'sk, and South Schmidt massifs, wt.%.

| | Sample 133 | | | Sample 138 | | | Sample 139 | |
| | Clinopyroxene harzburgite | | | Lherzolite | | | Wehrlite | |
Component	Ol	Opx	Cpx	Ol	Opx	Cpx	Ol	Cpx
SiO$_2$	39.30	55.65	54.00	39.88	55.74	52.64	40.02	54.41
TiO$_2$	N.d.	0.02	0.02	N.d.	0.05	0.19	N.d.	0.03
Al$_2$O$_3$	N.d.	0.86	0.96	N.d.	1.91	2.53	N.d.	1.68
Cr$_2$O$_3$	N.d.	0.16	0.18	N.d.	0.19	0.29	N.d.	0.30
FeO	17.74	11.84	3.99	16.52	9.66	4.22	13.15	2.92
MnO	0.25	0.30	0.10	0.25	0.25	0.15	0.27	0.10
MgO	42.69	30.68	16.95	43.96	31.65	16.40	46.34	17.92
CaO	0.02	0.41	23.68	0.02	0.59	23.46	0.03	22.51
Na$_2$O	N.d.	N.f.	0.11	N.d.	0.02	0.07	N.d.	0.04
K$_2$O	N.d.	N.f.	N.f.	N.d.	0.01	0.01	N.d.	N.f.
NiO	0.27	N.d.	N.d.	0.004	N.d.	N.d.	0.13	N.d.
H$_2$O	N.d.	N.d.	N.d.	N.d.	N.d.	N.d.	N.f.	N.d.
Total	100.27	99.92	99.98	100.53	100.21	99.95	99.93	99.90
Mg$^\#$	0.81	0.82	0.88	0.83	0.85	0.87	0.86	0.92
n	4	3	3	2	2	2	2	4

| | Sample 140 | | | Sample 155 | | | Sample 164 | | Sample 176 | |
| | Olivine websterites | | | | | | | | | |
Component	Ol	Opx	Cpx	Ol	Opx	Cpx	Ol	Opx	Ol	Opx
SiO$_2$	39.74	54.71	52.28	39.84	49.47	53.40	40.98	56.83	38.44	54.54
TiO$_2$	N.d.	0.02	0.08	N.d.	0.03	0.08	N.d.	0.01	N.d.	0.04
Al$_2$O$_3$	N.d.	2.09	2.41	N.d.	0.68	1.50	N.d.	0.89	N.d.	1.29
Cr$_2$O$_3$	N.d.	0.33	0.49	N.d.	0.12	0.45	N.d.	0.51	N.d.	0.16
FeO	17.17	11.57	4.30	15.68	12.38	3.67	9.50	6.07	24.06	16.23
MnO	0.29	0.28	0.14	0.19	0.24	0.12	0.13	0.14	0.33	0.35
MgO	43.20	30.82	16.41	44.37	36.89	16.30	49.23	34.19	37.49	27.00
CaO	0.02	0.49	23.58	0.01	0.30	23.33	0.02	0.87	0.02	1.09
Na$_2$O	N.d.	0.03	0.11	N.d.	0.01	0.13	N.d.	0.03	N.d.	0.02
K$_2$O	N.d.	0.01	N.f.	N.d.	0.01	N.f.	N.d.	0.01	N.d.	N.f.
NiO	0.17	N.d.	N.d.	0.27	N.f.	N.d.	0.34	N.d.	0.19	N.f.
H$_2$O	N.d.	N.d.	N.d.	N.d.	N.f.	N.d.	N.d.	N.d.	N.d.	N.d.
Total	100.49	100.32	99.79	100.37	100.13	98.98	100.20	99.55	100.53	100.71
Mg$^\#$	0.82	0.83	0.87	0.84	0.84	0.89	0.90	0.91	0.74	0.75
n	2	2	2	2	5	1	1	3	3	6

Table 8.11 Continued.

Component	Sample 176 Olivine websterite		Sample 152 Websterite		Sample 130 Gabbronorites			Sample 183	Sample 189 Amphibole-containing gabbronorite	
	Cpx	CrMt	Opx	Cpx	Opx	Cpx	Amph	Pl	Opx	Cpx
SiO_2	52.52	0.02	56.58	53.98	55.07	53.14	50.53	45.66	53.09	52.21
TiO_2	0.09	1.15	0.01	0.02	0.03	0.03	0.18	0.02	0.04	0.16
Al_2O_3	1.98	5.84	1.09	1.41	1.35	1.56	7.45	33.86	2.30	2.74
Cr_2O_3	0.30	12.66	0.33	0.76	0.07	0.08	0.14	0.01	0.07	0.23
FeO	7.65	77.68	6.63	2.12	14.24	5.19	7.67	0.48	15.71	6.04
MnO	0.18	0.25	0.17	0.08	0.33	0.14	0.11	N.f.	0.35	0.18
MgO	16.13	2.09	34.19	17.48	27.42	15.99	17.18	N.f.	28.53	15.33
CaO	20.73	N.f.	0.22	24.13	0.84	22.99	12.07	18.04	0.66	22.27
Na_2O	0.16	0.13	0.03	0.14	0.01	0.13	0.94	1.26	N.f.	0.23
K_2O	N.f.	0.01	0.01	0.01	0.01	N.f.	0.08	0.01	N.f.	N.f.
NiO	N.f.	N.d.	N.d.	N.d.	N.d.	N.d.	N.d.	N.d.	N.d.	N.d.
H_2O	N.f.	N.d.	N.d.	N.d.	N.d.	N.d.	2.00	N.d.	N.d.	N.d.
Total	99.74	99.81	99.22	100.10	99.34	99.25	98.34	99.34	100.75	99.39
$Mg^{\#}$	0.79	$Cr^{\#}$ (59.3)	0.90	0.94	0.77	0.85	0.20	An (88.7)	0.76	0.82
n	2	1	1	2	4	1	2	3	1	5

Component	Sample 189 Amphibole-containing gabbronorite		Sample 129 Gabbro		Sample 155-1 Anorthosite with Ol, Opx, and Cpx impurities			
	Pl	Amph	Cpx	Amph	Ol	Opx	Cpx	Pl
SiO_2	44.72	44.28	53.33	55.33	39.66	53.85	52.30	46.38
TiO_2	0.01	0.37	0.07	0.02	N.d.	0.14	0.47	0.02
Al_2O_3	34.88	12.61	2.33	3.00	N.d.	2.58	3.07	33.87
Cr_2O_3	N.f.	0.31	0.18	0.09	N.d.	0.34	0.33	N.f.
FeO	0.34	9.44	4.57	4.44	16.53	13.45	4.93	0.09
MnO	N.f.	0.09	0.14	0.16	0.24	0.32	0.14	0.02
MgO	N.f.	15.43	16.03	21.61	43.76	28.83	16.26	N.f.
CaO	18.48	12.08	22.89	12.70	0.01	0.69	21.89	17.42
Na_2O	0.80	2.04	0.14	0.39	N.d.	N.f.	0.24	1.38
K_2O	0.01	0.07	0.01	0.02	N.d.	0.001	0.007	N.f.
NiO	N.d.	N.d.	N.d.	N.d.	0.24	N.d.	N.d.	N.f.
H_2O	N.d.	2.00	N.d.	2.00	N.d.	N.d.	N.d.	N.f.
Total	99.25	98.72	99.65	99.76	100.45	100.20	99.64	99.18
$Mg^{\#}$	An (92.7)	0.74	0.86	0.10	0.82	0.79	0.85	An (87.5)
n	3	3	2	3	6	1	1	2

Note: Compiled after the data in Table 8.1. Massifs: Berezovka (samples 133, 138, 139, 140, 152, 130, 129, 155-1, and 155); Shel'ting (samples 176 and 183); Komsomol'sk (sample 189); South Schmidt (sample 164); n – the number of analyses of the same mineral grain performed to calculate the average contents of components.

Table 8.12 Distribution coefficients of components between coexisting minerals from rocks of the Berezovka, Shel'ting, Komsomol'sk, and South Schmidt massifs.

Component	Harzburgites						Lherzolites				Plagiolherzolite		
Sample	133			1610-2			138			154	142		
	Ol/Opx	Ol/Cpx	Opx/Cpx	Ol/Opx	Ol/CrSp	Opx/CrSp	Ol/Opx	Ol/Cpx	Opx/Cpx	Opx/Cpx	Ol/Opx	Ol/Cpx	Opx/Cpx
SiO_2	0.71	0.73	1.03	0.715	7953	11120	0.72	0.76	1.06	1.08	0.761	0.77	1.01
TiO_2	N.d.	N.d.	N.d.	0.588	0.103	0.176	N.d.	N.d.	0.26	N.d.	0.042	0.104	2.47
Al_2O_3	N.d.	N.d.	N.d.	0.0014	0.0001	0.090	N.d.	N.d.	0.75	0.803	0.035	0.012	0.341
Cr_2O_3	N.d.	N.d.	N.d.	0.225	0.003	0.015	N.d.	N.d.	0.64	N.d.	N.d.	0.095	N.d.
Fe_2O_3	N.d.	N.d.	N.d.	N.d.	N.d.	N.d.	N.d.	N.d.	N.d.	N.d.	N.d.	N.d.	0.379
FeO	1.45	4.45	2.97	1.51	0.341	0.225	1.71	3.92	2.29	2.19	0.634	3.67	5.79
MnO	0.82	2.45	3.00	0.943	0.553	0.587	1.00	1.72	1.72	N.d.	0.344	1.47	4.28
MgO	1.39	2.52	1.81	1.47	6.58	4.49	1.39	2.68	1.93	2.03	1.73	2.94	1.69
CaO	0.04	0.001	0.02	0.009	0.840	95.1	0.03	0.001	0.03	0.023	0.017	0.001	0.047
Na_2O	N.d.	N.d.	N.d.	1.50	2.00	1.33	N.d.	N.d.	0.231	N.d.	N.d.	0.407	N.d.
K_2O	N.d.	N.d.	N.d.	0.500	N.d.	N.d.	N.d.	N.d.	N.d.	N.d.	N.d.	1.09	N.d.
$Mg^{\#}$	0.99	0.92	0.93	0.99	N.d.	N.d.	0.97	0.95	0.98	0.99	1.18	0.99	0.84

Component	Plagiolherzolite			Plagiowehrlite	Olivine websterites								
Sample	142			136	145			134	137	1603-1	139	140	
	Ol/CrSp	Opx/CrSp	Cpx/CrSp	Cpx/CrSp	Opx/Cpx	Opx/CrSp	Cpx/CrSp	Opx/Cpx	Opx/Cpx	Opx/Cpx	Ol/Cpx	Ol/Opx	Ol/Cpx
SiO_2	957	1257	1239	N.d.	1.04	2171	2082	1.06	1.04	1.05	0.74	0.73	0.76
TiO_2	0.140	3.33	1.35	2.20	1.38	0.063	0.045	N.d.	N.d.	0.385	N.d.	N.d.	N.d.
Al_2O_3	0.001	0.022	0.064	0.074	0.877	0.169	0.193	0.821	0.833	0.778	N.d.	N.d.	N.d.
Cr_2O_3	0.003	N.d.	0.026	0.023	0.662	0.011	0.017	0.130	0.515	0.540	N.d.	N.d.	N.d.
Fe_2O_3	N.d.	N.d.	N.d.	N.d.	2.63	N.d.	N.d.	4.67	3.00	N.d.	N.d.	N.d.	N.d.
FeO	0.439	0.693	0.120	0.083	2.61	0.167	0.064	3.07	2.47	2.74	4.50	1.49	3.99
MnO	0.673	1.95	0.457	0.250	2.41	0.837	0.348	2.12	2.86	2.74	2.65	1.2	2.11
MgO	3.52	2.03	1.20	1.13	1.77	10.87	6.14	1.76	1.72	1.94	2.57	1.40	2.63
CaO	0.733	42.6	902	793	0.041	127	3100	0.060	0.043	0.021	0.001	0.04	0.001
Na_2O	1.79	N.d.	4.40	1.50	0.243	1.53	6.29	N.d.	N.d.	0.147	N.d.	N.d.	N.d.
K_2O	0.685	N.d.	0.630	N.d.	2.00	1.22	0.611	N.d.	N.d.	N.d.	N.d.	N.d.	N.d.
$Mg^{\#}$	N.d.	N.d.	N.d.	N.d.	1.01	N.d.	N.d.	0.93	0.94	0.96	0.94	0.99	0.94

Olivine websterites / Websterites / Olivine gabbronorite

Component	Olivine websterites								Websterites				Olivine gabbronorite
Sample	140	155			164*	176*			143	152	148-1		
	Opx/Cpx	Oll/Opx	Oll/Cpx	Opx/Cpx	Oll/Opx	Oll/Opx	Oll/Cpx	Opx/Cpx	Opx/Cpx	Opx/Cpx	Opx/Cpx	Opx/Amph	Cpx/Amph
SiO_2	1.05	0.81	0.75	0.93	0.70	0.70	0.73	1.04	1.02	0.96	1.03	1.09	1.06
TiO_2	N.d.	N.d.	N.d.	0.38	N.d.	N.d.	N.d.	0.55	0.659	0.67	0.838	0.667	0.795
Al_2O_3	N.d.	N.d.	N.d.	0.45	N.d.	N.d.	N.d.	0.65	0.642	0.78	0.895	0.219	0.245
Cr_2O_3	N.d.	N.d.	N.d.	0.27	N.d.	N.d.	N.d.	0.52	0.491	0.44	0.628	0.572	0.911
Fe_2O_3	N.d.	N.d.	N.d.	N.d.	N.d.	N.d.	N.d.	N.d.	N.d.	N.d.	N.d.	N.d.	N.d.
FeO	2.69	1.27	4.27	3.37	1.48	1.48	3.15	2.12	2.61	3.14	2.96	2.01	0.679
MnO	2.07	0.79	1.58	2.00	0.95	0.94	1.83	1.99	2.19	2.27	2.55	2.69	1.06
MgO	1.88	1.20	2.72	2.28	1.39	1.39	2.32	1.67	1.80	1.96	1.72	1.52	0.886
CaO	0.02	0.03	N.d.	0.01	0.02	0.02	N.d.	0.05	0.029	0.01	0.028	0.052	1.84
Na_2O	0.23	N.d.	N.d.	0.08	N.d.	N.d.	N.d.	0.11	0.112	0.21	0.281	0.042	0.148
K_2O	N.d.	N.d.	N.d.	N.d.	N.d.	N.d.	N.d.	3.33	1.17	N.d.	0.167	0.011	0.065
$Mg^\#$	0.95	1.00	0.94	0.94	0.98	0.98	0.93	0.95	0.93	0.96	0.91	0.90	0.99

Olivine gabbronorite / Olivine gabbro / Gabbronorites

Component	Olivine gabbronorite						Olivine gabbro				Gabbronorites		
Sample	148-1						155-3		155-5		128	131	131a
	Opx/CrSp	Cpx/CrSp	Amph/CrSp	Opx/Pl	Cpx/Pl	Amph/Pl	Oll/Cpx	Cpx/Pl	Oll/Cpx	Cpx/Pl	Cpx/Pl	Cpx/Pl	Cpx/Pl
SiO_2	919	891	844	1.22	1.19	1.12	0.74	1.15	0.73	1.16	1.20	1.25	1.27
TiO_2	0.030	0.036	0.045	N.d.	N.d.	N.d.	0.149	4.70	N.d.	19.29	N.d.	N.d.	N.d.
Al_2O_3	0.174	0.194	0.795	0.047	0.052	0.215	0.006	0.047	0.004	0.068	0.077	0.065	0.060
Cr_2O_3	0.008	0.013	0.014	7.73	12.31	13.51	0.228	19.0	0.171	10.36	N.d.	N.d.	N.d.
Fe_2O_3	N.d.	N.d.	N.d.	N.d.	N.d.	N.d.	N.d.	N.d.	N.d.	N.d.	N.d.	N.d.	N.d.
FeO	0.307	0.104	0.153	96.12	32.51	47.89	3.67	15.01	2.68	28.40	9.67	8.47	21.57
MnO	1.11	0.436	0.412	77.2	30.3	28.67	2.14	52.0	1.15	68.3	N.d.	N.d.	N.d.
MgO	12.17	7.08	7.99	N.d.	N.d.	N.d.	2.56	N.d.	2.16	N.d.	N.d.	N.d.	N.d.
CaO	4.81	169	92	0.034	1.21	0.657	0.001	1.17	0.001	0.976	1.20	20.88	45.70
Na_2O	N.d.	N.d.	N.d.	0.076	0.272	1.83	0.170	0.260	0.144	0.176	0.250	0.192	0.224
K_2O	0.200	1.20	18.50	0.077	0.462	7.12	N.d.	0.313	N.d.	0.333	N.d.	N.d.	N.d.
$Mg^\#$	N.d.	N.d.	N.d.	N.d.	N.d.	N.d.	0.93	N.d.	0.95	N.d.	N.d.	N.d.	N.d.

(Continued)

Table 8.12 Continued.

Gabbronorites

Component	132a Cpx/Pl	138a Opx/Cpx	141 Opx/Cpx	141 Cpx/Pl	144 Opx/Cpx	144 Opx/Pl	144 Cpx/Pl	146 Opx/Cpx	146 Opx/Pl	146 Cpx/Pl	149 Opx/Cpx	149 Cpx/Pl	151 Cpx/Pl
SiO$_2$	1.23	1.08	1.11	1.34	0.951	1.18	1.24	1.02	1.26	1.23	1.02	1.02	1.24
TiO$_2$	N.d.	0.500	N.d.	N.d.	3.33	N.d.	N.d.	N.d.	N.d.	N.d.	N.d.	N.d.	N.d.
Al$_2$O$_3$	0.062	0.917	0.719	0.064	1.05	0.075	0.071	0.515	0.035	0.069	1.17	0.080	0.052
Cr$_2$O$_3$	N.d.	N.d.	N.d.	N.d.	N.d.	N.d.	N.d.	N.d.	N.d.	N.d.	N.d.	N.d.	N.d.
Fe$_2$O$_3$	N.d.	N.d.	N.d.	N.d.	N.d.	N.d.	N.d.	N.d.	N.d.	N.d.	0.667	N.d.	2.07
FeO	30.67	2.01	5.75	115	2.64	52.42	19.83	2.02	16.82	3.21	0.725	8.33	23.67
MnO	N.d.	1.65	N.d.	N.d.	1.77	N.d.	N.d.	2.05	N.d.	N.d.	2.93	N.d.	N.d.
MgO	45.43	1.87	1.90	86.19	1.77	38.95	21.97	1.66	47.09	28.38	1.87	1.87	86.72
CaO	1.31	0.035	0.026	0.033	0.065	1.31	1.26	0.041	0.056	1.35	1.87	0.080	1.28
Na$_2$O	0.250	0.130	0.047	0.015	0.065	0.136	0.331	N.d.	0.226	0.226	0.080	N.d.	0.405
K$_2$O	N.d.	N.d.	N.d.	N.d.	N.d.	N.d.	N.d.	N.d.	N.d.	N.d.	N.d.	N.d.	N.d.
Mg$^\#$	N.d.	1.01	0.98	N.d.	0.92	N.d.	N.d.	0.92	N.d.	N.d.	0.99	N.d.	N.d.

	Amphibole-containing gabbronorites												Gabbronorites		Gabbro		Leucocratic gabbro		

Sample

Component	147 Opx/Cpx	147 Opx/Pl	147 Cpx/Pl	1607 Cpx/Pl	130 Opx/Cpx	130 Opx/Amph	130 Cpx/Amph	189* Opx/Cpx	189* Opx/Amph	189* Cpx/Amph	189* Cpx/Pl	189* Amph/Pl	129 Cpx/Amph	1612 Cpx/Pl	155-4 Cpx/Amph	155-4 Cpx/Pl	155-4 Amph/Pl
SiO$_2$	1.07	1.29	1.20	1.24	1.02	1.20	1.05	1.02	1.20	1.18	1.17	0.99	1.00	1.21	1.16	1.18	1.02
TiO$_2$	0.300	N.d.	N.d.	0.737	1.00	0.11	0.17	0.25	0.11	0.43	16.00	37.00	1.02	N.d.	0.275	1.49	5.42
Al$_2$O$_3$	0.522	0.028	0.053	0.035	0.87	0.18	0.21	0.84	0.18	0.22	0.08	0.36	0.437	0.059	0.148	0.077	0.522
Cr$_2$O$_3$	N.d.	N.d.	N.d.	3.49	0.91	0.23	0.57	0.30	0.23	0.74	N.d.	N.d.	1.87	30.0	0.875	N.d.	2.25
Fe$_2$O$_3$	N.d.	N.d.	N.d.	N.d.	N.d.	N.d.	N.d.	N.d.	N.d.	N.d.	N.d.	N.d.	N.d.	N.d.	N.d.	N.d.	N.d.
FeO	2.94	61.76	21.02	49.79	2.74	1.66	0.68	2.60	1.66	0.64	17.76	27.76	0.890	34.22	0.554	1.20	4.95
MnO	N.d.	N.d.	N.d.	11.06	2.34	3.89	1.27	1.94	3.89	2.00	N.d.	N.d.	1.10	15.72	1.20	1.00	2.74
MgO	1.90	81.89	43.09	N.d.	1.72	1.85	0.93	1.86	1.85	0.99	N.d.	N.d.	0.794	N.d.	0.897	1.00	0.836
CaO	0.021	0.029	1.39	1.24	0.04	0.05	1.91	0.03	0.05	1.84	1.21	0.65	1.83	1.13	1.96	1.16	1.11
Na$_2$O	N.d.	N.d.	N.d.	0.221	0.04	N.d.	0.14	N.d.	N.d.	0.11	0.29	2.55	0.195	0.351	0.108	0.450	0.591
K$_2$O	N.d.	N.d.	N.d.	0.682	N.d.	N.d.	N.d.	N.d.	N.d.	N.d.	N.d.	7.00	0.333	N.d.	0.131	1.28	4.16
Mg$^\#$	0.95	N.d.	N.d.	N.d.	0.91	1.03	4.23	0.93	1.03	1.10	N.d.	N.d.	0.99	N.d.	1.12	9.73	N.d.

Note: Compiled after the data in Table 8.11. 176* – Shel'ting massif. 189* – Komsomol'sk massif. 164* – South Schmidt massif.

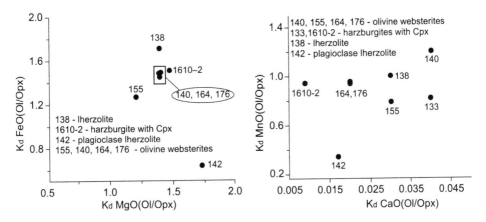

Figure 8.15 Variations in distribution coefficients of FeO, MgO, MnO, and CaO between coexisting olivines and orthopyroxenes in ultramafites and gabbroids (data from Table 8.12). 138 – lherzolite; 133, 1610-2 – harzburgite with Cpx; 142 – plagiolherzolite; 140, 155, 164, 176 – olivine websterites.

relationship between K_d(MgO) and K_d(FeO) and a direct relationship between K_d(MnO) and K_d(CaO) (Fig. 8.15).

K_d(*olivine–clinopyroxene*). The values of this coefficient for the mineral pairs from clinopyroxene-containing harzburgites, lherzolites, plagiolherzolites, olivine websterites, and olivine gabbro are in the following ranges: SiO_2 (0.73–0.77), FeO (2.68–4.45), MnO (1.15–2.65), MgO (2.16–2.94), and CaO (0.001). As evidenced from these values, K_d(olivine–clinopyroxene) for SiO_2 is nearly the same in all rocks: Clinopyroxene contains 3–4 times more of this component than olivine. The contents of FeO and MnO in olivine are 3–4 and 2–3 times higher, respectively, than those in clinopyroxene. In addition, clinopyroxene has three orders of magnitude higher contents of CaO than coexisting olivine.

K_d(*orthopyroxene–clinopyroxene*). The values of this coefficient for the mineral pairs from clinopyroxene-containing harzburgites, lherzolites, plagiolherzolites, olivine websterites, and gabbronorites are in the following ranges: SiO_2 (0.9–1.1), Al_2O_3 (0.3–1.2), FeO (2.0–5.8), MnO (1.7–4.3), MgO (1.7–2.3), and CaO (0.01–0.08). As evidenced from these values, the contents of FeO, MgO, and MnO in orthopyroxene are much higher than those in coexisting clinopyroxenes, SiO_2 is distributed almost equally between them, and the contents of Al_2O_3 in orthopyroxene from plagiolherzolites and some of the other two-pyroxene rocks are lower than those in coexisting clinopyroxene. In restite harzburgites and lherzolites, K_d(MgO) and K_d(FeO) for the orthopyroxene–clinopyroxene and olivine–orthopyroxene pairs are in inverse relationship (Fig. 8.16, 1). At the same time, K_d(MgO) and K_d(FeO) for the orthopyroxene–clinopyroxene pair from websterites and gabbronorites show a direct relationship (Fig. 8.16, 2, 3). The values of K_d(Al_2O_3) and K_d(CaO) are assumed to be in direct relationship in all studied two-pyroxene rocks (Fig. 8.17).

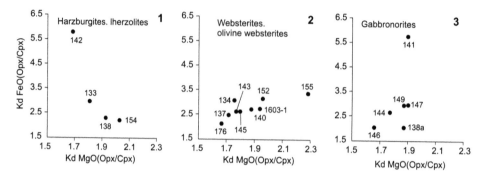

Figure 8.16 Variations in distribution coefficients of MgO and FeO between coexisting orthopyroxenes and clinopyroxenes in harzburgites, lherzolites, websterites, olivine websterites, and
gabbronorites (data from Table 8.12).

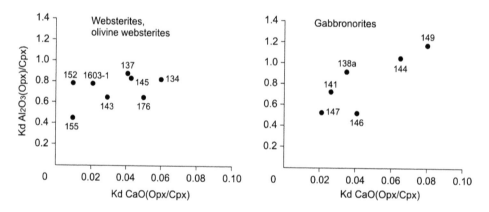

Figure 8.17 Variations in distribution coefficients of Al₂O₃ and CaO between coexisting orthopyroxenes and clinopyroxenes in websterites, olivine websterites, and gabbronorites (data from
Table 8.12).

K_d*(clinopyroxene–plagioclase).* The values of this coefficient for the mineral pairs
from gabronorites vary in the following ranges: SiO_2 (1.2–1.3), Al_2O_3 (0.05–0.09),
CaO (1.0–1.5), and Na_2O (0.1–0.4). Hence, SiO_2 and CaO are present mainly in
clinopyroxenes, whereas Al_2O_3 and Na_2O are concentrated mainly in plagioclases.
No clear relationship was revealed between $K_d(SiO_2)$ and $K_d(Al_2O_3)$ and between
$K_d(CaO)$ and $K_d(Na_2O)$ (Fig. 8.18).

The summarized data on the covariations in distribution coefficients of components between coexisting olivines, orthopyroxenes, clinopyroxenes, and plagioclases in
ultramafites and gabbroids of the Berezovka and other massifs are listed in Table 8.13.
They can be used as additional criteria for the mineral classification.

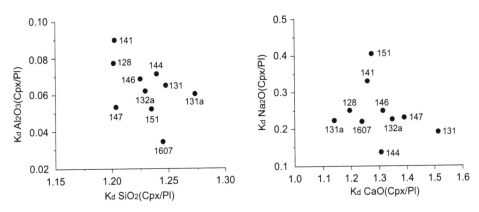

Figure 8.18 Variations in distribution coefficients of Al_2O_3 and SiO_2 and of CaO and Na_2O between coexisting clinopyroxenes and plagioclases in gabbronorites of the Berezovka massif (data from Table 8.12).

Table 8.13 Limited distribution coefficients of components (K_d) between coexisting olivines, orthopyroxenes, and clinopyroxenes from ultramafites and gabbroids of the Berezovka and other massifs.

$K_d(SiO_2)$	$K_d(Al_2O_3)$	$K_d(FeO)$	$K_d(MgO)$	$K_d(MnO)$	$K_d(CaO)$	$K_d(Na_2O)$
Olivine–orthopyroxene						
0.70–0.81	–	0.63–1.71	1.20–1.73	0.34–1.20	0.009–0.04	–
Orthopyroxene–clinopyroxene						
0.93–1.11	0.34–1.17	2.01–5.79	1.66–2.28	1.65–4.28	0.01–0.08	–
Clinopyroxene–plagioclase						
1.16–1.27	0.052–0.09	–	–	–	0.98–1.51	0.14–0.41

Note: Compiled after the data in Table 8.12.

8.3 ON THE TEMPERATURES OF CHEMICAL EQUILIBRIUM OF COEXISTING OLIVINES, ORTHOPYROXENES, AND CLINOPYROXENES IN ULTRAMAFIC RESTITES OF THE BEREZOVKA MASSIF

Data on the chemical composition of coexisting olivines, orthopyroxenes, and clinopyroxenes from lherzolite and clinopyroxene-containing harzburgite of the Berezovka massif were used to evaluate approximately the temperatures of the latest chemical equilibrium of the minerals during the cooling of these mantle restites (Table 8.14). The calculations were carried out using the Thermocalc-321 software including an internally consistent thermodynamic dataset [Powell and Holland, 1988; Holland and Powell, 1998]. This program permits calculation of the temperatures of chemical equilibrium of coexisting minerals if the corresponding pressures determined from geophysical data are specified. The boundary between the Earth's crust and upper

Table 8.14 Chemical composition (wt.%) of coexisting olivines, orthopyroxenes, and clinopyroxenes from clinopyroxene harzburgite and lherzolite of the Berezovka massif, based on which chemical equilibrium temperatures were calculated.

Mineral	SiO_2	TiO_2	Al_2O_3	Cr_2O_3	FeO	MnO	MgO	CaO	Na_2O	K_2O	NiO	H_2O	Total
Clinopyroxene harzburgite, sample 133													
Olivine	39.55	N.d.	N.d.	N.d.	17.46	0.25	42.25	0.02	N.d.	N.d.	0.31	N.d.	99.84
Orthopyroxene	55.65	0.02	0.86	0.16	11.84	0.30	30.68	0.41	N.d.	N.d.	N.d.	N.d.	99.92
Clinopyroxene	54.29	0.01	0.91	0.25	3.52	0.07	17.50	24.04	007	N.d.	N.d.	N.d.	100.66
Lherzolite, sample 138													
Olivine	39.88	N.d.	N.d.	N.d.	16.52	0.25	43.96	0.02	N.d.	N.d.	N.d.	N.d.	100.53
Orthopyroxene	50.00	0.08	3.03	0.38	12.23	0.14	30.88	1.43	N.d.	N.d.	0.06	N.d.	98.23
Clinopyroxene	52.64	0.19	2.53	0.29	4.22	0.15	16.40	23.43	0.07	N.d.	N.d.	N.d.	99.95

Note: Compiled after the data in Table 8.1.

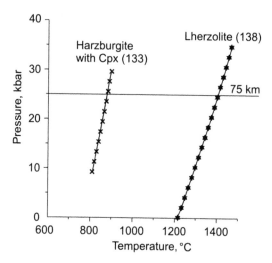

Figure 8.19 PT-diagram for chemical equilibrium of coexisting olivines, orthopyroxenes, and clinopyroxenes in clinopyroxene-containing harzburgite and lherzolite of the Berezovka massif. Calculations were made by A. Yu. Selyatitskii using the Thermocalc-321 software and invoking data from Table 8.14.

mantle beneath Sakhalin Island is located at depths of ~50 km [Filonenko, 2002]. Hence, the restite ultramafites of the Berezovka massif formed at depths no shallower than 50 km. On assumption of their formation in the depth range 55–65 km, this process should have occurred at ~18–22 kbar. These pressure values were set into the software for calculation of the temperatures of equilibrium of coexisting minerals in the Berezovka massif restites. The calculation results are presented on the diagram as two positively inclined lines (Fig. 8.19). If the results are valid, the latest chemical equilibrium of the coexisting minerals in lherzolite sample 138 and harzburgite sample 133 was established at ~1300°C and ~860°C, respectively. For comparison, the probable temperature of the sources of MORB is ~1380°C [Herzberg and O'Hara, 2002]. Moreover, the temperatures of equilibrium of minerals in spinel lherzolites of the Dinaridic ophiolite belt are 834–1070°C [Bazylev *et al.*, 2009].

8.4 ACCESSORY MINERALS

In studying the chemical composition of accessory minerals in the rocks of the considered mafic-ultramafic massifs, the main attention was focused on zircons and, less, Cr-spinels, whereas magnetites, ilmenites, sulfides, awaruites, and PGE minerals were examined on occasional grains (data on the PGE minerals are given below in a special chapter).

8.4.1 Zircons

Zircon is a highly informative accessory mineral of igneous and metamorphic rocks. It is often used for U–Pb isotope dating, which is finding increasing application for studying mafic-ultramafic complexes. In this section we present results of chemical analyses of the same zircon grains from rocks of the Berezovka massif as were used for U–Pb isotope dating and study of the morphologic and optical properties and trace-element composition of this mineral (see Chapter 10).

The general chemical composition of zircons was examined by X-ray spectroscopy. A total of 58 grains from eight rock varieties (gabbro-pyroxenites, olivine gabbronorites, other gabroids, and metavolcanics of the strata enclosing the Berezovka massif) were studied. In addition to major components (SiO_2, ZrO_2, and HfO_2), the contents of impurities UO_2, ThO_2, Ce_2O_3, Yb_2O_3, and Y_2O_3 were determined (Table 8.15). In general, all the grains were rather homogeneous in the contents of major components but inhomogeneous in the contents of impurity elements (Fig. 8.20).

The contents of SiO_2 in the zircons are 32.2–33.1 wt.%, with the grains from gabbroids being somewhat poorer in this component than the grains from gabbropyroxenites (Fig. 8.20, 1). The contents of ZrO_2 also vary over a narrow range of values (64.0–66.2 wt.%) (Fig. 8.20, 1–6). These estimates are close to those obtained earlier for a more representative collection of zircons: The average SiO_2 content is ~33 wt.%, and the average ZrO_2 content is 67 wt.% [Lesnov, 2009a]. The ZrO_2 contents seem to be in inverse relationship with the Yb_2O_3 and Y_2O_3 contents (Fig. 8.20, 6, 7). The contents of HfO_2 in the studied zircons vary from 0.01 to 2.15 wt.%. Elevated contents of this component are observed mostly in zircons from gabbroids (Fig. 8.20, 2). The ZrO_2/SiO_2 ratio varies over a wide range of values (29.9–78.9); moreover, it shows significant variations in different zircon grains from the same sample.

The contents of trace elements in the zircons, measured by X-ray spectroscopy, are in the following ranges (wt.%): UO_2 (0.002–0.287), ThO_2 (0.004–0.288), Ce_2O_3 (0.004–0.132), Yb_2O_3 (0.007–0.307), and Y_2O_3 (0.009–0.868). As follows from the diagram (Fig. 8.20, 10), there is a direct relationship between UO_2 and ThO_2 contents and between ZrO_2 and ThO_2 contents (Fig. 8.20, 12). Note that zircons from gabbropyroxenite (sample 1610-2) are richer in ThO_2 than those from gabbroids (samples 1596-1A, 1596-6, and 1595) (Fig. 8.21). The UO_2/ThO_2 ratio in all studied zircons varies from 0.02 to 5.71; moreover, it shows wide variations in the mineral grains even within some rock samples. Thus, zircons from ultramafic and gabbroid rocks of the Berezovka massif show a rather homogeneous chemical composition but differ in contents of some impurity elements and in ZrO_2/SiO_2 and UO_2/ThO_2 ratios.

Table 8.15 Chemical composition of zircons from rocks of the Berezovka massif, wt.%.

Sample 1	Grain 2	Rock 3	SiO$_2$ 4	ZrO$_2$ 5	HfO$_2$ 6	UO$_2$ 7	ThO$_2$ 8	Ce$_2$O$_3$ 9	Yb$_2$O$_3$ 10	Y$_2$O$_3$ 11	Total 12	ZrO$_2$/HfO$_2$ 13	UO$_2$/ThO$_2$ 14
1607	4.1	Gabbro-pyroxenite	32.70	65.52	1.16	0.054	0.073	N.f.	0.148	0.139	99.79	56.6	0.74
-"-	5.1	-"-	32.66	64.77	1.56	0.091	0.021	0.004	0.019	0.068	99.19	41.6	4.33
-"-	6.1	-"-	32.36	65.12	1.87	0.024	N.f.	0.036	N.f.	0.063	99.47	34.9	N.d.
-"-	7.1	-"-	32.47	65.41	1.72	0.226	N.f.	N.f.	N.f.	N.f.	99.83	38.1	N.d.
-"-	8.1	-"-	32.99	65.26	1.55	N.f.	0.025	0.008	0.177	0.229	100.24	42.2	N.d.
-"-	9.1	-"-	32.53	65.70	1.85	0.08	0.014	N.f.	N.f.	N.f.	100.17	35.5	5.71
-"-	10.1	-"-	32.97	64.54	1.87	N.f.	N.f.	N.f.	N.f.	N.f.	99.38	34.4	N.d.
1610-2	1.1	Gabbro-pyroxenite	32.90	64.36	1.60	0.042	0.096	0.056	N.f.	0.121	99.18	40.2	0.44
-"-	2.1	-"-	32.57	64.87	1.22	0.111	0.073	0.016	0.137	0.403	99.40	53.0	1.52
-"-	3.1	-"-	32.50	64.52	1.14	0.011	0.066	0.08	0.263	0.550	99.13	56.6	0.17
-"-	4.1	-"-	32.67	64.95	1.28	0.072	N.f.	0.064	0.104	0.339	99.48	50.6	N.d.
-"-	5.1	-"-	32.56	64.87	1.13	0.140	0.142	0.040	0.106	0.240	99.23	57.5	0.99
-"-	6.1	-"-	32.30	65.09	1.18	0.173	0.130	0.068	0.166	0.522	99.63	55.1	1.33
-"-	9.1	-"-	32.62	65.36	1.39	0.018	N.f.	N.f.	0.007	0.047	99.44	47.1	N.d.
-"-	10.1	-"-	32.64	65.16	1.80	0.058	N.f.	N.f.	N.f.	0.034	99.69	36.1	N.d.
-"-	11.1	-"-	32.80	65.14	1.16	0.112	N.f.	0.008	N.f.	0.037	99.26	56.0	N.d.
1604	1.1	Olivine gabbronorite	32.69	65.82	1.25	0.022	0.105	N.f.	N.f.	0.058	99.95	52.8	0.21
-"-	2.1	-"-	32.43	64.52	1.37	0.287	0.288	0.127	0.080	0.292	99.39	47.2	1.00
-"-	3.1	-"-	32.61	65.53	1.37	0.062	N.f.	0.044	0.045	0.139	99.80	47.8	N.d.
-"-	4.1a	-"-	32.70	66.21	1.36	N.f.	0.022	0.091	N.f.	0.038	100.42	48.9	N.d.
-"-	4.1b	-"-	33.01	65.48	1.48	N.f.	0.017	N.f.	N.f.	N.f.	99.99	44.3	N.d.
-"-	5.1	-"-	32.39	65.17	1.82	0.210	0.106	N.f.	N.f.	0.203	99.90	35.9	1.98
-"-	6.1	-"-	32.65	65.03	2.09	0.046	0.052	N.f.	0.069	0.094	100.00	31.1	0.89
-"-	7.1	-"-	32.67	64.89	2.03	0.132	N.f.	N.f.	0.039	N.f.	99.76	31.9	N.d.
-"-	8.1	-"-	32.47	65.08	1.76	0.003	0.027	0.099	N.f.	N.f.	99.44	36.9	0.11

Sample	No.	Rock type											
1596-A1	1.1	Gabbroid	32.76	65.18	1.74	N.f.	0.044	0.024	N.f.	0.009	99.76	37.6	N.d.
	2.1	-"-	32.26	65.25	2.04	0.194	N.f.	0.004	0.028	0.165	99.94	32.0	N.d.
	3.1	-"-	33.11	65.27	1.45	N.f.	0.023	N.f.	0.100	0.297	100.25	44.9	N.d.
	4.1	-"-	32.64	65.26	1.31	0.108	0.020	N.f.	N.f.	N.f.	99.34	36.3	5.40
	4.2	-"-	32.60	64.89	1.79	0.053	0.055	0.080	N.f.	0.037	99.51	49.7	0.96
1596-6	1.1	Gabbroid	32.44	64.72	1.62	N.f.	0.008	N.f.	0.014	0.063	98.87	40.0	N.d.
	2.2	-"-	32.52	65.68	1.18	0.101	0.080	N.f.	N.f.	0.071	99.63	55.5	1.26
	3.1	-"-	32.43	65.76	1.64	N.f.	N.f.	N.f.	N.f.	N.f.	99.83	40.0	N.d.
	5.1	-"-	32.32	65.78	1.77	0.058	0.020	N.f.	0.031	0.069	100.05	37.2	2.90
1595	1.1	Gabbroid	32.29	65.90	0.86	0.045	0.016	0.132	0.126	0.320	99.69	78.9	2.81
	2.1	-"-	32.58	65.23	1.78	0.082	N.f.	0.060	0.026	0.023	99.78	36.7	N.d.
	3.1	-"-	32.46	64.26	2.15	N.f.	0.031	0.008	0.015	N.f.	98.92	29.9	N.d.
	4.1	-"-	32.55	65.08	1.42	0.079	0.038	0.056	0.077	0.093	99.39	45.8	2.08
	5.1	-"-	32.57	64.28	1.88	0.067	0.021	0.024	0.132	0.173	99.15	34.2	3.19
	6.1a	-"-	32.42	65.49	1.47	0.035	0.037	0.028	N.f.	N.f.	99.48	44.5	0.95
	6.1b	-"-	32.53	65.37	1.27	0.068	0.042	N.f.	0.038	0.049	99.37	51.5	1.62
1655	1.1	Gabbro-diorite	32.32	64.87	1.26	0.014	0.027	0.016	0.099	0.302	98.91	51.4	0.52
	2.1	-"-	32.45	64.00	1.88	0.0244	N.f.	N.f.	0.058	0.246	98.66	34.0	N.d.
	3.1	-"-	32.55	64.65	1.19	0.042	0.015	N.f.	0.128	0.363	98.94	54.3	2.80
	5.1	-"-	32.58	65.26	1.26	0.008	0.073	N.f.	0.090	0.224	99.50	51.6	0.11
	6.1	-"-	32.40	64.68	1.29	0.135	0.098	0.012	0.199	0.455	99.27	50.3	1.38
	7.1	-"-	32.19	64.07	1.20	0.037	0.108	0.008	0.307	0.868	98.79	53.5	0.34
	8.1	-"-	32.55	65.50	1.09	0.104	N.f.	0.068	0.075	0.233	99.62	60.0	N.d.
	9.1	-"-	32.35	65.93	1.11	N.f.	0.004	N.f.	0.007	0.148	99.55	59.6	N.d.
	10.1	-"-	32.52	65.28	0.01	0.032	0.045	N.f.	0.019	0.184	99.09	64.5	0.71
	11.1	-"-	32.31	64.34	1.27	0.122	0.111	N.f.	0.257	0.714	99.12	50.8	1.38
156	1.1	Trachyandesite	32.59	64.64	1.89	0.023	N.f.	N.f.	0.073	0.078	99.29	34.2	N.d.
	1.2	-"-	32.48	64.97	1.75	N.f.	0.071	N.f.	N.f.	0.141	99.41	37.0	N.d.
	2.1	-"-	32.47	64.71	1.44	0.002	0.097	N.f.	0.164	0.164	99.05	44.9	0.02
	3.1	-"-	32.63	65.33	1.22	N.f.	0.006	0.012	N.f.	0.031	99.23	53.6	N.d.
	4.1	-"-	32.85	65.37	1.46	0.111	0.059	0.016	N.f.	0.061	99.93	44.9	1.88
	5.1	-"-	32.83	65.56	1.39	N.f.	N.f.	N.f.	0.030	0.051	99.86	47.1	N.d.
	6.1	-"-	32.74	65.21	1.57	0.023	N.f.	N.f.	0.012	0.050	99.61	44.9	N.d.

Note: Compiled after the data in Table 8.1.

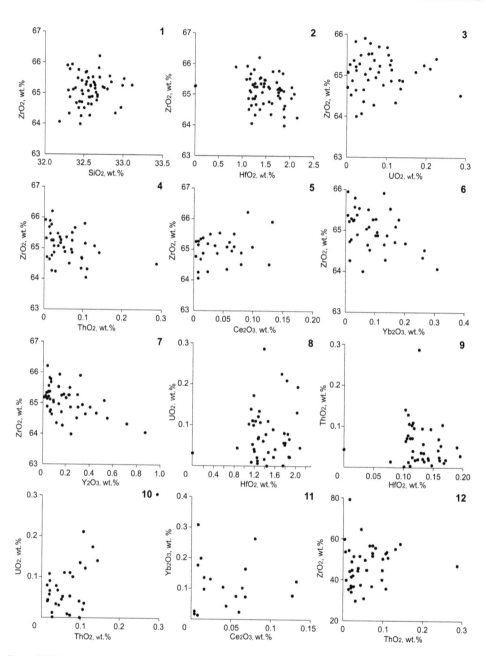

Figure 8.20 Variations in contents of ZrO_2, SiO_2, HfO_2, UO_2, ThO_2, Ce_2O_3, Yb_2O_3, and Y_2O_3 in zircons from ultramafites and gabbroids of the Berezovka massif (data from Table 8.15).

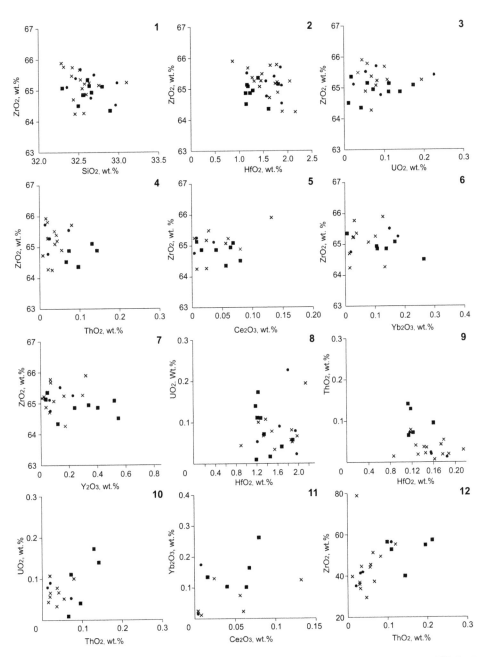

Figure 8.21 Variations in contents of SiO₂, ZrO₂, HfO₂, UO₂, ThO₂, Ce₂O₃, Yb₂O₃, and Y₂O₃ in zircons from gabbro-pyroxenites (sample 1607 (points) and sample 1610-2 (squares)) and gabbroids (samples 1596-1A, 1596-6, and 1595 (crosses)) of the Berezovka massif (data from Table 8.15).

8.4.2 Cr-spinels

In the Sakhalin mafic-ultramafic massifs, accessory Cr-spinel is present mainly in harzburgites, lherzolites, dunites, wehrlites, and, more seldom, websterites, orthopyroxenites, olivine gabbro, and olivine gabbronorites and usually amounts to no more than 5%. Its grains vary from euhedral to xenomorphous and from few tenths of mm to few mm in size. Usually they are uniformly distributed throughout the rocks. Seldom, the grains form bunching or jet-like clusters, sometimes passing into chromitite phenocrysts. Some Cr-spinel grains are present as inclusions in olivine, orthopyroxene, or clinopyroxene grains.

According to X-ray spectral data, the contents of Cr_2O_3 in the studied Cr-spinels vary from $\sim$54 wt.% in high-Cr varieties to $\sim$12 wt.% in low-Cr ones (Table 8.16). The contents of other components also vary over wide ranges of values (wt.%): Al_2O_3 (5.84–44.82), MgO (2.09–16.85), FeO (15.03–77.68), TiO_2 (0.030–1.48), and MnO (0.144–0.433). Analysis of several mineral grains from lherzolite of the South Schmidt massif showed that their Cr_2O_3 and Al_2O_3 contents are nearly the same. The arrangement of figurative points on the composition diagram points to an inverse relationship between the FeO and MgO contents of these Cr-spinel samples (Fig. 8.22).

The $Cr^{\#}$ value of the analyzed Cr-spinels varies from 26–29% in the grains from lherzolites and harzburgites to 71–82% in the grains from olivine websterites and orthopyroxenites. The $Mg^{\#}$ value also considerably varies: from 70–71% (websterites) to 7.0–7.9% (olivine websterites, olivine–clinopyroxene orthopyroxenites, and olivine gabbro). We assume an inverse relationship between $Mg^{\#}$ and $Cr^{\#}$ (Fig. 8.23). In most of the Cr-spinel grains, MnO dominates over TiO_2, and TiO_2/MnO is within 0.11–0.68. The only exception is Cr-spinel grains from olivine gabbro (sample 148-1) and olivine orthopyroxenite (sample 176), in which TiO_2 dominates over MnO and TiO_2/MnO is 4.66 and 4.40, respectively.

Using the obtained $Cr^{\#}$ and $Mg^{\#}$ values for Cr-spinels from ultramafic restites of the Berezovka and South Schmidt massifs, we approximately estimated the degrees of partial melting of the upper-mantle ultramafic sources during the formation of these restites. The calculations were made by the equation based on the results of physical experiments on the partial melting of peridotites and on the calibration scale combined with the $Cr^{\#}$–$Mg^{\#}$ diagram [Lesnov et al., 2008]:

$$D = 0.426 \times Cr^{\#} + 1.538,$$

where D is the degree of partial melting of the ultramafic protolith.

The available $Cr^{\#}$ and $Mg^{\#}$ values for the studied Cr-spinels and the results of the above calculations suggest that the South Schmidt massif lherzolites (sample 163) formed at $\sim$15% partial melting of the upper-mantle source, and the Berezovka massif harzburgites (sample 1610-2), at $\sim$30% partial melting.

8.4.3 Magnetites

Study of the chemical composition of occasional grains of magnetites, Ti-magnetites, and ilmenites from rocks of the Berezovka and Komsomol'sk massifs showed that

Table 8.16 Chemical composition of Cr-spinels from rocks of the Berezovka, Shel'ting, and South Schmidt massifs, wt.%.

Grain	Rock	SiO_2	TiO_2	Al_2O_3	Cr_2O_3	FeO	MnO	MgO	CaO	Na_2O	K_2O	Total	$Mg^\#$, %	$Cr^\#$, %
Berezovka massif														
1610-2	Harz	0.005	0.149	18.27	37.16	35.65	0.325	8.08	0.007	0.006	N.f.	99.65	58	29
1610-2a	Harz	N.f.	0.187	17.24	32.25	42.92	0.433	6.09	0.018	0.015	N.f.	99.15	56	29
1596	Harz	N.f.	0.34	11.22	30.75	51.16	0.43	4.68	0.01	0.03	N.d.	98.62	14	65
142	Pl Lherz	0.042	0.011	38.10	24.77	20.27	0.223	14.20	0.025	0.022	0.018	97.68	56	30
142a	Pl Lherz	0.043	0.046	37.81	25.49	21.06	0.217	13.77	N.f.	0.049	N.f.	98.49	54	31
136	Pl Wehrl	N.d.	0.05	32.17	29.23	26.11	0.32	13.23	0.03	0.04	N.d.	101.19	53	38
145	Ol Web	0.019	1.30	7.69	18.58	60.35	0.399	2.68	0.011	0.019	0.004	91.05	7.3	71
145a	Ol Web	0.032	1.48	8.44	21.05	58.25	0.390	2.82	0.004	0.014	0.014	92.48	7.9	71
148-1	Ol Gb	0.059	1.46	9.54	28.75	51.13	0.339	2.25	0.137	N.f.	0.005	93.67	7.3	67
Shel'ting massif														
174	Ol Enst	0.014	0.105	7.65	50.56	35.10	0.355	5.83	0.006	N.f.	N.f.	99.62	23	82
174a	Ol Enst	0.039	0.241	11.33	42.90	37.96	0.357	6.58	N.f.	N.f.	N.f.	99.41	24	72
176	Cpx–Ol Enst	0.023	1.15	5.84	12.66	77.68	0.25	2.09	N.d.	0.13	0.008	99.81	7.0	59
South Schmidt massif														
163	Lherz	0.014	0.039	44.82	23.25	15.03	0.144	16.85	0.010	0.037	0.002	100.20	67	26
163a	Lherz	N.f.	0.055	42.27	25.85	16.32	0.182	15.64	0.015	0.006	0.003	100.34	63	29
161	Lherz	0.022	0.033	22.67	43.58	19.85	0.291	11.76	0.005	0.044	0.004	98.28	51	56
161a	Lherz	0.028	0.030	24.15	41.38	20.50	0.269	11.79	0.012	N.f.	N.f.	98.13	51	54
161b	Lherz	0.022	0.054	22.51	44.22	17.41	0.226	13.22	N.f.	N.f.	0.009	97.65	58	57
161c	Lherz	0.026	0.052	23.16	42.48	19.34	0.278	11.96	0.011	0.045	N.f.	97.37	52	55
161d	Lherz	0.043	0.038	23.29	45.87	19.59	0.266	11.77	0.016	0.033	N.f.	100.91	52	57
161e	Lherz	0.103	0.048	23.03	43.59	20.16	0.284	12.25	0.027	0.065	0.020	99.54	52	56
161f	Lherz	0.033	0.034	22.95	45.62	19.57	0.265	12.11	0.011	0.018	0.006	100.60	52	57
161g	Lherz	0.035	0.048	23.05	46.41	17.09	0.258	13.06	N.f.	0.011	0.005	99.94	58	58
161h	Lherz	0.041	0.042	24.10	45.16	19.19	0.254	11.70	0.018	0.016	0.003	100.50	52	56
162	Web	0.017	0.079	15.36	54.80	19.23	0.287	10.61	0.002	0.010	0.004	100.40	70	50
162a	Web	N.f.	0.071	14.97	54.71	20.08	0.304	10.28	0.018	N.f.	N.f.	100.43	71	48

Note: Compiled after the data in Table 8.1.

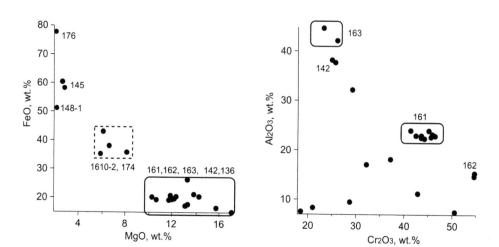

Figure 8.22 Variations in contents of MgO, FeO, Cr$_2$O$_3$, and Al$_2$O$_3$ in accessory Cr-spinels from rocks of the Berezovka, Shel'ting, and South Schmidt massifs (data from Table 8.16).

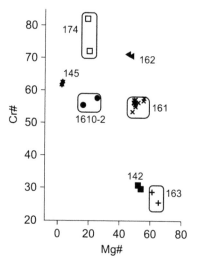

Figure 8.23 Variations in the values of Cr$^{\#}$ = 100 · Cr/(Cr +Al) and Mg$^{\#}$ = 100 · Mg/(Mg + Fe) (formula coefficients) in accessory Cr-spinels from rocks of the Berezovka (samples 142, 145, and 1610-2), Shel'ting (sample 174), and South Schmidt (samples 161, 162, and 163) massifs (data from Table 8.16).

the content of FeO decreases in the series from magnetites in websterite to those in gabbro and, then, to Ti-magnetite in gabbro-pegmatite and to ilmenite in harzburgite (Table 8.17, Fig. 8.24). It was established that the content of MgO in magnetite from websterites is lower than that in magnetite from gabbro. The magnetites and ilmenites

Table 8.17 Chemical composition of magnetites, Ti-magnetites, and ilmenites from rocks of the Berezovka and Komsomol'sk massifs, wt.%.

Sample	Rock	SiO_2	TiO_2	Al_2O_3	Cr_2O_3	FeO	MnO	MgO	CaO	Na_2O	K_2O	Total
Berezovka massif												
143	Web	0.036	0.437	0.413	2.70	90.28	0.153	0.078	0.153	0.105	N.f.	94.36
143a	Web	0.010	0.371	0.316	1.96	91.13	0.090	0.020	0.090	0.108	0.005	94.10
143b	Web	0.071	0.425	0.439	2.78	90.15	0.153	0.011	0.153	0.067	N.f.	94.25
155-3	Gb	0.051	0.581	0.855	3.35	88.82	0.175	0.479	0.056	0.040	0.005	94.41
155-3a	Gb	0.064	0.249	0.597	2.30	89.22	0.171	0.564	0.046	0.027	0.007	93.25
159b	Harz	0.04	46.54	0.49	1.86	40.52	0.23	9.10	N.f.	0.02	N.d.	98.90
Komsomol'sk massif												
192	Gb-pegm	0.009	2.21	4.17	0.319	86.50	0.232	1.62	0.001	0.029	N.f.	95.09
192a	Gb-pegm	1.15	2.55	3.04	0.068	88.56	0.144	1.86	0.001	0.016	0.006	97.40

Note: Compiled after the data in Table 8.1. Magnetites (samples 143, 143a, 143b, 153-3, and 153-3a); Ti-magnetites (samples 192 and 192a); and ilmenite (sample 159b).

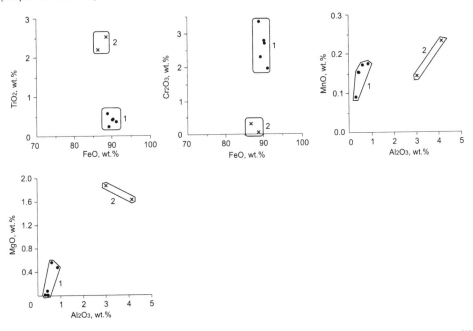

Figure 8.24 Variations in contents of FeO, TiO_2, Cr_2O_3, MnO, MgO, and Al_2O_3 in magnetites (1) and Ti-magnetites (2) from rocks of the Berezovka and Komsomol'sk massifs (data from Table 8.17).

are characterized by nearly the same contents of Cr_2O_3. The Ti-magnetites have lower contents of this impurity but higher contents of Al_2O_3 than the magnetites.

* * *

In olivines from ultramafites and gabbroids of the studied massifs, $Mg^{\#}$ varies from 0.73 (chrysolites) to 0.91 (ferroan forsterites); in orthopyroxenes from ultramafites it increases in passing from bronzites (olivine orthopyroxenites) to ferroan enstatites (websterites). In orthopyroxenes from gabbronorites, $Mg^#$ varies over a narrow range of values close to that in bronzite. In clinopyroxenes, $Mg^#$ is within 0.81–0.96; the maximum values are revealed in the mineral grains from lherzolites and websterites of the South Schmidt massif, and the grains from plagiolherzolites and plagiowehrlites of the Berezovka massif are characterized by slightly lower $Mg^#$. Clinopyroxenes from gabbronorites and gabbro have higher contents of Na_2O than those from websterites, wehrlites, lherzolites, and clinopyroxene-containing harzburgites. The composition of plagioclases from gabbroids of the Berezovka massif varies from bytownite (gab-bronorite) to anorthite (gabbro and leucocratic gabbro). In the lump of parallel-banded wehrlite–gabbroid rock of the Berezovka massif, the plagioclase grains localized along the profile orthogonal to the rock banding have nearly constant contents of CaO and Na_2O, i.e., the latter do not depend on the quantitative mineral composition of the bands. This confirms that the rock lacks "latent foliation", which is usually of "cumu-lative" genesis. Amphiboles from gabbroids of the Berezovka and Komsomol'sk massifs are pargasites and scarcer edenites.

As seen from the calculated distribution coefficients of major chemical components between rock-forming minerals from ultramafites, $K_d(FeO)$ and $K_d(MgO)$ are more than unity for the olivine–orthopyroxene, olivine–clinopyroxene, and orthopyroxene–clinopyroxene pairs. The contents of SiO_2 in zircons from gabbroids and ultramafites of the Berezovka massif are within 32–33 wt.%; the mineral grains from gabbroids are poorer in silica than those from gabbro-pyroxenites. The contents of ZrO_2 vary from 64 to 66 wt.%, and the contents of HfO_2, from 0.01 to 2.15 wt.%; the zircons from gabbroids are richer in this component than those from the other rocks. The ZrO_2/SiO_2 ratio in the zircons varies from 30 to 79. The mineral shows a direct relationship between UO_2 and ThO_2 contents and between ZrO_2/SiO_2 and ThO_2 content.

Cr-spinels from the studied rocks are divided into high-Cr ($Cr_2O_3 = 54$ wt.%) and low-Cr ($Cr_2O_3 = 12$ wt.%) varieties. The $Cr^#$ values in Cr-spinels from lherzolites and harzburgites are lower than those in the mineral from olivine websterites and orthopyroxenites. The $Mg^#$ values in the Cr-spinels vary from 7.0 to 71%. Estimations based on $Cr^#$ in the mineral show that restite lherzolites of the South Schmidt massif formed at ~15% partial melting of the upper-mantle source, and harzburgites of the Berezovka massif, at ~30% partial melting.

Thus, the rock-forming and accessory minerals from rocks of the Sakhalin mafic-ultramafic massifs show wide variations in chemical composition, which suggests their crystallization in different conditions. The studies revealed significant differences between the chemical compositions of similar rock-forming and accessory minerals in rocks of different petrographic types: from restite ultramafites and hybrid ultramafites to orthomagmatic and hybrid gabbroids.

Chapter 9

Distribution of rare earth elements in rock-forming minerals from mafic-ultramafic massifs of the East Sakhalin ophiolite association

In petrological research into igneous rocks, including mafic-ultramafic ones, more and more attention is paid to studies of the regularities of distribution of REE and other trace elements between rock-forming and accessory minerals, using different analytical methods [Lesnov, 2000a]. Such investigations were first carried out for orthopyroxene, clinopyroxene, and plagioclase samples from various rocks of mafic-ultramafic massifs of the East Sakhalin ophiolite association.

The contents of REE in rock-forming minerals from rocks of the Sakhalin massifs were measured in the laboratories of the Analytical Center of the Sobolev Institute of Geology and Mineralogy (Novosibirsk), using two methods: radiochemical neutron activation analysis (RNAA) and mass-spectrometric analysis with inductively coupled plasma and laser ablation (LA ICP-MS). The characteristics of both methods are described in Section 6.3.1.

With RNAA, the contents of ten REE in minerals were measured: La, Ce, Nd, Sm, Eu, Gd, Tb, Tm, Yb, and Lu. Their detection limits were as follows (ppm): La (0.005), Ce (0.05), Nd (0.05), Sm (0.0001), Eu (0.01), Gd (0.01), Tb (0.001), Tm (0.001), Yb (0.005), and Lu (0.001). The LA ICP-MS method ensured much lower detection limits of these REE (ppm): La (0.0010), Ce (0.0011), Pr (0.0005), Nd (0.0008), Sm (0.0004), Eu (0.0003), Gd (0.0003), Tb (0.0002), Dy (0.0002), Ho (0.0002), Er (0.0002), Tm (0.0001), Yb (0.0002), and Lu (0.0001). The latter method has a crucial advantage: It permits a parallel analysis of all coexisting minerals in each rock sample. The results of performed mineral analyses for REE are presented in Table 9.1.

9.1 ORTHOPYROXENES

The REE patterns of orthopyroxenes were studied on a collection of their grains from lherzolites, websterites, and olivine and olivine-free gabbronorites of the Berezovka, Shel'ting, Komsomol'sk, and South Schmidt massifs. The study was performed mainly by RNAA. This method permits determination of contents of only 10 REE; therefore, the total contents of REE impurities in these samples were not calculated (Table 9.2). In orthopyroxene grains from websterite (sample 143), analyzed by LA ICP-MS, the total contents of REE were 3.52 and 4.51 ppm. According to the summarized data, the total contents of REE in orthopyroxenes from the mafic-ultramafic rocks vary in the following ranges (ppm): spinel lherzolites from xenoliths – 0.5–5.2, lherzolites from the massifs – 0.2–2.8, harzburgites from the massifs – 0.8–1.4, gabbronorites – 4–42, and norites – 0.5–2.4. The $(La/Yb)_n$ value in orthopyroxenes from these rocks ranges from 0.12 to 1.83 [Lesnov, 2001a, 2007b].

Table 9.1 Contents of REE in coexisting minerals from rocks of the Berezovka, Shel'ting, Komsomol'sk, and South Schmidt massifs, ppm.

Sample	Rock	Mineral	La	Ce	Pr	Nd	Sm	Eu	Gd	Tb	Dy	Ho	Er	Tm	Yb	Lu	Total	(La/Yb)n	(Eu/Eu*)n
1	2	3	4	5	6	7	8	9	10	11	12	13	14	15	16	17	18	19	20
									Berezovka massif										
154	Lherz	Opx	0.130	0.310	N.d.	0.150	0.03	0.120	0.040	0.008	N.d.	N.d.	N.d.	0.005	0.048	0.008	N.d.	1.83	10.55
"	"	Cpx	0.070	0.280	N.d.	0.260	0.110	0.070	0.250	0.047	N.d.	N.d.	N.d.	0.025	0.190	0.028	N.d.	0.25	1.24
134	Web	Opx	0.100	0.270	N.d.	0.140	0.040	0.047	0.100	0.025	N.d.	N.d.	N.d.	0.023	0.200	0.032	N.d.	0.34	2.17
"	"	Cpx	0.110	0.360	N.d.	0.290	0.100	0.070	0.300	0.058	N.d.	N.d.	N.d.	0.045	0.330	0.046	N.d.	0.22	1.14
142	Wehrl	Cpx*	5.195	9.403	1.013	3.822	0.670	0.144	0.697	0.102	0.550	0.111	0.407	0.049	0.468	0.058	22.69	7.49	0.64
143	Web	Opx*	0.089	0.526	0.113	0.654	0.340	0.052	0.327	0.044	0.502	0.100	0.353	0.020	0.355	0.040	3.52	0.17	0.47
"	"	Opx*	0.145	0.721	0.141	0.995	0.406	0.071	0.414	0.068	0.628	0.128	0.365	0.033	0.361	0.035	4.51	0.27	0.52
"	"	Cpx*	1.408	2.957	0.326	1.51	0.437	0.101	0.518	0.077	0.586	0.215	0.454	0.039	0.462	0.040	9.13	2.06	0.65
"	"	Cpx*	1.823	3.865	0.529	2.70	0.690	0.137	0.677	0.234	0.919	0.179	0.538	0.047	0.637	0.055	12.88	1.93	0.61
148-1	Gbn	Pl*	0.398	0.063	0.36	0.243	0.123	0.171	0.029	0.076	0.066	0.224	0.034	0.304	0.047	0.004	2.30	5.72	6.25
132a	Ol Gbn	Opx	0.060	0.130	N.d.	0.150	0.040	0.020	0.100	0.025	N.d.	N.d.	N.d.	0.015	0.095	0.020	N.d.	0.43	0.92
"	"	Cpx	0.330	0.570	N.d.	0.460	0.160	0.150	0.420	0.080	0.430	N.d.	N.d.	0.050	0.220	0.040	N.d.	1.01	1.67
"	"	Pl	N.d.	0.100	N.d.	0.080	0.030	0.018	N.d.	N.d.	N.d.	N.d.	N.d.	N.d.	N.d.	N.d.	N.d.	N.d.	N.d.
138a	Ol Gbn	Opx	0.040	0.130	N.d.	0.100	0.027	0.015	0.110	0.025	N.d.	N.d.	N.d.	0.012	0.090	0.014	N.d.	0.30	0.73
"	"	Cpx	0.110	0.370	N.d.	0.360	0.180	0.090	0.330	0.060	N.d.	N.d.	N.d.	0.050	0.430	0.060	N.d.	0.17	1.11
"	"	Pl	0.041	N.d.	N.d.	0.060	0.015	0.016	N.d.	N.d.	N.d.	N.d.	N.d.	N.d.	N.d.	N.d.	N.d.	N.d.	N.d.
132	Ol Gbn	Pl	0.036	0.020	N.d.	0.051	0.010	0.017	N.d.	N.d.	N.d.	N.d.	N.d.	N.d.	N.d.	N.d.	N.d.	N.d.	N.d.
128	Gbn	Pl	0.180	0.350	N.d.	0.140	0.025	0.075	N.d.	0.007	N.d.	N.d.	N.d.	0.001	0.006	N.d.	N.d.	20.25	N.d.
130	Gbn	Pl	0.060	0.140	N.d.	0.110	0.020	0.050	N.d.	0.003	N.d.	N.d.	N.d.	0.002	0.009	N.d.	N.d.	4.50	N.d.
131	Gbn	Opx	0.110	0.310	N.d.	0.160	0.060	0.013	0.070	0.015	N.d.	N.d.	N.d.	0.020	0.090	0.028	N.d.	0.82	0.61
"	"	Cpx	0.240	0.530	N.d.	0.960	0.270	0.160	0.370	0.080	0.700	N.d.	N.d.	0.070	0.500	0.100	N.d.	0.32	1.55
"	"	Pl	0.044	N.d.	N.d.	0.072	0.012	0.058	N.d.	0.005	N.d.	N.d.	N.d.	0.002	0.010	N.d.	N.d.	2.97	N.d.
131a	Gbn	Cpx*	0.776	1.41	0.157	0.584	0.079	0.017	0.128	0.012	0.123	0.019	0.084	N.d.	0.092	N.d.	N.d.	5.69	0.50
"	"	Cpx*	1.39	2.30	0.274	1.14	0.209	0.039	0.256	0.020	0.179	0.030	0.105	0.008	0.116	0.006	6.1	8.09	0.52
"	"	Cpx*	1.42	2.59	0.265	1.09	0.180	0.037	0.196	0.018	0.189	0.022	0.111	0.002	0.102	0.007	6.2	9.40	0.60
"	"	Pl	0.075	0.120	N.d.	N.d.	0.018	0.056	0.002	0.002	N.d.	N.d.	N.d.	0.002	0.009	N.d.	N.d.	5.62	0.60
141	Gbn	Opx	0.150	0.250	N.d.	0.200	0.050	0.013	0.070	0.017	N.d.	N.d.	N.d.	0.015	0.170	0.023	N.d.	0.60	0.67
"	"	Cpx	0.310	0.750	N.d.	0.500	0.140	0.060	0.200	0.044	N.d.	N.d.	N.d.	0.039	N.d.	0.040	N.d.	N.d.	1.10
"	"	Pl	0.020	0.140	N.d.	0.070	0.027	0.031	N.d.	0.004	N.d.	N.d.	N.d.	N.d.	0.002	N.d.	N.d.	6.75	N.d.

Sample	Rock	Min.																	
144	Gbn	Opx	0.570	1.460	N.d.	0.930	0.240	0.044	0.300	0.050	N.d.	N.d.	N.d.	0.040	0.240	0.034	N.d.	1.60	0.50
"	"	Cpx	0.320	1.280	N.d.	1.440	0.620	0.450	2.160	0.310	N.d.	N.d.	N.d.	0.100	0.600	0.100	N.d.	0.36	1.06
"	"	Pl	0.030	N.d.	N.d.	0.040	0.017	0.050	N.d.	N.d.	N.d.	N.d.	N.d.	0.001	0.010	N.d.	N.d.	2.02	N.d.
146	Gbn	Pl	N.d.	0.050	N.d.	N.d.	0.010	0.047	N.d.	0.005	N.d.	N.d.	N.d.	0.002	0.013	N.d.	N.d.	N.d.	N.d.
147	Gbn	Opx	0.060	0.490	N.d.	0.460	0.240	0.060	0.250	0.050	N.d.	N.d.	N.d.	0.033	0.260	0.055	N.d.	0.16	0.86
"	"	Cpx	0.960	1.410	N.d.	0.880	0.220	0.140	0.260	0.048	N.d.	N.d.	N.d.	0.035	0.230	0.025	N.d.	2.82	1.79
"	"	Pl	0.026	0.057	N.d.	0.052	0.013	0.050	N.d.	0.007	N.d.	N.d.	N.d.	0.003	0.015	N.d.	N.d.	1.17	N.d.
151	Gbn	Pl	0.024	0.081	N.d.	0.065	0.015	0.056	N.d.	0.005	N.d.	N.d.	N.d.	0.040	0.020	N.d.	N.d.	0.81	N.d.
148	Gb-pegm	Pl	0.018	0.061	N.d.	0.050	0.020	0.040	N.d.	0.005	N.d.	N.d.	N.d.	0.002	0.018	N.d.	N.d.	0.67	N.d.

Shel'ting massif

Sample	Rock	Min.																	
173	Gbn	Opx	0.120	0.600	N.d.	0.760	0.320	0.110	0.510	0.130	N.d.	N.d.	N.d.	0.094	0.680	0.100	N.d.	0.12	0.83
"	"	Cpx	0.380	1.370	N.d.	1.610	0.810	0.660	3.160	0.710	N.d.	N.d.	N.d.	0.400	2.340	0.360	N.d.	0.11	1.10

Komsomol'sk massif

Sample	Rock	Min.																	
190	Ol Gbn	Opx	0.140	0.340	N.d.	0.190	0.044	0.010	0.066	0.014	N.d.	N.d.	N.d.	0.020	0.170	0.033	N.d.	0.56	0.57
"	"	Cpx	0.290	0.800	N.d.	0.610	0.220	0.100	0.440	0.045	N.d.	N.d.	N.d.	0.100	0.680	0.120	N.d.	0.29	0.96

South Schmidt massif

Sample	Rock	Min.																	
160	Lherz	Opx	0.120	0.290	N.d.	0.170	0.039	0.040	0.170	0.034	N.d.	N.d.	N.d.	0.029	0.250	0.041	N.d.	0.32	1.27
162	Web	Opx	0.022	0.080	N.d.	0.050	0.012	0.008	0.024	0.06	N.d.	N.d.	N.d.	0.004	0.036	0.004	N.d.	0.41	1.41
"	"	Cpx	0.062	0.130	N.d.	0.055	0.008	0.011	0.052	0.010	N.d.	N.d.	N.d.	0.009	0.064	0.011	N.d.	0.65	1.24

Note: Analyses were carried out at the Analytical Center of the Sobolev Institute of Geology and Mineralogy (Novosibirsk); * – LA ICP-MS method (analyst V.S. Paleskii); other analyses – RNAA method (analyst V.A. Kovaleva). Rocks: Lherz – lherzolite, Web – websterite, Ol Gbn – olivine gabbronorite, Gbn – gabbronorite, Gb-pegm – gabbro-pegmatite. N.d. – No data.

Table 9.2 Contents of REE in orthopyroxenes from rocks of the Berezovka, Shel'ting, Komsomol'sk, and South Schmidt massifs, ppm.

Sample	Rock	La	Ce	Pr	Nd	Sm	Eu	Gd	Tb	Dy	Ho	Er	Tm	Yb	Lu	$(La/Yb)_n$	$(La/Sm)_n$	$(Eu/Eu^*)_n$
								Berezovka massif										
154	Lherz	0.130	0.310	N.d.	0.150	0.03	0.120	0.040	0.008	N.d.	N.d.	N.d.	0.005	0.048	0.008	1.83	2.73	10.55
134	Web	0.100	0.270	N.d.	0.140	0.040	0.047	0.100	0.025	N.d.	N.d.	N.d.	0.023	0.200	0.032	0.34	1.57	2.16
143*	Web	0.089	0.526	0.113	0.654	0.340	0.052	0.327	0.044	0.502	0.100	0.353	0.020	0.355	0.040	0.17	0.16	0.47
-"-	-"-	0.145	0.721	0.141	0.995	0.406	0.071	0.414	0.068	0.628	0.128	0.365	0.033	0.361	0.035	0.27	0.22	0.52
132a	Ol Gbn	0.060	0.130	N.d.	0.150	0.040	0.020	0.100	0.025	N.d.	N.d.	N.d.	0.015	0.095	0.020	0.43	0.94	0.92
138a	Ol Gbn	0.040	0.130	N.d.	0.100	0.027	0.015	0.110	0.025	N.d.	N.d.	N.d.	0.012	0.090	0.014	0.30	0.93	0.72
131	Gbn	0.110	0.310	N.d.	0.160	0.060	0.013	0.070	0.015	N.d.	N.d.	N.d.	0.020	0.090	0.028	0.82	1.15	0.61
141	Gbn	0.150	0.250	N.d.	0.200	0.050	0.013	0.070	0.017	N.d.	N.d.	N.d.	0.015	0.170	0.023	0.60	1.89	0.67
144	Gbn	0.570	1.460	N.d.	0.930	0.240	0.044	0.300	0.050	N.d.	N.d.	N.d.	0.040	0.240	0.034	1.60	1.50	0.50
147	Gbn	0.060	0.490	N.d.	0.460	0.240	0.060	0.250	0.050	N.d.	N.d.	N.d.	0.033	0.260	0.055	0.16	0.16	0.74
								Shel'ting massif										
173	Gbn	0.120	0.600	N.d.	0.760	0.320	0.110	0.510	0.130	N.d.	N.d.	N.d.	0.094	0.680	0.100	0.12	0.24	0.83
								Komsomol'sk massif										
190	Ol Gbn	0.140	0.340	N.d.	0.190	0.044	0.010	0.066	0.014	N.d.	N.d.	N.d.	0.020	0.170	0.033	0.56	2.00	0.57
								South Schmidt massif										
160	Lherz	0.120	0.290	N.d.	0.170	0.039	0.040	0.170	0.034	N.d.	N.d.	N.d.	0.029	0.250	0.041	0.32	1.94	1.27
162	Web	0.022	0.080	N.d.	0.050	0.012	0.008	0.024	0.006	N.d.	N.d.	N.d.	0.004	0.036	0.004	0.41	1.15	1.41

Note: Compiled after the data in Table 9.1.

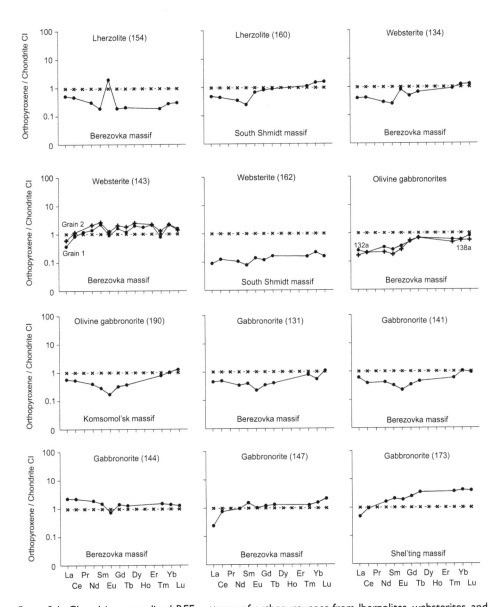

Figure 9.1 Chondrite-normalized REE patterns of orthopyroxenes from lherzolites, websterites, and olivine and olivine-free gabbronorites from the Berezovka, Shel'ting, Komsomol'sk, and South Schmidt massifs (data from Table 9.2).
Here and in Figs. 9.2–9.5, the REE patterns are chondrite CI-normalized [Evensen *et al.*, 1978].

The chondrite-normalized REE patterns of most of the studied orthopyroxenes from the Sakhalin massifs are positively sloped (Fig. 9.1). The $(La/Sm)_n$ value is within 0.16–2.73, and $(Eu/Eu^*)_n$ is predominantly within 0.50–0.92. Correspondingly, the patterns show moderate or weak negative Eu anomalies. A significant domination of

HREE over LREE is one of typomorphic features of orthopyroxenes [Lesnov, 2007b]. Some orthopyroxenes, including those from the Sakhalin massifs, are abnormally enriched in LREE; therefore, their REE patterns are negatively sloped. The LREE enrichment of these orhopyroxenes is probably due to the minor amounts of structural (isomorphic) REE impurities and the varying amount of nonstructural REE impurities. The latter are localized in the grain microcracks and, probably, in fluid microinclusions [Lesnov, 2007b].

REE isomorphism in orthopyroxenes. The available data suggest that the major mechanism of isomorphous incorporation of REE into orthopyroxene structure is their heterovalent ion substitution for ions of one or several mineral framework elements [Lesnov, 2007b, 2011b]. The main ion undergoing this substitution is Ca^{2+} because its radius is 1.12 Å, i.e., is within the interval of the radii of trivalent REE ions, varying from 1.16 Å for La^{3+} to 0.977 Å for Lu^{3+}. The ionic radii of other framework elements in orthopyroxenes, such as Mg^{2+} and Fe^{2+}, are beyond the above interval; therefore, substitution of these ions by REE ions is unlikely. The probable scheme of isomorphous incorporation of REE into orthopyroxene structure is as follows: $3Ca^{2+} \rightarrow 2REE^{3+} + vacancy$. The appearance of vacancies in the mineral structure must reduce its physical strength. This, in turn, must limit isomorphous incorporation of REE ions into the structure under high-pressure mineral crystallization. Another reason for limited REE isomorphism in orthopyroxene structure might be the relatively low content of Ca^{2+}. Heavy REE are predominant in orthopyroxene structure, probably because the radii of their trivalent ions are closer to the ionic radius of Ca^{2+} than the radii of trivalent LREE ions.

Correlation between REE and fluid components in orthopyroxenes. Fluid components (CO, H_2, N_2, CH_4, CO_2, H_2O, etc.) play a crucial role in crystallization of minerals from magmatic melts and in fractionation of REE between solid phases and their parental melts [Balashov, 1976; Green, 1994; Mysen, 1983; Neruchev et al., 1997; Roedder, 1987]. This was taken into account during the study of the relationship between the contents of REE and fluid components in orthopyroxenes from gabbronorites, websterites, and lherzolites of the Berezovka, Shel'ting, Komsomol'sk, and South Schmidt massifs [Lesnov et al., 1998b]. The REE contents were measured in orthopyroxene monofractions by RNAA, and the fluid contents, by chromatography. The results showed that the contents of HREE in orthopyroxenes increased with the concentration of reduced CO gas and DR (degree of fluid reduction) values ($DR = 100(CO + H_2 + CH_4)/(CO + H_2 + CH_4 + CO_2 + H_2O)$). The contents of some REE decreased with an increase in the total content of fluid components released from orthopyroxene during their decrepitation at 900°C.

9.2 CLINOPYROXENES

This mineral is known to be an important concentrator of REE in ultramafic and gabbroid rocks. Its REE composition was studied on a large collection of grains from lherzolites, wehrlites, websterites, and olivine and olivine-free gabbronorites of the Berezovka, Shel'ting, Komsomol'sk, and South Schmidt massifs (Table 9.3). The total REE contents in these clinopyroxene grains examined by LA ICP-MS are as

Table 9.3 Contents of REE in clinopyroxenes from rocks of the Berezovka, Shel'ting, Komsomol'sk, and South Schmidt massifs, ppm.

Sample	Rock	La	Ce	Pr	Nd	Sm	Eu	Gd	Tb	Dy	Ho	Er	Tm	Yb	Lu	$(La/Yb)_n$	$(La/Sm)_n$	$(Eu/Eu*)_n$
								Berezovka massif										
154	Lherz	0.070	0.280	N.d.	0.260	0.110	0.070	0.250	0.047	N.d.	N.d.	N.d.	0.025	0.190	0.028	0.25	0.40	1.25
134	Web	0.110	0.360	N.d.	0.290	0.100	0.070	0.300	0.058	N.d.	N.d.	N.d.	0.045	0.330	0.046	0.22	0.69	1.14
142*	Wehrl	5.195	9.403	1.013	3.822	0.670	0.144	0.697	0.102	0.550	0.111	0.407	0.049	0.468	0.058	7.49	4.88	0.64
143*	Web	1.408	2.957	0.326	1.51	0.437	0.101	0.518	0.077	0.586	0.215	0.454	0.039	0.462	0.040	2.06	2.03	0.65
-"-	-"-	1.823	3.865	0.529	2.70	0.690	0.137	0.677	0.076	0.919	0.179	0.538	0.047	0.637	0.055	1.93	1.66	0.61
132a	Ol Gbn	0.330	0.570	N.d.	0.460	0.160	0.150	0.420	0.080	0.430	N.d.	N.d.	0.050	0.220	0.040	1.01	1.30	1.67
138a	Ol Gbn	0.110	0.370	N.d.	0.360	0.180	0.090	0.330	0.060	N.d.	N.d.	N.d.	0.050	0.430	0.060	0.17	0.38	1.11
131	Gbn	0.240	0.530	N.d.	0.960	0.270	0.160	0.370	0.080	0.700	N.d.	N.d.	0.070	0.500	0.100	0.32	0.56	1.55
131a*	Gbn	0.776	1.41	0.157	0.584	0.079	0.017	0.128	0.012	0.123	0.019	0.084	N.d.	0.092	N.d.	5.69	6.18	0.51
-"-	-"-	1.39	2.30	0.274	1.14	0.209	0.039	0.256	0.020	0.179	0.030	0.105	0.008	0.116	0.006	8.09	4.19	0.52
-"-	-"-	1.42	2.59	0.265	1.09	0.180	0.037	0.196	0.018	0.189	0.022	0.111	0.002	0.102	0.007	9.40	4.97	0.60
141	Gbn	0.310	0.750	N.d.	0.500	0.140	0.060	0.200	0.044	N.d.	N.d.	N.d.	0.039	N.d.	0.040	N.d.	1.39	1.10
144	Gbn	0.320	1.280	N.d.	1.440	0.620	0.450	2.160	0.310	N.d.	N.d.	N.d.	0.100	0.600	0.100	0.36	0.32	1.06
147	Gbn	0.960	1.410	N.d.	0.880	0.220	0.140	0.260	0.048	N.d.	N.d.	N.d.	0.035	0.230	0.025	2.82	2.75	1.79
								Shel'ting massif										
173	Gbn	0.380	1.370	N.d.	1.610	0.810	0.660	3.160	0.710	N.d.	N.d.	N.d.	0.400	2.340	0.360	0.11	0.30	1.10
								Komsomol'sk massif										
190	Ol Gbn	0.290	0.800	N.d.	0.610	0.220	0.100	0.440	0.045	N.d.	N.d.	N.d.	0.100	0.680	0.120	0.29	0.83	0.96
								South Schmidt massif										
162	Web	0.062	0.130	N.d.	0.055	0.008	0.011	0.052	0.010	N.d.	N.d.	N.d.	0.009	0.064	0.011	0.65	4.88	1.24

Note: Compiled after the data in Table 9.1.

follows (ppm): Cpx from wehrlites (sample 142) – 22.7, Cpx from websterites (sample 143) – 9.1–12.9, and Cpx from gabbronorites (sample 131a) – 6.1–6.2. According to available data, the total REE contents in clinopyroxenes from ultramafic, gabbroid, and some other rocks vary over wider ranges of values [Lesnov, 2001b, 2007b]: Cpx from meteorites – 14–20 ppm, Cpx from lunar mafic rocks – 12–323 ppm, Cpx from lherzolite xenoliths in alkali basalts – up to 20 ppm, Cpx from lherzolite xenoliths in alkali basalts of Shavaryn-Tsaram Volcano (Mongolia) – 22–39 ppm, and Cpx from lherzolites of the Troodos massif (Cyprus) – 2.0–3.5 ppm.

The studied clinopyroxenes from rocks of the Sakhalin mafic-ultramafic massifs are characterized by a more intense REE fractionation as compared with the coexisting orthopyroxenes, which is due to the higher LREE contents; $(La/Yb)_n = 1.3–9.4$, and $(La/Sm)_n$ is usually also much higher than 1.

The database on REE contents in clinopyroxenes from rocks of many mafic-ultramafic massifs, including the Sakhalin ones, was used during computer-aided multiparametric geochemical discrimination of this mineral [Lesnov et al., 2011]. The computations showed statistically significant differences between the REE compositions of clinopyroxenes from the following rock pairs: (1) ultramafites from massifs–ultramafites from deep-seated xenoliths in alkali basalts; (2) clinopyroxene-containing harzburgites–lherzolites from massifs; (3) clinopyroxene-containing harzburgites–plagioclase-containing lherzolites from massifs; and (4) clinopyroxene-containing harzburgites–garnet lherzolites from massifs.

The studied clinopyroxenes also show certain differences in the location on composition diagrams and in the shape of their chondrite-normalized REE patterns (Fig. 9.2). Based on these features, the clinopyroxenes can be separated into two groups: (1) clinopyroxenes whose REE patterns lie close to the line marking accumulation of REE in chondrite CI and (2) clinopyroxenes whose REE patterns lie mostly above or below the line of REE contents in chondrite CI. The REE pattern of clinopyroxene from lherzolite (sample 154, Berezovka massif) shows that its contents of MREE and HREE are close to the chondrite ones, but the sample is depleted in LREE, whose contents are noticeably lower than those in the chondrite (Fig. 9.2, 1). Higher REE fractionation and depletion in LREE relative to HREE were observed in clinopyroxenes from lherzolites of the Lizard, Balmuccia, and Newfoundland massifs and from the same rocks dragged from the Indian Ocean [Lesnov, 2007b].

The REE pattern of clinopyroxene from gabbronorite of the Shel'ting massif (sample 173) (Fig. 9.2, 12) shows its noticeable depletion in LREE relative to HREE (though the total content of these impurities is rather high), which indicates intense REE fractionation. The REE patterns of clinopyroxene from gabbronorite of the Berezovka massif (sample 131a) are of significantly different shape, because the LREE contents are much higher and the contents of MREE and, particularly, HREE are, on the contrary, somewhat lower than those in the chondrite (Fig. 9.2, 9). Note that the REE patterns of clinopyroxenes from lherzolites, wehrlites, and websterites are complicated by weak negative Eu anomalies, whereas the REE patterns of this mineral from gabbronorites show weak positive Eu anomalies.

Of special interest are clinopyroxenes whose REE patterns point to anomalous enrichment in LREE, first of all, La and Ce. This enrichment is best seen in the patterns of clinopyroxenes from wehrlite (Fig. 9.2, 2), websterite (Fig. 9.2, 4), gabbronorite (Fig. 9.2, 11), and some other rocks. It is probably due to the varying contents of

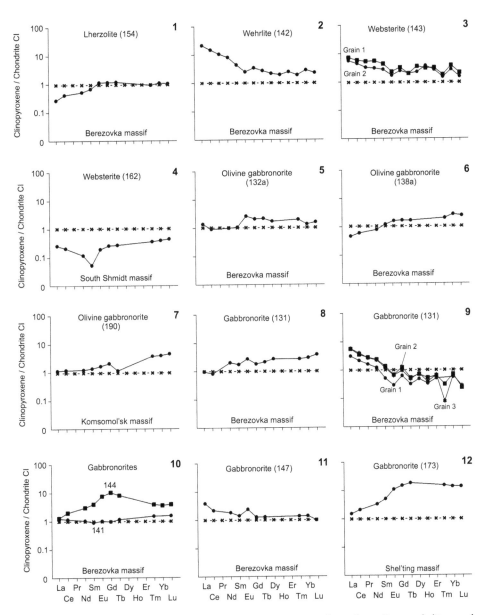

Figure 9.2 Chondrite-normalized REE patterns of clinopyroxenes from lherzolites, wehrlites, websterites, and olivine and olivine-free gabbronorites from the Berezovka, Shel'ting,
Komsomol'sk, and South Schmidt massifs (data from Table 9.3).

nonstructural LREE impurities in the grain microcracks. Taking into account this
fact, one should ignore their contents when considering the endogenous formation of
clinopyroxenes. They can be of interest only in the context of epigenetic changes of
the mineral.

REE isomorphism in clinopyroxenes. The crystal structure of clinopyroxenes can accumulate more or less significant amounts of isomorphic impurities of different elements, including REE. Therefore, clinopyroxenes, as a widespread mineral phase of igneous and some other rocks, is usually considered their important concentrator of REE. Accumulation of isomorphic REE impurities in clinopyroxene during its formation was controlled by the influence of different factors, such as the crystallochemical properties of the mineral, the composition of parental melts, the redox regime, and the sequence of phase crystallization.

On discussion of the possible mechanisms of isomorphic incorporation of REE ions into clinopyroxene structure, one should take into account the fact that ions of the same framework and impurity elements can occupy different positions in the mineral structure. The variants of isomorphic substitution in clinopyroxenes can be judged from the revealed negative and positive correlations between REE and some framework elements. For example, we established a direct relationship between the contents of Fe, Ti, and Na and the contents of Tm, Yb, and Lu and an inverse relationship between the contents of some REE and the contents of Ca and Mg.

During the study of REE isomorphism in orthopyroxenes we noted that Ca^{2+} might be the best (as regards crystallochemistry) "candidate" for isomorphic substitution by trivalent REE ions in silicates. The ionic radii and charge balance between substituting and substitutable ions and the correlations between the contents of REE and framework elements suggest the following schemes of isomorphic substitution in clinopyroxenes [Lesnov, 2007b, 2011b]:

1) $3Ca^{2+} \rightarrow REE^{3+} + Fe^{3+}$;
2) $2Ca^{2+} \rightarrow REE^{3+} + Na^{+}$;
3) $3Mg^{2+} \rightarrow Eu^{2+} + Ti^{4+}$.

9.3 PLAGIOCLASES

According to available data, plagioclases can accumulate limited amounts of REE (usually no more than few ppm) in the structure. However, plagioclase from lunar gabbroids was found to have abnormally high contents of REE impurities (15–50 ppm) [Lesnov, 2000d, 2001c, 2007b]. Chondrite-normalized contents of La in plagioclases are almost always higher than those of Yb; $(La/Yb)_n = 2$–20. A crucial typomorphic feature of plagioclases is relative Eu enrichment and related strong positive Eu anomalies in REE patterns.

The REE composition of plagioclases from rocks of the Sakhalin massifs was studied on a small collection of their samples from gabbronorites, olivine gabbronorites, and gabbro-pegmatites of the Berezovka massif (Table 9.4, Fig. 9.3). The obtained data (though not on all REE) show that the examined plagioclases have a REE composition typical of this mineral. They are depleted in REE, with the chondrite-normalized contents of LREE being more or less higher than those of HREE, and have abnormally high Eu contents. Using the analysis results for these plagioclases and the earlier published distribution coefficients of REE in the plagioclase/melt system [Pietruszka and Garcia, 1999], we calculated the model REE composition of the parental melts of the host gabbroids. The calculations showed that the total REE content in the parental

Table 9.4 Contents of REE in plagioclases from rocks of the Berezovka massif, ppm.

Sample	Rock	La	Ce	Pr	Nd	Sm	Eu	Gd	Tb	Dy	Ho	Er	Tm	Yb	Lu	(La/Yb)n
132a	Ol Gbn	N.d.	0.100	N.d.	0.080	0.030	0.018	N.d.	N.d.	N.d.	N.d.	N.d.	N.d.	N.d.	N.d.	N.d.
138a	Ol Gbn	0.041	N.d.	N.d.	0.060	0.015	0.016	N.d.	N.d.	N.d.	N.d.	N.d.	N.d.	N.d.	N.d.	N.d.
132	Ol Gb	0.036	0.020	N.d.	0.051	0.010	0.017	N.d.	N.d.	N.d.	N.d.	N.d.	N.d.	N.d.	N.d.	N.d.
128	Gbn	0.180	0.350	N.d.	0.140	0.025	0.075	N.d.	0.007	N.d.	N.d.	N.d.	0.001	0.006	N.d.	20.25
130	Gbn	0.060	0.140	N.d.	0.110	0.020	0.050	N.d.	0.003	N.d.	N.d.	N.d.	0.002	0.009	N.d.	4.50
131	Gbn	0.044	N.d.	N.d.	0.072	0.012	0.058	N.d.	0.005	N.d.	N.d.	N.d.	0.002	0.010	N.d.	2.97
131a	Gbn	0.075	0.120	N.d.	N.d.	0.018	0.056	N.d.	0.002	N.d.	N.d.	N.d.	0.002	0.009	N.d.	5.62
141	Gbn	0.020	0.140	N.d.	0.070	0.027	0.031	N.d.	0.004	N.d.	N.d.	N.d.	N.d.	0.002	N.d.	6.75
144	Gbn	0.030	N.d.	N.d.	0.040	0.017	0.050	N.d.	N.d.	N.d.	N.d.	N.d.	0.001	0.010	N.d.	2.02
146	Gbn	N.d.	0.050	N.d.	N.d.	0.010	0.047	N.d.	0.005	N.d.	N.d.	N.d.	0.002	0.013	N.d.	N.d.
147	Gbn	0.026	0.057	N.d.	0.052	0.013	0.050	N.d.	0.007	N.d.	N.d.	N.d.	0.003	0.015	N.d.	1.17
151	Gbn	0.024	0.081	N.d.	0.065	0.015	0.056	N.d.	0.005	N.d.	N.d.	N.d.	0.004	0.020	N.d.	0.81
148-1*	Gbn	0.398	0.063	N.d.	0.243	0.123	0.171	N.d.	0.234	N.d.	N.d.	N.d.	0.304	0.047	N.d.	5.72
148	Gb-pegm	0.018	0.061	N.d.	0.050	0.020	0.040	N.d.	0.005	N.d.	N.d.	N.d.	0.002	0.018	N.d.	0.67

Note: Analyses were carried out by RNAA method (rock designations follow Table 9.1). Plagioclase from sample 148-1*, additionally analyzed by LA ICP-MS method, contains (ppm): Nd (0.36), Gd (0.029), Dy (0.066), Ho (0.224), Er (0.034), total REE (2.30), (Eu/Eu*)$_n$ = 6.25.

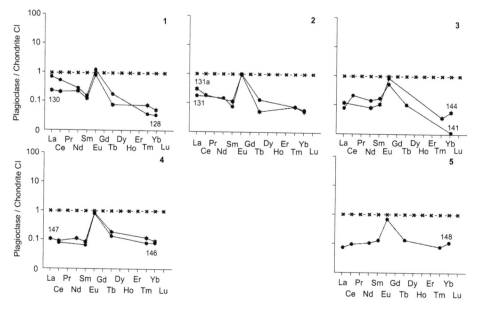

Figure 9.3 Chondrite-normalized REE patterns of plagioclases from gabbronorites (*1–4*) and gabbro-pegmatites (*5*) from the Berezovka massif (data from Table 9.4).

melt was higher than that in chondrite CI but lower than the REE content in N-MORB (especially the contents of Eu and other MREE) [Lesnov, 2000c].

Let us consider one more aspect concerned with the REE composition of plagioclases. Though the studied plagioclase structures contain much Ca, whose bivalent crystals are the most appropriate (as regards crystallochemistry) for substitution by REE ions, this mineral has generally low contents of REE impurities. The reason for this discordance calls for additional study. We can only put forward some hypotheses. Probably, the REE depletion of the plagioclases was additionally due to their crystallization in plutonic gabbroids and ultramafic rocks at the late stages of melt cooling, often, after the completion of clinopyroxene crystallization. Since clinopyroxene structure is favorable for isomorphic incorporation of REE ions, we can admit that the parental melts were seriously depleted in these elements by the beginning of plagioclase crystallization. Therefore, REE could not accumulate in large amounts in the plagioclase structure.

Taking into account possible combinations of the sizes of ionic radii and charges of substituting and substitutable ions, we suggest that REE isomorphism in the plagioclase structure proceeded by one or several of the following schemes [Lesnov, 2007b]:

1) $[2Ca^{2+}] \rightarrow [La^{3+} + Na^{+}]$;
2) $[3Ca^{2+}] \rightarrow [La^{3+} + Na^{+} + Eu^{2+}]$;
3) $[3Ca^{2+}] \rightarrow [La^{3+} + Sr^{2+} + Na^{+}]$;
4) $[Ca^{2+} + Si^{4+}] \rightarrow [REE^{3+} + Al^{3+}]$.

Table 9.5 Distribution coefficients of REE (K_d) between coexisting clinopyroxenes (Cpx), and orthopyroxenes (Opx) from rocks of the Berezovka, Shel'ting, Komsomol'sk, and South Schmidt massifs.

Rock											
Lherz	Web	Web	Web	Ol Gbn	Ol Gbn	Ol Gbn	Gbn	Gbn	Gbn	Gbn	Gbn
Sample											
154	134	143	162	132a	138a	190	131	141	144	147	173

Element	Distribution coefficient (K_d = Cpx/Opx)										
La	0.54	1.10	13.81	2.82	5.50	2.75	2.07	2.18	2.07	0.56	16.0	3.17
Ce	0.90	1.33	5.47	1.63	4.38	2.85	2.35	1.71	3.00	0.88	2.88	2.28
Pr	N.d.	N.d.	3.37	N.d.	N.d.	N.d.	N.d.	N.d.	N.d.	N.d.	N.d.	N.d.
Nd	1.73	2.07	2.55	1.10	3.07	3.60	3.21	6.00	2.50	1.55	1.91	2.12
Sm	3.67	2.50	1.51	0.67	4.00	6.67	5.00	4.50	2.80	2.58	0.92	2.53
Eu	0.58	1.49	1.93	1.38	7.50	6.00	10.0	12.31	4.62	10.23	2.33	6.00
Gd	6.25	3.00	1.61	2.17	4.20	3.00	6.67	5.29	2.86	7.20	1.04	6.20
Tb	5.88	2.32	1.37	0.17	3.20	2.40	3.21	5.33	2.59	6.20	0.96	5.46
Dy	N.d.	N.d.	1.33	N.d.	N.d.	N.d.	N.d.	N.d.	N.d.	N.d.	N.d.	N.d.
Ho	N.d.	N.d.	1.73	N.d.	N.d.	N.d.	N.d.	N.d.	N.d.	N.d.	N.d.	N.d.
Er	N.d.	N.d.	1.38	N.d.	N.d.	N.d.	N.d.	N.d.	N.d.	N.d.	N.d.	N.d.
Tm	5.00	1.96	1.62	2.25	3.33	4.17	5.00	3.50	2.60	2.50	1.06	4.26
Yb	3.96	1.65	1.53	1.78	2.32	4.78	4.00	5.56	N.d.	2.50	0.88	3.44
Lu	3.50	1.44	1.27	2.75	2.00	4.29	3.64	3.57	1.74	2.94	0.45	3.60

Note: Distribution coefficients were calculated from the data of Table 9.1. Lherz – lherzolite, Web – websterite, Ol Gbn – olivine gabbronorite, Gbn – gabbronorite. Massifs: Berezovka (131, 132a, 134, 138a, 141, 143, 144, 147, 154); Shel'ting (173); Komsomol'sk (190); South Schmidt (162).

9.4 ON THE DISTRIBUTION COEFFICIENTS OF RARE EARTH ELEMENTS AMONG COEXISTING CLINOPYROXENES, ORTHOPYROXENES, AND PLAGIOCLASES

Study of the regularities of distribution of REE and other impurity elements among coexisting minerals is one of the topical lines in modern geochemistry of igneous rocks [Zharikov and Yaroshevskii, 2003]. Of special interest are data on the REE distribution between clinopyroxene and coexisting phases in rocks of different compositions and genesis, because this mineral is the major concentrator of many impurity elements [Lesnov and Gora, 1996, 1997, 1998a,b; Lesnov, 2000b, 2007b].

Using a database on the REE compositions of rock-forming minerals from lherzolites, websterites, and olivine and olivine-free gabbronorites of the Berezovka and other Sakhalin massifs, we calculated the distribution coefficients (K_d) of REE between coexisting orthopyroxenes and clinopyroxenes (Table 9.5) and between coexisting clinopyroxenes and plagioclases (Table 9.6). The estimated K_d(Cpx–Opx) values testify to the highly irregular distribution of REE between these minerals. For example,

Table 9.6 Distribution coefficients of REE (K_d) between coexisting clinopyroxenes (Cpx), and plagioclases (Pl) from rocks of the Berezovka massif.

	Rock						
	Olivine gabbronorites		Gabbronorites				
	Sample						
	132a	*138a*	*131*	*131a*	*141*	*144*	*147*
Element	Distribution coefficient (K_d = Cpx/Pl)						
La	N.d.	2.68	5.45	15.9	15.5	10.6	36.9
Ce	5.70	N.d.	N.d.	17.5	5.36	N.d.	24.7
Pr	N.d.	N.d.	N.d.	N.d.	N.d.	N.d.	N.d.
Nd	5.75	6.00	13.3	N.d.	7.14	36.0	16.9
Sm	5.33	12.0	22.5	8.67	5.19	36.5	16.9
Eu	8.33	5.63	2.76	0.55	1.94	9.00	2.80
Gd	N.d.	N.d.	N.d.	N.d.	N.d.	N.d.	N.d.
Tb	N.d.	N.d.	16.0	8.33	11.0	N.d.	6.86
Dy	N.d.	N.d.	N.d.	N.d.	N.d.	N.d.	N.d.
Ho	N.d.	N.d.	N.d.	N.d.	N.d.	N.d.	N.d.
Er	N.d.	N.d.	N.d.	N.d.	N.d.	N.d.	N.d.
Tm	N.d.	N.d.	35.0	2.50	N.d.	100.0	11.7
Yb	N.d.	N.d.	50.0	11.5	N.d.	60.0	15.3
Lu	N.d.	N.d.	N.d.	N.d.	N.d.	N.d.	N.d.

Note: Distribution coefficients were calculated from the data in Table 9.1.

K_d(Cpx–Opx) of La and Eu in lherzolite (sample 154) does not exceed unity, i.e., these elements accumulate in clinopyroxenes, whereas K_d(Cpx–Opx) of other REE is much higher than unity, meaning that the elements accumulate in orthopyroxenes. In websterites (samples 134, 143, and 162), K_d(Cpx–Opx) of all REE is >1, i.e., the elements accumulate mostly in clinopyroxenes. In olivine gabbronorites (samples 132a, 138a, and 190) and in gabbronorites (samples 190, 131, 141, and 173), K_d(Cpx–Opx) of all REE is ≫1, which indicates still higher fractionation of the elements toward their accumulation in clinopyroxenes. In gabbronorites (samples 144 and 173), most of REE are concentrated in clinopyroxenes, though in the first sample La and Ce occur mostly in orthopyroxene (Fig. 9.4). The estimated K_d(Cpx–Opx) values in ultramafites and gabbroids agree with the earlier established regularity of the preferable accumulation of REE in clinopyroxenes rather than coexisting orthopyroxenes. The deviation from this regularity for elements with K_d(Cpx–Opx) < 1 is, most likely, due to their epigenetic redistribution or (which is doubtful) analytical errors.

Omitting detailed discussion of the K_d(Cpx–Opx) estimates for REE in gabbronorites of the Berezovka massif (Fig. 9.5), note that these elements accumulated mostly in clinopyroxenes and that the fractionation of Eu between the minerals was less contrasting as compared with the other REE. In general, the available estimates of distribution coefficients of REE among coexisting clinopyroxenes, orhtopyroxenes,

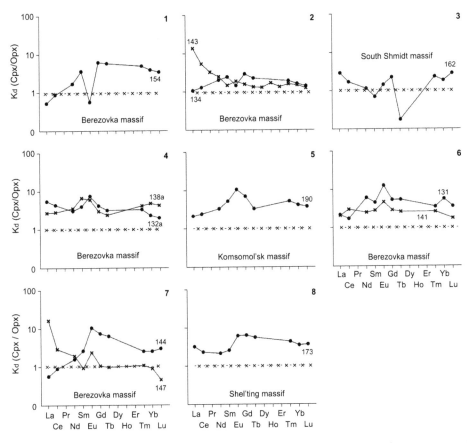

Figure 9.4 Variations of the REE distribution coefficients between coexisting orthopyroxenes and clinopyroxenes from lherzolites (*1*), websterites (*2, 3*), and from olivine (*4*) and olivine-free (*5–8*) gabbronorites from the Berezovka, Shel'ting, Komsomol'sk, and South Schmidt massifs (data from Table 9.5).

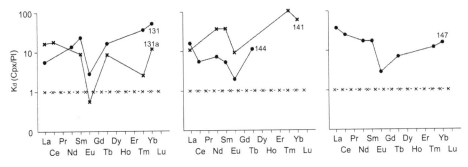

Figure 9.5 Variations of the REE distribution coefficients between coexisting clinopyroxenes and plagioclases from gabbronorites from the Berezovka massif (data from Table 9.6).

and plagioclases suggest that these minerals are in geochemical disequilibrium in rocks of the studied massifs.

* * *

New data on REE distribution in orthopyroxenes, clinopyroxenes, and plagioclases from mafic-ultramafic rocks of the Sakhalin massifs have been obtained. Orthopyroxenes are generally depleted in REE; their REE patterns are positively sloped, $(La/Sm)_n = 0.16–2.73$ (mostly, > 1), and $(Eu/Eu^*)_n = 0.50–0.92$. Correspondingly, the REE patterns show weak negative Eu anomalies. In orthopyroxenes, most of REE ions probably substitute Ca^{2+}. Clinopyroxenes are enriched in REE. The shapes of their REE patterns and $(La/Yb)_n$ values point to different intensities of element fractionation. The clinopyroxenes are divided into two groups according to their REE patterns: (1) minerals with REE patterns localized near the line of REE contents in chondrite CI and (2) minerals with REE patterns lying above or below this line. Some of the analyzed clinopyroxenes are abnormally enriched in LREE, which is probably due to the varying contents of their nonstructural impurities in the grain microcracks. In clinopyroxenes, as in orthopyroxenes, Ca^{2+} seems to be the best "candidate" for isomorphic substitution by REE ions. Plagioclases are generally poor in REE, especially HREE. Their REE patterns are negatively sloped and complicated by positive Eu anomalies of different intensities. The calculated distribution coefficients of REE among coexisting clinopyroxenes, orthopyroxenes, and plagioclases in the massif rocks suggest that these minerals did not reach a geochemical equilibrium during crystallization.

Chapter 10

Isotope-geochronological and geochemical systematization of zircons from rocks of the Berezovka mafic-ultramafic massif

As was shown in the previous chapters, among the problems related to geologic structure and formation conditions of mafic-ultramafic massifs, spatiotemporal relationship between the ultramafic and gabbroid bodies composing such massifs and their dating remain a hotly discussed subject. For quite a long time, these issues have been dealt with mainly on the basis of structure–geological and petrographical research methods. As geochemical and especially isotope techniques were applied far more rarely, it has not always been possible to establish with certainty age relations in ultramafic and gabbroid bodies composing such massifs. In recent years, greater possibilities for solving such problems have been provided with the development of isotope methods for dating rocks and minerals, specifically, the U–Pb method for analysis of individual zircon grains with the use of SHRIMP II multicollector secondary-ion mass-spectrometers. Afterward, along with the age determinations made on zircons from granitoids [Rudnev *et al.*, 2012], alkaline igneous rocks [Vrublevskii *et al.*, 2014], and metamorphic rocks [Kaulina, 2010; Turkina *et al.*, 2014], the zircons from gabbroids [Krasnobaev *et al.*, 2007; Bortnikov *et al.*, 2008; Tsukanov and Skolotnev, 2010; Skolotnev *et al.*, 2010; Ledneva *et al.*, 2012; etc.], ultramafic rocks [Knauf, 2008; Fershtater *et al.*, 2009; Malitch *et al.*, 2009; Lesnov *et al.*, 2010a; Badanina and Malich, 2012; Oh *et al.*, 2012; Ronkin *et al.*, 2013; etc.] and chromitites [Savelieva *et al.*, 2006, 2007] began to be more widely used for such estimates.

The isotope analysis of zircons with the use of the SHRIMP II was for the first time applied for dating of this mineral from the Berezovka mafic-ultramafic massif.

10.1 DESCRIPTION OF THE COLLECTION OF ZIRCONS FROM ROCKS OF THE BEREZOVKA MASSIF

For isotope-geochronological studies (zircon dating) of the Berezovka massif, V.G. Gal'versen (Sakhalin Geological-Prospecting Expedition 2007–2008) took 22 samples (each being up to several kilograms in weight) from its rocks and those in its vicinity. Eighteen of them were pyroxenites, gabbro-pyroxenites, melanocratic olivine gabbronorites, and some other types of gabbroids as well as gabbro-diorites, diorites, and quartz diorites; the other four were metavolcanics of the enclosing strata. In the Mineralogical Laboratory of the Sakhalin Geological-Prospecting Expedition, zircon monofractions containing about 450 grains were extracted from these samples with the necessary precautions taken to prevent their contamination with foreign materials. These monofractions were delivered to the Center of Isotopic Research of the Russian Geological Research Institute (VSEGEI) (St. Petersburg), where about 190 best grains selected for the study were implanted into the epoxy preparations. About 200 isotopic age determinations of zircon were thereby carried out, and their oscillatory zoning was studied using cathodoluminescence (CL) images. Then 160 LA ICP-MS analyses of zircon grains from the same preparations were carried out to measure the contents of 26 impurity elements, including REE, as well as 40 analyses to determine the Re and Mo contents and more than 140 analyses to evaluate the contents of Hf, U, Ce, and Yb in two zircon crystals, along their cross sections.

10.2 METHODS FOR ISOTOPE DATING AND TRACE-ELEMENT ANALYSIS OF ZIRCONS

The isotope-geochronological studies of zircons were carried out with a SHRIMP II high-resolution secondary-ion mass-spectrometer. From five to 12 zircon grains were analyzed in each of the samples, with the analyses performed at several points in some of them. Zircons from the TEMORA and 91500 standards were used as reference samples. Subsequent to isotope measurements, the study of morphology and internal structure of zircon grains was carried out, with their microphotographs obtained with a CamScan MX2500 scanning electron microscope, in optical and cathodoluminescence modes. Isotope measurements and processing of digital data were made, following the techniques accepted at the VSEGEI Center of Isotopic Research (Schuth *et al.*, 2012). The trace-element composition of zircons was studied by LA ICP-MS with an Element mass spectrometer (Thermo Finnigan MAT) coupled with a UP-213, Nd:YAG laser ablation system (New Wave Research). The NIST-612 (USGS) glass was used as a reference sample (geochemical standard). The morphology and elemental composition of zircons were also studied using a LEO 1430VP scanning electron microscope and EPMA probe with a Camebax-Micro spectrometer.

10.3 MORPHOLOGY AND OPTICAL PROPERTIES OF ZIRCONS

In the studied collection, zircon grains differ in size and morphology, CL intensity, and kind of oscillatory zoning. The grains range from 50 to 400 μm in length, with those 50–200 μm long prevailing among them. Their length-to-width ratio varies from 1 to 4, with most of its values lying between 1.0 and 2.5. By their morphology, the grains are divided into the following varieties: (1) short-prismatic crystals with well-developed faces, edges, and pyramidal apices; (2) long-prismatic crystals with well-developed faces, edges, and pyramidal apices; (3) prismatic crystals with slightly resorbed faces and edges; (4) prismatic crystals with strongly resorbed faces and edges; and (5) extremely intensely resorbed ovoid grains totally or almost totally lacking crystal faces.

By CL intensity, zircon grains are classified into several types featured by (1) intense CL, (2) moderate CL, (3) reduced CL, (4) extremely low CL, and (5) almost zero luminescence.

The oscillatory zoning patterns of grains also differ, dividing them into (1) crystals with fine and mostly regular zoning concordant with their faceting; (2) crystals with clear fine zoning and dark "cores"; (3) crystals with coarse irregular zoning, often discordant with their faceting; (4) crystals with coarse irregular zoning and dark cores; (5) crystals with sectorial or spotted zoning; and (6) crystals totally or nearly totally lacking oscillatory zoning.

The observed differences in zircon grains served as proxy indicators in their isotope-geochronological and geochemical systematization.

10.4   ISOTOPIC AGE OF ZIRCONS

Within the studied collection of zircon grains, their isotopic age varies in the range from ~3100 to ~20 Ma (Tables 10.1, 10.2). The histograms of the occurrence of zircon ages both in the studied collection (Fig. 10.1, *a*) and within the set of samples not older than 200 Ma (Fig. 10.1, *b*) have a polymodal shape, which is indicative of the polychronous formation of zircons and the host rocks. The histograms show that the most intense maxima represent the interval of ~170–150 Ma. It is remarkable that a wide scatter of ages is observed not only within the entire studied collection of zircon grains but also in the grains from many individual rock samples (Fig. 10.2). Earlier, the author published preliminary results of the isotope-geochronological studies of zircons from the Berezovka massif [Lesnov *et al.*, 2010a; Lesnov, 2012a].

By the cumulative isotopic ages, zircons, in the first approximation, were divided into six age populations (Ma): (1) ~3100-990, (2) ~790-410, (3) ~395-210, (4) ~200-100, (5) ~90-65, and (6) ~30-20. Zircons from the oldest population (~17% of the entire collection) occurred mainly in hybrid ultramafites (pyroxenites and gabbro-pyroxenites) and in several hybrid gabbroids (melanocratic olivine gabbronorites), while they are more rare in other gabbroid varieties. The greatest number of zircon grains (~40% of the entire collection) is represented by the varieties with age ranging in the interval of ~190-140 Ma. These zircons occur mainly in orthomagmatic gabbroids and more rarely in hybrid amphibole gabbro, gabbro-diorites, and diorites.

As is shown in optical microphotographs, the grain morphology of the analyzed zircons is inhomogeneous, which is remarkably noticeable when grains of ancient zircons (Fig. 10.3, *A*) and of those dated ~190-140 Ma are compared (Fig. 10.3, *B*). The largest number of ancient grains tends to have rounded shape, down to totally lacking crystal faces, and rough surfaces. Such zircons are the most common in pyroxenites (sample 1597) and gabbro-pyroxenites (samples 1607, 1610-2, and 1606-1), regarded as hybrid ultramafites. These grains are thought to have undergone more or less intense resorption caused by the action of later generated mafic melts and their fluids. Unlike the intensely resorbed grains of ancient zircons, almost all grains dated at ~190-140 Ma have fairly distinct crystal faces.

Also, zircons from various age populations differ in CL imaging and in peculiarities of oscillatory zoning (Fig. 10.4). Most of zircon grains from the oldest populations are characterized by extremely low (even zero) CL and often lack oscillatory zoning. In the grains with preserved oscillatory zoning, the latter is irregular or spotted and the zones are often discordant with the crystal faces. The overwhelming majority of zircon grains dated at ~190-140 Ma shows moderate to intense CL and usually has a clear regular oscillatory zoning concordant with the crystal faces.

Table 10.1 Characteristics of the collection of zircon crystals from the Berezovka massif, studied by the U–Pb isotope method.

Sample	Rock	Total number of crystals in sample	Number of age determinations of crystal by U–Pb method	Distribution of isotope dating in age population, Ma					
				3100-990	790-410	395-210	200-100	99-65	30-20
1597	Pyroxenite	20	10	7	3	–	–	–	–
1607	Gabbro-pyroxenite	25	10	3	–	6	1	–	–
1610-2	Gabbro-pyroxenite	27	15	–	1	5	8	–	–
1606-1	Gabbro-pyroxenite	24	10	4	5	–	1	–	–
1604	Melanocratic olivine gabbronorite	35	11	2	–	2	4	3	–
2622	Gabbro	13	9	4	1	–	2	2	–
1595	Gabbroid	12	9	1	3	4	1	–	–
1596-A1	Gabbroid	4	6	2	4	–	–	–	–
1596-6	Gabbroid	11	8	–	3	2	3	–	–
1596-4	Gabbroid	25	11	1	3	2	2	3	–
10	Gabbroid	11	10	5	2	1	2	–	–
1655	Gabbro-diorite	35	11	–	1	–	10	–	–
1658	Gabbro-diorite	35	10	–	–	–	10	–	–
1652	Gabbro-diorite	30	10	–	–	–	10	–	–
2612-2	Gabbro-diorite	40	10	–	2	–	3	4	–
160	Diorite	43	9	–	1	1	2	5	–
197	Diorite	38	10	–	–	–	10	–	–
2612	Quartz diorite	14	10	–	–	2	3	5	–
156	Trachyandesibasalt	9	9	5	–	–	2	1	1
233	Andesite	20	10	–	2	2	5	1	–
181	Trachyandesite	30	11	1	2	2	5	–	–
1655-1	Metatuff latite	5	4	–	2	–	2	–	9
	Total	456	213	36	31	28	83	24	10

Note: Isotope-geochronological studies of zircons were carried out at the Center of Isotopic Research of VSEGEI (St. Petersburg) with a SHRIMP II high-resolution secondary-ion mass-spectrometer.

Table 10.2 Results of isotope-geochronological, geochemical, morphological, and optical studies of zircons from rocks of Berezovka massif

Consecutive numbers 1	Rocks 2	Samples 3	Crystals 4	U, ppm 5	Th, ppm 6	206Pb, ppm 7	232Th/238U 8	238U/206Pb 9	207Pb*/206Pb 10	207Pb*/235U 11	206Pb*/238U 12	Crystal habit 13	Length of crystals, μm (length-to-width ratio) 14	Cathodoluminescence intensity 15	Oscillatory zoning 16	Age, Ma (206Pb/238U) 17	Age, Ma (207Pb/206Pb) 18
1	Pyroxenite	1597	1.1	318	60	102	0.19	2.671	0.12914	6.627	0.3741	Ovoid	239 (1.10)	V-weak	N-zon	2048 ± 14	2077 ± 10
2	Pyroxenite	1597	2.1	329	339	32.4	1.07	8.712	0.06649	0.988	0.1142	Ovoid	N.d.	Moder.	Brok	697.2 ± 5.9	700 ± 70
3	Pyroxenite	1597	3.1	49	18	26.3	0.37	1.617	0.2356	19.87	0.6164	Ovoid	208 (1.37)	Weak	N-zon	3096 ± 34	3078 ± 15
4	Pyroxenite	1597	4.1	79	104	11.1	1.36	6.109	0.0768	1.666	0.1631	Ovoid	250 (1.20)	Moder.	N-zon	974 ± 12	1043 ± 55
5	Pyroxenite	1597	5.1	172	62	55	0.37	2.694	0.13012	6.554	0.3703	Ovoid	208 (1.72)	Moder.	Brok-Cor	2031 ± 16	2076 ± 15
6	Pyroxenite	1597	5.2	182	29	58.3	0.17	2.678	0.12724	6.54	0.3733	Ovoid	208 (1.72)	Moder.	Brok	2045 ± 16	2058 ± 13
7	Pyroxenite	1597	6.1	120	49	43.3	0.42	2.373	0.1369	7.95	0.4213	Ovoid	2.90 (1.75)	Moder.	Good-C	2267 ± 23	2188 ± 30
8	Pyroxenite	1597	7.1	53	37	4.79	0.72	9.47	0.0638	0.929	0.1056	Ovoid	241 (1.20)	Moder.	Good-C	647.3 ± 9.8	735 ± 70
9	Pyroxenite	1597	8.1	148	36	37.9	0.25	3.351	0.097	3.976	0.2983	Ovoid	200 (1.62)	Weak	Good	1683 ± 17	1561 ± 25
10	Pyroxenite	1597	9.1	41	24	3.92	0.62	9	0.0718	0.87	0.1092	Ovoid	250 (130)	V-weak	Brok	668 ± 12	527 ± 250
11	Gabbro-pyroxenite	1607	1.1	290	3	14.5	0.01	17.15	0.0555	0.439	0.0583	Res-SI	N.d.	Weak	Brok-Cor	365 ± 7.0	401 ± 43
12	Gabbro-pyroxenite	1607	1.2	216	23	10.7	0.11	17.33	0.0529	0.421	0.0577	S-Prism	N.d.	Weak	Brok	362 ± 7.0	326 ± 47
13	Gabbro-pyroxenite	1607	4.1	603	429	21.1	0.74	24.54	0.0514	0.286	0.0407	S-Prism	N.d.	V-weak	N-zon	257 ± 4.9	240 ± 68
14	Gabbro-pyroxenite	1607	5.1	701	128	122	0.19	4.93	0.20505	5.72	0.203	Res-SI	120 (1.90)	V-weak	Brok	1190 ± 21	2865 ± 6
15	Gabbro-pyroxenite	1607	6.1	384	33	18.7	0.09	17.59	0.0551	0.425	0.0568	Res-SI	91 (3.71)	Weak	Brok-Cor	356 ± 6.9	381 ± 45
16	Gabbro-pyroxenite	1607	7.1	330	16	16.4	0.05	17.31	0.05465	0.419	0.0576	Res-SI	83 (1.57)	V-weak	Brok-Cor	361 ± 7.0	320 ± 54
17	Gabbro-pyroxenite	1607	8.1	565	323	21	0.59	23.1	0.05092	0.301	0.0433	Res-St	105 (2.63)	V-weak	Brok-Cor	273 ± 5.2	214 ± 37
18	Gabbro-pyroxenite	1607	9.1	119	86	3.35	0.75	30.51	0.1204	0.453	0.032	Res-SI	N.d.	Moder.	Brok-Cor	N.d.	1672 ± 150
19	Gabbro-pyroxenite	1607	10.1	898	497	36.8	0.57	20.95	0.1329	0.866	0.0477	Res-SI	N.d.	Moder.	Brok-Cor	N.d.	2123 ± 88
20	Gabbro-pyroxenite	1610-2	1.1	899	584	27.5	0.67	28.13	0.0504	0.2472	0.03556	S-Prism	74 (1.50)	V-weak	Good-C	225 ± 2.2	214 ± 63
21	Gabbro-pyroxenite	1610-2	2.1	423	202	9.17	0.49	39.57	0.047	0.1639	0.2527	S-Prism	N.d.	V-weak	N-zon	161 ± 2.2	51 ± 120
22	Gabbro-pyroxenite	1610-2	3.1	505	183	10.7	0.37	41.05	0.0495	0.166	0.02435	S-Prism	N.d.	V-weak	Brok	155 ± 2.2	170 ± 240

(Continued)

Table 10.2 Continued.

Consecutive numbers 1	Rocks 2	Samples 3	Crystals 4	U, ppm 5	Th, ppm 6	206Pb, ppm 7	232Th/238U 8	238U/206Pb 9	207Pb*/206Pb 10	207Pb*/235U 11	206Pb*/238U 12	Crystal habit 13	Length of crystals, μm (length-to-width ratio) 14	Cathodoluminescence intensity 15	Oscillatory zoning 16	Age, Ma (206Pb/238U) 17	Age, Ma (207Pb/206Pb) 18
23	Gabbro-pyroxenite	1610-2	4.1	219	78	4.72	0.37	39.87	0.0491	0.1699	0.02508	S-Prism	N.d.	Intense	Good	160±2.6	154±120
24	Gabbro-pyroxenite	1610-2	5.1	446	842	11.3	1.95	34.17	0.0499	0.201	0.02926	S-Prism	71 (1.14)	Moder.	Good-C	186±2.7	189±320
25	Gabbro-pyroxenite	1610-2	6.1	745	342	16	0.47	40.31	0.052	0.178	0.0248	S-Prism	108 (2.16)	Intense	Brok	158±1.7	286±150
26	Gabbro-pyroxenite	1610-2	7.1	63	41	1.98	0.67	27.54	0.0533	0.267	0.03632	S-Prism	N.d.	Intense	Brok	230±5.8	340±160
27	Gabbro-pyroxenite	1610-2	8.1	415	233	11.8	0.58	30.26	0.0502	0.229	0.03305	S-Prism	N.d.	V-weak	Brok	210±2.5	204±100
28	Gabbro-pyroxenite	1610-2	9.1	239	80	5.16	0.34	40.03	0.0495	0.171	0.02498	S-Prism	N.d.	V-weak	Brok	159±2.4	172±160
29	Gabbro-pyroxenite	1610-2	10.1	237	3	12.7	0.01	16.09	0.0546	0.468	0.06215	S-Prism	78 (1.72)	Moder.	Brok-Cor	389±4.3	397±88
30	Gabbro-pyroxenite	1610-2	11.1	94	47	7.2	0.51	11.22	0.0602	0.74	0.0891	Res-Sl	N.d.	Weak	Brok-Cor	550±8	611±99
31	Gabbro-pyroxenite	1610-2	12.1	326	112	7.17	0.35	39.34	0.0503	0.176	0.02542	Res-Sl	132 (3.00)	Weak	Brok-Cor	162±2	209±210
32	Gabbro-pyroxenite	1610-2	13.1	216	179	11.7	0.86	15.82	0.0556	0.485	0.06322	Res-Sl	138 (2.10)	Moder.	Brok-Cor	395±5	437±89
33	Gabbro-pyroxenite	1610-2	14.1	440	135	61.9	0.32	6.11	0.0735	1.659	0.1637	Res-Sl	147 (2.10)	Ex-weak	Brok	977±8.4	1028±31
34	Gabbro-pyroxenite	1606-1	1.1	154	73	13.5	0.49	9.8	0.066	0.857	0.1015	Res-St	200 (1.26)	Ex-weak	N-zon	623±6.2	649±76
35	Gabbro-pyroxenite	1606-1	2.1	129	148	10.9	1.18	10.23	0.0602	0.812	0.0978	Res-St	N.d.	Ex-weak	Brok-Cor	601.4±7.3	611±45
36	Gabbro-pyroxenite	1606-1	3.1	43	30	6.62	0.73	5.532	0.0802	1.794	0.179	Ovoid	222 (1.67)	V-weak	Brok-Cor	1061±15	1005±88
37	Gabbro-pyroxenite	1606-1	4.1	136	87	28.6	0.67	4.075	0.0911	3.018	0.2449	Ovoid	N.d.	Weak	Brok	1412±15	1413±23
38	Gabbro-pyroxenite	1606-1	5.1	83	72	7.33	0.89	9.77	0.0631	0.84	0.1019	Ovoid	206 (1.41)	Weak	Brok	626.6±7.9	595±100

No.	Rock type	Sample															
39	Gabbro-pyroxenite	1606-1	6.1	113	38	23	0.35	4.238	0.0879	2.859	0.236	Ovoid	N.d.	V-weak	Brok	1366±15	1379±27
40	Gabbro-pyroxenite	1606-1	7.1	57	50	18.7	0.9	2.617	0.1348	6.91	0.3805	Ovoid	215 (1.08)	V-weak	Good-C	2079±23	2121±27
41	Gabbro-pyroxenite	1606-1	8.1	84	62	7.04	0.76	10.25	0.0632	0.802	0.0971	Res-St	240 (1.85)	V-weak	Good-C	597.5±7.5	601±81
42	Gabbro-pyroxenite	1606-1	9.1	268	102	5.62	0.39	40.93	0.056	0.173	0.0243	Res-Sl	382 (1.83)	Moder.	Brok	154.7±2	273±210
43	Gabbro-pyroxenite	1606-1	10.1	102	55	10.3	0.56	8.45	0.0675	0.998	0.1174	Res-St	N.d.	Weak	Good	715.8±8.7	662±190
44	Melanocratic olivine gabbronorite	1604	1.1	360	123	7.77	0.35	39.8	0.0497	0.1722	0.02512	L-Prism	N.d.	Weak	Good	160±3.3	N.d.
45	Melanocratic olivine gabbronorite	1604	2.1	707	525	28.8	0.77	21.1	0.05148	0.3349	0.04737	S-Prism	N.d.	Ex-weak	N-zon	298±5.7	N.d.
46	Melanocratic olivine gabbronorite	1604	3.1	379	188	8.24	0.51	39.51	0.0493	0.1719	0.02531	S-Prism	N.d.	V-weak	N-zon	161±3.3	N.d.
47	Melanocratic olivine gabbronorite	1604	4.1	191	114	2.45	0.62	76.4	0.0424	0.075	0.01291	S-Prism	171 (1.50)	Intense	Good-C	83±2.1	N.d.
48	Melanocratic olivine gabbronorite	1604	4.2	139	111	1.36	0.82	84.8	0.0424	0.068	0.01165	S-Prism	N.d.	Intense	Good-C	75±2.0	N.d.
49	Melanocratic olivine gabbronorite	1604	4.3	186	106	1.97	0.59	81.2	0.0471	0.0792	0.01221	S-Prism	N.d.	Intense	Good-C	78±1.8	N.d.
50	Melanocratic olivine gabbronorite	1604	5.1	1281	810	482	0.65	2.282	0.1293	7.8	0.4381	Res-Sl	88 (2.00)	V-weak	Brok	N.d.	2086±14
51	Melanocratic olivine gabbronorite	1604	5.2	1381	661	386	0.53	2.898	0.126	5.99	0.345	Res-Sl	N.d.	V-weak	Brok	N.d.	2043±15
52	Melanocratic olivine gabbronorite	1604	6.1	166	32	3.67	0.2	36.85	0.0728	0.198	0.02522	S-Prism	70 (2.20)	Weak	Brok	161±3.8	N.d.
53	Melanocratic olivine gabbronorite	1604	7.1	985	52	49	0.05	17.27	0.05291	0.434	0.058	S-Prism	N.d.	Ex-weak	N-zon	364±6.8	N.d.
54	Melanocratic olivine gabbronorite	1604	8.1	86	85	2.17	1.02	34.18	0.0548	0.219	0.02901	S-Prism	60 (1.67)	Weak	Brok	184±4.6	N.d.

(Continued)

Table 10.2 Continued.

Consecutive numbers 1	Rocks 2	Samples 3	Crystals 4	U, ppm 5	Th, ppm 6	206Pb, ppm 7	232Th/238U 8	238U/206Pb 9	207Pb/206Pb 10	207Pb*/235U 11	206Pb*/238U 12	Crystal habit 13	Length of crystals, μm (length-to-width ratio) 14	Cathodoluminescence intensity 15	Oscillatory zoning 16	Age, Ma (206Pb/238U) 17	Age, Ma (207Pb/206Pb) 18
55	Gabbro	2622	1.1	276	159	50.5	0.59	4.698	0.08303	2.407	0.2126	Ovoid	232 (1.29)	V-weak	Brok-Cor	1243 ± 19	1248 ± 23
56	Gabbro	2622	2.1	900	505	17.8	0.58	43.5	0.05098	0.1568	0.02295	Res-St	N.d.	V-weak	Brok	146 ± 2.4	N.d.
57	Gabbro	2622	3.1	536	169	6.96	0.33	66.1	0.0504	0.1016	0.0151	S-Prism	N.d.	V-weak	Brok	97 ± 1.7	N.d.
58	Gabbro	2622	4.1	777	329	113	0.44	5.92	0.07502	1.733	0.1688	Res-St	N.d.	Ex-weak	N-zon	1006 ± 15	1054 ± 14
59	Gabbro	2622	5.1	352	226	3.55	0.66	85.4	0.0517	0.0763	0.01165	S-Prism	N.d.	Moder.	Brok	75 ± 1.4	N.d.
60	Gabbro	2622	5.2	368	128	3.77	0.36	84	0.06	0.082	0.01176	S-Prism	N.d.	Moder.	Brok	75 ± 1.5	N.d.
61	Gabbro	2622	6.1	296	197	20.6	0.69	12.35	0.0571	0.645	0.081	Res-St	N.d.	V-weak	N-zon	502 ± 8	N.d.
61	Gabbro	2622	7.1	52	11	1.11	0.22	40.4	0.0631	0.152	0.02421	S-Prism	172 (1.19)	Weak	Brok-Cor	154 ± 5	N.d.
62	Gabbro	2622	8.1	115	59	31.9	0.53	3.1	0.1184	5.2	0.3221	Ovoid	121 (1.17)	Ex-weak	Brok-Cor	1800 ± 28	1913 ± 37
63	Gabbro	2622	9.1	29	7	8.96	0.24	2.758	0.1202	5.92	0.362	Res-St	176 (2.43)	Weak	Brok-Cor	1991 ± 42	1936 ± 41
64	Gabbroid	1595	1.1	212	304	32.5	1.48	5.6	0.0793	1.95	0.1784	Ovoid	N.d.	V-weak	Brok	1058 ± 24	1179 ± 59
65	Gabbroid	1595	2.1	174	13	9.4	0.07	16.04	0.056	0.481	0.0623	S-Prism	82 (3.10)	Moder.	Brok-Cor	390 ± 6.3	451 ± 180
66	Gabbroid	1595	3.1	183	4	9.9	0.02	16.09	0.0523	0.448	0.06211	S-Prism	90 (2.20)	Moder.	Brok-Cor	389 ± 5.4	299 ± 220
67	Gabbroid	1595	4.1	308	140	6.41	0.47	41.42	0.0494	0.1643	0.02414	S-Prism	90 (2.20)	Moder.	Brok	154 ± 2.1	166 ± 140
68	Gabbroid	1595	5.1	221	2	9.69	0.01	19.68	0.0498	0.349	0.0508	Res-Sl	95 (1.80)	Weak	Brok-Cor	320 ± 4.1	186 ± 120
69	Gabbroid	1596-A1	1.1	328	165	18.6	0.52	15.21	0.0556	0.504	0.06574	Res-Sl	100 (1.67)	Weak	Brok-Cor	410 ± 3.5	436 ± 44
70	Gabbroid	1596-A1	2.1	584	412	36.2	0.73	13.93	0.0567	0.561	0.07176	Res-St	76 (2.47)	Moder.	Brok-Cor	447 ± 4.5	480 ± 100
71	Gabbroid	1596-A1	3.1	57	31	5.09	0.56	9.74	0.0609	0.86	0.1025	Ovoid	120 (1.22)	Moder.	Brok	629 ± 15	635 ± 340
72	Gabbroid	1596-A1	3.2	109	165	9.52	1.56	9.86	0.0616	0.862	0.1014	Ovoid	120 (1.22)	Moder.	Brok	623 ± 10	661 ± 80
73	Gabbroid	1596-A1	4.1	458	82	144	0.18	2.727	0.1244	6.29	0.3667	Res-St	110 (1.38)	Weak	Brok	2014 ± 17	2020 ± 42
74	Gabbroid	1596-A1	4.2	249	55	69.2	0.23	3.1	0.1244	5.402	0.3226	Res-St	110 (1.38)	Weak	Brok	1802 ± 17	1978 ± 19
75	Gabbroid	1596-6	1.1	212	110	18.8	0.54	9.69	0.06	0.854	0.1032	Res-St	N.d.	Ex-weak	Brok	633 ± 12	N.d.
76	Gabbroid	1596-6	2.1	478	234	9.21	0.51	44.54	0.0488	0.147	0.02241	S-Prism	122 (2.15)	Moder.	Good	143 ± 2.8	N.d.
77	Gabbroid	1596-6	2.2	468	231	9.02	0.51	44.54	0.0526	0.1519	0.02236	S-Prism	122 (2.15)	Moder.	Good	143 ± 2.8	N.d.
78	Gabbroid	1596-6	3.1	314	14	15.6	0.05	17.35	0.0554	0.422	0.0575	Ovoid	92 (1.38)	Moder.	Brok	360 ± 7.0	N.d.
79	Gabbroid	1596-6	5.1	520	108	33.2	0.22	13.43	0.0586	0.582	0.0743	Res-St	N.d.	Moder.	N.d.	462 ± 9.2	N.d.
80	Gabbroid	1596-4	6.1	303	7	16.4	0.02	15.82	0.05896	0.487	0.063	Res-St	74 (1.83)	Moder.	Brok	394 ± 7.7	N.d.
81	Gabbroid	1596-4	1.1	524	512	12.7	1.01	35.45	0.0531	0.189	0.02806	S-Prism	167 (3.57)	Weak	Brok	178 ± 3.1	N.d.
82	Gabbroid	1596-4	2.1	178	113	2.49	0.65	61.5	0.066	0.123	0.01604	S-Prism	N.d.	Moder.	Brok	103 ± 2.2	N.d.
83	Gabbroid	1596-4	3.1	910	295	31.6	0.34	24.71	0.0516	0.28	0.0404	S-Prism	N.d.	Ex-weak	N-zon	255 ± 4.0	N.d.
84	Gabbroid	1596-4	4.1	252	183	5.58	0.75	38.77	0.0542	0.181	0.02569	S-Prism	171 (2.33)	Moder.	Good	164 ± 3.3	N.d.
85	Gabbroid	1596-4	5.1	498	495	4.95	1.03	86.4	0.0519	0.074	0.0115	S-Prism	N.d.	Weak	Brok	74 ± 1.4	N.d.

No.	Rock	Sample	Spot	U	Th	Th/U	Pb*	206Pb/238U err	207Pb/235U	207Pb/206Pb	Shape	CL	Zoning	Quality	Age1	Age2
86	Gabbroid	1596-4	6.1	79	89	1.04	1.17	65.1	0.0652	0.096	0.015	Res-Sl	132 (1.23)	Moder.	Brok-Cor	96 ± 3.1	N.d.
87	Gabbroid	1596-4	7.1	221	179	5.22	0.84	36.35	0.0537	0.182	0.02732	S-Prism	N.d.	Moder.	Brok-Cor	174 ± 3.2	N.d.
88	Gabbroid	1596-4	8.1	638	128	0.21	0.21	16.15	0.055	0.457	0.0618	Res-Sl	N.d.	Ex-weak	Brok	387 ± 6.1	N.d.
89	Gabbroid	1596-4	9.1	535	148	146	0.29	3.139	0.11498	4.912	0.3175	Res-Sl	N.d.	Ex-weak	N-zon	1778 ± 25	N.d.
90	Gabbroid	1596-4	10.1	994	507	9.57	0.53	89.2	0.0487	0.0707	0.01117	S-Prism	N.d.	V-weak	Brok-Cor	72 ± 1.3	N.d.
91	Gabbroid	1596-4	11.1	228	117	5.05	0.53	38.76	0.0515	0.1723	0.02571	S-Prism	167 (1.60)	Moder.	Good	164 ± 3.2	N.d.
92	Gabbroid	10	1.1	72	32	26	0.46	2.392	0.1415	8.04	0.417	Ovoid	152 (1.00)	Moder.	Good-C	2247 ± 34	2226 ± 31
93	Gabbroid	10	1.2	188	8	61.8	0.05	2.605	0.12792	6.73	0.3835	Ovoid	124 (1.30)	Moder.	Good-C	2093 ± 29	2060 ± 15
94	Gabbroid	10	1.3	76	34	24.7	0.46	2.632	0.1327	6.78	0.3784	Ovoid	N.d.	Moder.	Good-C	2069 ± 31	2097 ± 38
95	Gabbroid	10	2.1	376	231	9.3	0.63	35	0.0604	0.207	0.0283	Res-Sl	N.d.	Moder.	Brok-Cor	180 ± 3	N.d.
96	Gabbroid	10	2.2	483	308	11.9	0.66	35.03	0.0561	0.199	0.02836	Res-Sl	N.d.	Moder.	Brok-Cor	180 ± 3	N.d.
97	Gabbroid	10	3.1	2.33	108	60.6	0.48	3.301	0.1045	4.364	0.3029	Res-Sl	N.d.	Ex-weak	N-zon	1706 ± 24	1705 ± 18
98	Gabbroid	10	4.1	137	108	22.4	0.82	5.248	0.0802	2.033	0.1899	S-Prism	N.d.	V-weak	Brok	1121 ± 19	N.d.
99	Gabbroid	10	5.1	1068	120	55.4	0.12	16.56	0.0924	0.432	0.05746	S-Prism	109 (1.94)	Ex-weak	Good	360 ± 6	N.d.
100	Gabbroid	10	6.1	286	84	18.9	0.3	13.01	0.061	0.571	0.0762	S-Prism	N.d.	Weak	Brok	474 ± 8	N.d.
101	Gabbroid	10	6.2	436	131	27.5	0.31	13.63	0.05839	0.558	0.0731	S-Prism	N.d.	Weak	Brok	455 ± 7	N.d.
102	Gabbro-diorite	1655	1.1	241	95	5.46	0.41	37.91	0.052	0.178	0.02629	L-Prism	N.d.	Weak	Brok	167 ± 3.6	N.d.
103	Gabbro-diorite	1655	2.1	608	212	12.4	0.36	42.1	0.0502	0.1645	0.02377	L-Prism	N.d.	V-weak	Brok	151 ± 4.4	N.d.
104	Gabbro-diorite	1655	3.1	413	203	8.91	0.51	39.78	0.0493	0.1707	0.02514	L-Prism	107 (2.65)	V-weak	Brok	160 ± 3.2	N.d.
105	Gabbro-diorite	1655	4.1	305	138	6.33	0.47	41.3	0.0529	0.158	0.02405	L-Prism	N.d.	Weak	Brok	153 ± 4.2	N.d.
106	Gabbro-diorite	1655	5.1	288	152	6.41	0.55	38.6	0.0523	0.164	0.02571	L-Prism	176 (4.62)	Moder.	Brok	164 ± 3.5	N.d.
107	Gabbro-diorite	1655	6.1	322	178	7.34	0.57	37.7	0.0539	0.178	0.02636	L-Prism	N.d.	Weak	Brok	168 ± 3.5	N.d.
108	Gabbro-diorite	1655	7.1	1163	850	25.7	0.76	38.82	0.04859	0.1681	0.02572	L-Prism	136 (3.53)	V-weak	Brok	164 ± 3.2	N.d.
109	Gabbro-diorite	1655	8.1	601	172	12.8	0.3	40.24	0.049	0.1678	0.02485	L-Prism	N.d.	V-weak	Brok	158 ± 3.1	N.d.
110	Gabbro-diorite	1655	9.1	790	424	17.1	0.56	39.85	0.0488	0.1689	0.0251	S-Prism	100 (1.25)	Weak	Brok	160 ± 3.1	N.d.
111	Gabbro-diorite	1655	10.1	129	53	14.5	0.43	7.64	0.0661	1.194	0.1309	L-Prism	N.d.	Weak	Brok	793 ± 16	N.d.
112	Gabbro-diorite	1655	11.1	645	410	13.2	0.66	42.04	0.0483	0.1577	0.02378	L-Prism	125 (4.00)	Weak	Brok	152 ± 3.0	N.d.
113	Gabbro-diorite	1658	1.1	228	79	4.7	0.36	41.64	0.0481	0.1633	0.02405	L-Prism	300 (1.92)	Moder.	Brok	153 ± 2.2	N.d.
114	Gabbro-diorite	1658	2.1	214	71	4.51	0.34	40.68	0.0524	0.1721	0.2453	S-Prism	N.d.	Weak	Brok	156 ± 2.3	N.d.
115	Gabbro-diorite	1658	3.1	148	28	3.21	0.2	39.59	0.0547	0.173	0.02512	S-Prism	N.d.	Moder.	Brok	160 ± 2.5	N.d.
116	Gabbro-diorite	1658	4.1	389	146	8.45	0.39	39.54	0.0499	0.1669	0.02523	S-Prism	N.d.	Weak	Brok	161 ± 2.1	N.d.
117	Gabbro-diorite	1658	5.1	158	46	3.49	0.3	38.94	0.511	0.1735	0.02562	S-Prism	N.d.	Moder.	Brok	163 ± 2.5	N.d.
118	Gabbro-diorite	1658	6.1	111	23	2.36	0.22	40.47	0.0554	0.17	0.02455	L-Prism	N.d.	Moder.	Brok	156 ± 2.8	N.d.
119	Gabbro-diorite	1658	6.1	111	23	2.36	0.22	40.47	0.0554	0.17	0.02455	L-Prism	N.d.	Moder.	Brok	156 ± 2.8	N.d.
120	Gabbro-diorite	1658	7.1	260	102	5.65	0.41	39.49	0.0515	0.1733	0.02526	L-Prism	N.d.	Moder.	Brok	161 ± 2.3	N.d.

(Continued)

Table 10.2 Continued.

Consecutive numbers 1	Rocks 2	Samples 3	Crystals 4	U, ppm 5	Th, ppm 6	206Pb, ppm 7	232Th/238U 8	238U/206Pb 9	207Pb*/206Pb 10	207Pb*/235U 11	206Pb*/238U 12	Crystal habit 13	Length of crystals, μm (length-to-width ratio) 14	Cathodoluminescence intensity 15	Oscillatory zoning 16	Age, Ma (206Pb/238U) 17	Age, Ma (207Pb/206Pb) 18
121	Gabbro-diorite	1658	8.1	233	83	5.04	0.37	39.70	0.0508	0.1653	0.02509	S-Prism	N.d.	Moder.	Brok	160 ± 2.4	N.d.
122	Gabbro-diorite	1658	9.1	243	93	5.25	0.4	39.73	0.0528	0.1658	0.02502	L-Prism	N.d.	Moder.	Brok	159 ± 2.4	N.d.
123	Gabbro-diorite	1658	10.1	205	67	4.44	0.34	39.58	0.0512	0.1681	0.02518	S-Prism	N.d.	Moder.	Brok	160 ± 2.5	N.d.
124	Gabbro-diorite	1652	1.1	501	188	10.4	0.39	41.59	0.0474	0.1572	0.02404	S-Prism	N.d.	Weak	Brok	153 ± 2.6	N.d.
125	Gabbro-diorite	1652	2.1	598	483	12.3	0.83	41.81	0.047	0.155	0.02391	S-Prism	N.d.	Weak	Brok	152 ± 2.6	N.d.
126	Gabbro-diorite	1652	3.1	590	206	12.3	0.36	41.28	0.0495	0.1652	0.02423	S-Prism	N.d.	Weak	Brok	154 ± 2.6	N.d.
127	Gabbro-diorite	1652	4.1	229	71	4.8	0.32	41.57	0.0531	0.176	0.02405	S-Prism	N.d.	Moder.	Brok	153 ± 2.9	N.d.
128	Gabbro-diorite	1652	5.1	610	214	12.6	0.36	41.73	0.0493	0.1628	0.02396	S-Prism	N.d.	V-weak	Brok	153 ± 2.5	N.d.
129	Gabbro-diorite	1652	6.1	424	231	8.8	0.56	41.93	0.0476	0.157	0.02384	S-Prism	133 (1.60)	V-weak	Brok	152 ± 2.7	N.d.
130	Gabbro-diorite	1652	7.1	616	382	13	0.64	40.71	0.0482	0.1633	0.02456	S-Prism	200 (1.50)	Weak	Brok	156 ± 2.7	N.d.
131	Gabbro-diorite	1652	8.1	299	90	6.2	0.31	41.96	0.0483	0.159	0.02383	S-Prism	N.d.	Weak	Brok	152 ± 3.1	N.d.
132	Gabbro-diorite	1652	9.1	405	172	8.27	0.44	42.32	0.0525	0.171	0.02363	S-Prism	N.d.	Weak	Brok	151 ± 2.8	N.d.
133	Gabbro-diorite	1652	10.1	597	231	12.1	0.4	42.44	0.0511	0.1659	0.02356	S-Prism	187 (1.28)	V-weak	Brok	150 ± 2.6	N.d.
134	Gabbro-diorite	2612-2	1.1	303	282	6.73	0.96	38.74	0.0509	0.1742	0.02575	L-Prism	143 (2.15)	Weak	Brok	164 ± 3.0	N.d.
135	Gabbro-diorite	2612-2	2.1	290	176	3.66	0.63	68.1	0.0538	0.0982	0.0146	Res-SI	N.d.	Moder.	Good-C	93 ± 1.8	N.d.
136	Gabbro-diorite	2612-2	3.1	371	138	39.3	0.39	8.1	0.6592	1.104	0.1233	S-Prism	N.d.	Weak	Brok	750 ± 12	N.d.
137	Gabbro-diorite	2612-2	4.1	635	218	7.37	0.35	74.1	0.051	0.0862	0.01342	S-Prism	190 (1.90)	Moder.	Good-C	86 ± 1.6	N.d.
138	Gabbro-diorite	2612-2	5.1	108	113	4.95	10.8	18.76	0.0571	0.384	0.053	S-Prism	200 (1.62)	Weak	Brok-Cor	333 ± 6.3	N.d.
139	Gabbro-diorite	2612-2	6.1	183	215	3.23	1.22	48.57	0.0515	0.1346	0.02049	S-Prism	143 (1.36)	Weak	Brok	131 ± 2.6	N.d.
140	Gabbro-diorite	2612-2	7.1	466	263	5.45	0.58	73.5	0.0488	0.0916	0.0136	S-Prism	143 (1.76)	Weak	Brok-Cor	87 ± 1.6	N.d.
141	Gabbro-diorite	2612-2	8.1	194	99	14.3	0.53	11.63	0.0568	0.679	0.086	S-Prism	N.d.	V-weak	Brok	532 ± 9.4	N.d.
142	Gabbro-diorite	2612-2	9.1	225	90	3.97	0.42	48.72	0.0529	0.138	0.02042	S-Prism	207 (2.07)	Moder.	Good-C	130 ± 2.6	N.d.
143	Gabbro-diorite	2612-2	10.1	364	237	4.43	0.67	70.7	0.0536	0.0928	0.01405	S-Prism	N.d.	Moder.	Brok-Cor	90 ± 1.8	N.d.
144	Diorite	160	1.1	143	163	1.4	1.18	87.3	0.0561	0.069	0.01128	S-Prism	N.d.	Weak	Brok	72 ± 1.9	N.d.
145	Diorite	160	2.1	328	259	8.45	0.82	34.54	0.0499	0.1993	0.02895	S-Prism	N.d.	Weak	Brok	184 ± 3.4	N.d.
146	Diorite	160	3.1	358	272	12.1	0.78	25.54	0.05136	0.2686	0.03908	S-Prism	156 (1.25)	Weak	Good-C	247 ± 4.3	N.d.
147	Diorite	160	4.1	107	112	1.08	1.08	84.7	0.0619	0.071	0.01154	S-Prism	156 (1.25)	Weak	Brok	74 ± 2.0	N.d.
148	Diorite	160	5.1	365	307	22.6	0.87	13.86	0.05731	0.566	0.0721	S-Prism	162 (1.62)	Weak	Brok	449 ± 7.7	N.d.
149	Diorite	160	6.1	794	528	9.44	0.69	72.2	0.0495	0.0906	0.1381	S-Prism	N.d.	Weak	Good-C	88 ± 1.6	N.d.
150	Diorite	160	6.2	646	406	7.62	0.65	72.8	0.0499	0.0910	0.0137	S-Prism	N.d.	Weak	Good-C	88 ± 1.6	N.d.
151	Diorite	160	7.1	177	82	1.87	0.48	81.1	0.0552	0.072	0.01214	S-Prism	N.d.	Moder.	Good-C	78 ± 1.8	N.d.
152	Diorite	160	8.1	231	60	2.38	0.27	83.3	0.0478	0.0791	0.01201	S-Prism	186 (1.86)	Moder.	Good-C	77 ± 1.6	N.d.
153	Diorite	160	9.1	224	97	5.16	0.44	37.36	0.0502	0.1853	0.02677	L-Prism	148 (2.71)	Weak	Good-C	170 ± 3.3	N.d.

ID	Rock														
154	Diorite	1.1	44	11	0.98	0.26	38.7	0.0569	0.16	0.02547	S-Prism	N.d.	Weak	Sec	162 ± 4.7	N.d.
155	Diorite	2.1	58	14	1.33	0.24	37.05	0.0524	0.152	0.02662	S-Prism	333 (1.47)	Weak	Sec	169 ± 4.0	N.d.
156	Diorite	3.1	78	22	1.7	0.29	39.57	0.0538	0.167	0.02509	L-Prism	356 (2.33)	Weak	Sec	160 ± 3.2	N.d.
157	Diorite	4.1	84	26	1.87	0.32	38.49	0.0544	0.176	0.02582	S-Prism	313 (1.72)	Weak	Sec	164 ± 3.5	N.d.
158	Diorite	5.1	102	35	2.22	0.35	39.62	0.0571	0.153	0.02485	S-Prism	269 (1.65)	Weak	Sec	158 ± 3.2	N.d.
159	Diorite	6.1	79	21	1.8	0.27	37.62	0.06	0.187	0.0263	S-Prism	338 (1.50)	Weak	Sec	167 ± 3.5	N.d.
160	Diorite	7.1	186	52	4	0.29	39.86	0.0504	0.165	0.02501	S-Prism	247 (1.53)	Moder.	Sec	159 ± 2.7	N.d.
161	Diorite	8.1	42	9	0.92	0.22	39.3	0.0602	0.163	0.02502	S-Prism	347 (1.53)	Moder.	Sec	159 ± 4.2	N.d.
162	Diorite	9.1	50	10	1.14	0.21	38.12	0.0574	0.154	0.2577	S-Prism	252 (1.50)	Moder.	Sec	164 ± 4.6	N.d.
163	Diorite	10.1	70	22	1.4	0.32	41.48	0.0527	0.159	0.2397	S-Prism	N.d.	Moder.	Brok	153 ± 3.2	N.d.
164	Quartz diorite 2612	1.1	54	21	0.604	0.4	76.5	0.0711	0.087	0.01271	S-Prism	179 (1.23)	Weak	Sec	81 ± 3.0	N.d.
165	Quartz diorite 2612	2.1	311	211	6.85	0.7	39.03	0.0519	0.1763	0.02556	L-Prism	188 (2.25)	Weak	Good-C	163 ± 3.0	N.d.
166	Quartz diorite 2612	3.1	137	46	1.43	0.35	82.5	0.0519	0.071	0.01199	S-Prism	175 (1.34)	Weak	Good-C	77 ± 1.9	N.d.
167	Quartz diorite 2612	4.1	538	256	13.7	0.49	33.77	0.05112	0.2018	0.029955	S-Prism	N.d.	V-weak	Good-C	188 ± 3.2	N.d.
168	Quartz diorite 2612	5.1	121	83	2.73	0.71	37.97	0.0536	0.177	0.02618	S-Prism	133 (1.45)	Weak	Sec	167 ± 3.5	N.d.
169	Quartz diorite 2612	6.1	832	1322	8.83	1.64	80.9	0.0492	0.0815	0.01234	L-Prism	225 (2.25)	Moder.	Good-C	79 ± 1.4	N.d.
170	Quartz diorite 2612	7.1	86	38	0.885	0.46	83.1	0.0786	0.09	0.01168	S-Prism	150 (1.43)	Moder.	Brok	75 ± 2.3	N.d.
171	Quartz diorite 2612	8.1	71	47	3.17	0.68	19.29	0.0555	0.365	0.516	S-Prism	N.d.	V-weak	Brok	324 ± 6.7	N.d.
172	Quartz diorite 2612	9.1	178	80	1.85	0.46	82.6	0.0552	0.0795	0.01199	S-Prism	N.d.	Weak	Good-C	77 ± 1.9	N.d.
173	Quartz diorite 2612	10.1	690	376	24.7	24.7	24	0.05238	0.2956	0.04161	S-Prism	N.d.	V-weak	Good	263 ± 4.4	N.d.
174	Trachyandesite 156	1.1	424	60	61.6	0.15	5.93	0.07298	1.697	0.1686	L-Prism	158 (2.50)	V-weak	Good-C	1004 ± 17	1013 ± 17
175	Trachyandesite 156	1.2	264	27	37.7	0.11	6.02	0.07247	1.6661	0.1662	L-Prism	158 (2.50)	V-weak	Good-C	991 ± 19	999 ± 25
176	Trachyandesite 156	2.1	608	270	2.48	0.46	213	0.0461	0.0299	0.0047	L-Prism	117 (2.60)	V-weak	Good-C	30 ± 0.67	N.d.
177	Trachyandesite 156	2.2	550	269	2.11	0.50	226.2	0.0419	0.0255	0.00442	L-Prism	117 (2.60)	V-weak	Good-C	28 ± 0.62	N.d.
178	Trachyandesite 156	2.3	1050	692	3.92	0.68	231.2	0.0477	0.0284	0.004326	L-Prism	117 (2.60)	V-weak	Good-C	28 ± 0.55	N.d.
179	Trachyandesite 156	3.1	74	18	1.43	0.25	45.1	0.0442	0.135	0.02215	S-Prism	142 (1.52)	Intense	Brok	141 ± 3.6	N.d.
180	Trachyandesite 156	4.1	1340	548	14	0.42	82.3	0.048	0.0804	0.01215	S-Prism	114 (2.00)	Moder.	Brok	78 ± 1.4	N.d.
181	Trachyandesite 156	5.1	222	84	76.4	0.39	2.498	0.12853	7.09	0.4003	Res-SI	150 (1.83)	Moder.	Brok-Cor	2170 ± 39	2078 ± 10
182	Trachyandesite 156	5.2	900	2	254	N.d.	3.039	0.1149	5.22	0.3291	Res-SI	150 (1.83)	Moder.	Brok-Cor	1834 ± 33	1879 ± 28
183	Trachyandesite 156	5.3	568	287	170	0.52	2.872	2.872	6.17	0.3482	Res-SI	150 (1.83)	Moder.	Brok-Cor	1926 ± 31	2079 ± 10
184	Trachyandesite 156	6.1	69	17	1.42	0.25	42.1	0.053	0.174	0.02373	S-Prism	84 (1.23)	Moder.	Brok-Cor	151 ± 3.9	N.d.
185	Andesite 233	1.1	386	28	5.2	0.08	64.3	0.0495	0.106	0.01556	S-Prism	100 (2.00)	Intense	Brok	100 ± 1.7	170 ± 260
186	Andesite 233	2.1	157	198	3.56	1.3	38.28	0.0498	0.179	0.02612	L-Prism	102 (2.50)	Moder.	Brok	166 ± 3.3	186 ± 310
187	Andesite 233	3.1	767	422	18.5	0.57	35.79	0.0516	0.199	0.02794	S-Prism	74 (2.00)	Good	Good	178 ± 1.7	269 ± 120
188	Andesite 233	4.1	252	184	4.57	4.57	49.3	0.041	0.115	0.02029	S-Prism	N.d.	Intense	Brok	130 ± 2.9	N.d.
189	Andesite 233	5.1	512	276	37.3	0.56	11.84	0.0592	0.689	0.08441	S-Prism	86 (2.00)	Ex-weak	Brok-Cor	522 ± 4.6	573 ± 100
190	Andesite 233	6.1	541	177	9.49	0.34	49.59	0.0474	0.132	0.02016	L-Prism	171 (2.67)	Intense	Brok-Cor	129 ± 1.8	70 ± 340

(Continued)

Table 10.2 Continued.

Consecutive numbers 1	Rocks 2	Samples 3	Crystals 4	U, ppm 5	Th, ppm 6	206Pb, ppm 7	232Th/238U 8	238U/206Pb 9	207Pb*/206Pb 10	207Pb*/235U 11	206Pb*/238U 12	Crystal habit 13	Length of crystals, μm (length-to-width ratio) 14	Cathodoluminescence intensity 15	Oscillatory zoning 16	Age, Ma (206Pb/238U) 17	Age, Ma (207Pb/206Pb) 18
191	Andesite	233	7.1	157	42	1.38	0.28	99.4	0.0479	0.066	0.01006	S-Prism	N.d.	Moder.	Good-C	85 ± 1.7	93 ± 400
192	Andesite	233	8.1	959	287	61.2	0.31	13.47	0.05561	0.5691	0.01006	S-Prism	N.d.	Weak	Brok	462 ± 3.5	437 ± 32
193	Andesite	233	9.1	732	276	22.5	0.39	27.92	0.0503	0.2485	0.03582	S-Prism	N.d.	V-weak	Brok-Cor	227 ± 2.4	210 ± 78
194	Andesite	233	10.1	484	480	26	1.02	16	0.0545	0.469	0.06248	S-Prism	N.d.	Intense	Brok-Cor	391 ± 3.8	390 ± 61
195	Trachyandesibasalt	181	1.1	984	1692	2.83	1.78	299.7	0.0466	0.0214	0.003337	S-Prism	N.d.	Ex-weak	N-zon	21 ± 0.43	N.d.
196	Trachyandesibasalt	181	2.1	231	145	0.714	0.65	286.2	0.051	0.0246	0.00349	S-Prism	127 (1.44)	Weak	Brok-Cor	22 ± 0.64	N.d.
197	Trachyandesibasalt	181	3.1	371	255	1.09	0.71	295.3	0.0433	0.0202	0.003386	S-Prism	N.d.	Weak	Brok-Cor	22 ± 0.51	N.d.
198	Trachyandesibasalt	181	4.1	293	205	0.819	0.72	312.8	0.035	0.0152	0.003197	S-Prism	98 (1.36)	Moder.	Brok	21 ± 0.58	N.d.
199	Trachyandesibasalt	181	5.1	218	140	0.578	0.66	331	0.057	0.0237	0.003022	S-Prism	83 (1.36)	Moder.	Brok-Cor	19 ± 0.63	N.d.
200	Trachyandesibasalt	181	6.1	248	164	0.775	0.68	280.6	0.046	0.0226	0.00356	S-Prism	N.d.	V-weak	N-zon	23 ± 0.68	N.d.
201	Trachyandesibasalt	181	7.1	194	75	0.539	0.4	318.5	0.046	0.02	0.003139	S-Prism	174 (1.72)	Moder.	Brok-Cor	20 ± 0.62	N.d.
202	Trachyandesibasalt	181	8.1	177	84	10.1	0.49	15.16	0.0551	0.501	0.0659	S-Prism	N.d.	V-weak	N-zon	412 ± 7.6	N.d.
202	Trachyandesibasalt	181	9.1	315	114	54.8	0.37	4.941	0.07973	2.225	0.2024	Res-St	N.d.	Ex-weak	N-zon	1188 ± 20	1190 ± 25
203	Trachyandesibasalt	181	10.1	246	167	0.732	0.7	299.5	0.04	0.0183	0.00334	S-Prism	N.d.	Weak	Brok-Cor	21 ± 0.69	N.d.
204	Trachyandesibasalt	181	11.1	196	106	0.558	0.56	313	0.48	0.021	0.0032	S-Prism	N.d.	Moder.	Brok	21 ± 0.81	N.d.
205	Latite metatuff	1655-1	1.1	230	111	4.8	0.5	41.18	0.0482	0.1615	0.02428	S-Prism	100 (2.00)	Moder.	Brok	154.6 ± 3.1	N.d.
206	Latite metatuff	1655-1	1.2	295	161	6.58	0.56	38.63	0.0483	0.1723	0.02588	S-Prism	130 (1.63)	Moder.	Brok	164.7 ± 3.2	N.d.
207	Latite metatuff	1655-1	2.1	423	152	27.9	0.37	13.03	0.05755	0.609	0.0768	Res-SI	N.d.	Ex-weak	Brok	476.5 ± 8.5	N.d.
208	Latite metatuff	1655-1	2.2	493	174	33.9	0.36	12.49	0.05696	0.629	0.0801	Res-SI	N.d.	Ex-weak	Brok	496 ± 8.8	N.d.

Note: Isotope-geochronological study of zircons was made at the Center of Isotopic Research of VSEGEI (St. Petersburg) with a SHRIMP II mass spectrometer. The crystal habit: S-Prism – short-prismatic crystals and their fragments with nonresorbed faces and edges; L-Prism – long-prismatic crystals (ovoids); Res-SI – prismatic crystals with slightly resorbed faces and edges; Res-St – prismatic crystals and their fragments with strongly resorbed faces. The intensity of cathodoluminescence: Intense; Moder – moderate; Weak; V-weak – very weak; Ex-weak – extremely weak. Oscillatory zoning: Good – fine rhythmic zoning parallel to the edges of the prism and pyramid; Good-C – fine rhythmic zoning parallel to the edges of the prism and pyramid with the presence of a darker "core"; Brok – broken, rough zoning, often nonparallel to faces; Brok-Cor – broken with the presence of a darker "core"; Sec – sectorial zoning; N-zon – no zoning. Pb* – radiogenic lead. N.d. – no data.

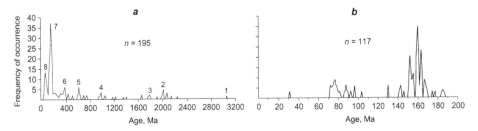

Figure 10.1   Histogram of the occurrence of the ^{206}Pb/^{238}U isotopic ages of zircons from ultramafic and mafic rocks of the Berezovka massif (according to Table 10.2).
a – Total number of analyzed grains; *b* – analyses of young population ages (<200 Ma).

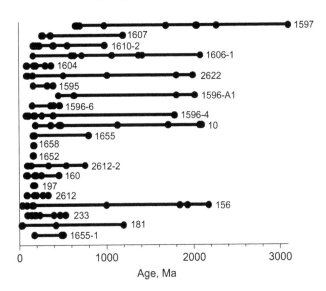

Figure 10.2   Ranges of the isotope dates of zircons from rock samples of the Berezovka massif. The samples are numbered according to their numbering in Table 10.2.

On the ^{206}Pb/^{238}U – ^{207}Pb/^{206}Pb concordia diagrams, the centers of ellipsoids of the confidence intervals of isotope ratios for most of zircon grains from the oldest population lie on the concordia or near it; sometimes they are localized on the discordia, with the dispersion of ^{207}Pb/^{235}U values usually exceeding that of ^{206}Pb/^{238}U (Fig. 10.5). For most of zircon grains from the ~190-140 Ma age populations, the centers of the above-mentioned ellipsoids also lie on the concordia. These diagrams evidence that the estimated isotopic ages of zircons prove highly reliable.

The calculations performed by V.V. Khlestov revealed a significant positive correlation between the size of zircon grains, on the one hand, and their isotopic ages, on the other. This suggests that under the subsolidus conditions, the U–Pb isotope system of smaller zircon grains was affected more strongly by the isotope disturbance and "rejuvenation", compared to such systems in larger mineral grains. This may have been the reason why coarse crystals were subjected to a lesser "rejuvenation" than finer zircon grains.

a

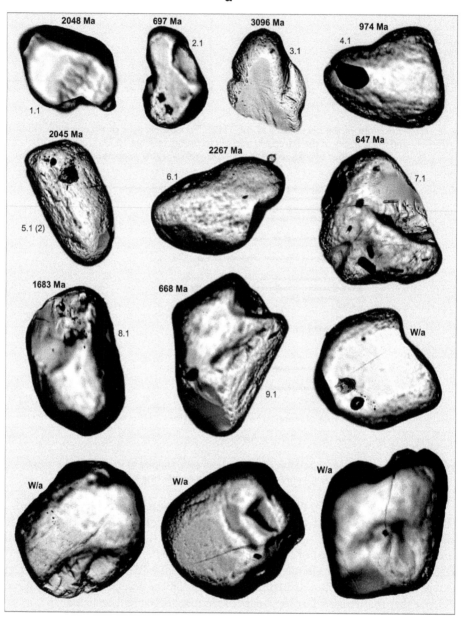

Figure 10.3 Microphotographs of grains of polychronous zircons from rocks of the Berezovka massif (taken with a SHRIMP II mass-spectrometer in optical mode).
a – Relict ovoid grains from the ancient and intermediate age populations (sample 1597, pyroxenite), *b* – syngenetic prismatic crystals from the young age population (sample 1658, gabbro-diorite). Here and in Fig. 10.4, the crystals are numbered according to the numbers of their respective analyses in Tables 10.2 and 10.3, and their age is shown. W/a – Without analysis.

b

Figure 10.3 Continued.

Results of the isotope studies of zircons from gabbroids and ultramafic rocks of the Berezovka massif permitted the following conclusions: (1) the isotopic age comprises a very wide time interval within the representative collection of zircon grains (i.e., the zircons are polychronous); (2) grains of various age are often present in one sample of rocks; (3) the analyzed zircon grain collection consists of several groups of age

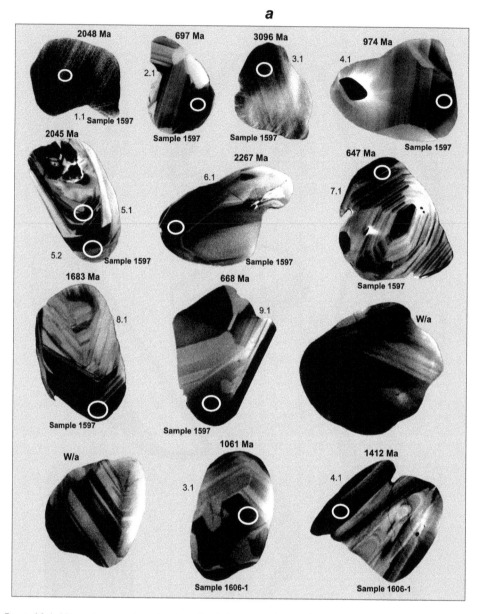

Figure 10.4 Microphotographs of grains of polychronous zircons from rocks of the Berezovka massif (taken with a SHRIMP II mass-spectrometer in CL mode).
a – Relict ovoid grains from the ancient and intermediate age populations (sample 1597, pyroxenite; sample 1606-1, gabbro-pyroxenite), *b* – syngenetic prismatic crystals from young age populations (samples 1658 and 2612-2, gabbro-diorites). White circles mark the position of the probe during the analysis. W/a – Without analysis.

Figure 10.4 Continued.

populations; (4) the oldest zircons occur predominantly in hybrid ultramafic rocks and are rare in hybrid and orthomagmatic gabbroids; (5) most ancient zircons are characterized by ovoid shapes due to intensely resorbed grains, different nature of oscillatory zoning, and low CL intensity; (6) the presence of grains from intermediate age populations (~790-200 Ma) in the studied collection of zircons is likely due to their more or less significant "rejuvenation" caused by actions of later generated mafic melt and its fluids; (7) most of zircon grains dated at ~190-140 Ma have clear crystal faces and distinct oscillatory zoning; (8) the 170-150 Ma zircon grains with clear crystal faces, most commonly occurring in the collection, are likely to have crystallized from the mafic melt that formed the gabbroid intrusion and thus determine the time of its emplacement.

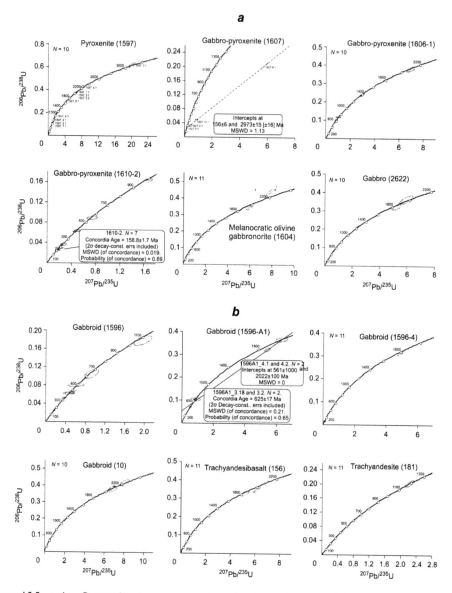

Figure 10.5 a–d – Concordia diagrams, based on the determinations of the isotopic age of polychronous zircons from rocks of the Berezovka massif (pyroxenite, gabbro-pyroxenites,
different gabbroids, gabbro-diorites, diorites, quartz diorite, and other rocks) (data from
Table 10.2).

10.5   GEOCHEMISTRY OF ZIRCONS

In recent years, voluminous data on the regularities of REE and trace-element distribution in zircons from different igneous rocks, including gabbroid and ultramafic rocks,

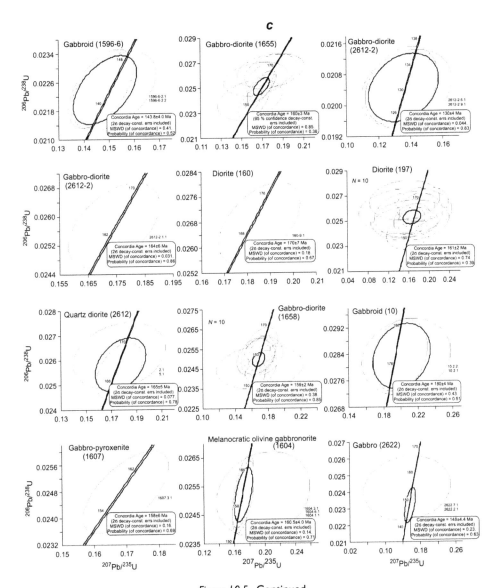

Figure 10.5 Continued.

have been summarized in [Hoskin and Schaltegger, 2003]; however, the data on the zircons from the ultramafic and gabbroid rocks are very limited. In this context, of interest are new LA ICP-MS data on the trace element composition of zircons from the ultramafic rocks and gabbroids of the Berezovka massif, partly published earlier [Lesnov, 2009a, 2011c; Lesnov, 2012b].

Rare earth elements (REE). The total contents of REE in the studied zircons vary from 42 to 10,906 ppm (Table 10.3). Most grains are characterized by specific intense fractionation of REE, which is expressed in chondrite-normalized HREE essentially

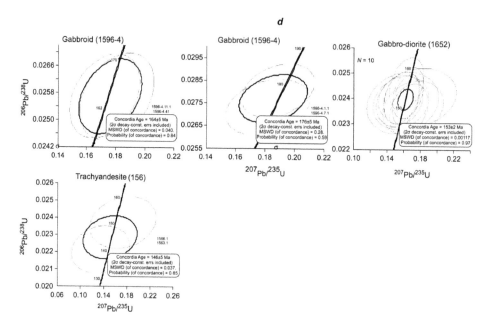

Figure 10.5 Continued.

dominating over LREE and extremely low parameter $(La/Yb)_n$. All but a few analyzed grains of zircons show chondrite-normalized REE patterns typical of magmatic zircons and nearly the same in shape (Fig. 10.6, *a*, *b*). Almost all of them are complicated by intense positive Ce anomalies and weaker negative Eu anomalies. Besides the predominant zircon grains with their typical REE patterns, the studied collection contains a few mineral grains with REE patterns differing both in the position on the diagrams and in shape (Fig. 10.6, *c*). The diagrams usually show weak positive Ce anomalies and negative Eu anomalies or even lack them. Such REE patterns testify to nonuniform enrichment of the grains in LREE, as well as the nonuniform depletion in HREE. Similar anomalous REE patterns were observed in zircons of different age populations but most often in grains of intermediate populations dated at the interval from ~990 to ~190 Ma. The above differences in REE patterns are likely due to the epigenetic redistribution of these elements during recrystallization of grains and subsequent infiltration of fluids along the zircon grain microcracks caused by plastic deformations.

It is likely that this REE redistribution in zircons from ancient population proceeded synchronously with the above disturbance of their U–Pb isotope systems resulting in "rejuvenation" and "transition" from oldest to intermediate populations. This hypothesis also follows from the negative correlation between the isotopic age of zircons calculated from $^{206}Pb/^{238}U$ and the contents of La, Ce, Yb, on one hand, and total REE, $(Ce/Ce^*)_n$, and $(Eu/Eu^*)_n$, on the other (Fig. 10.7). Therefore, note that T.F. Zinger *et al.* [2010] revealed a correlation between plastic deformations of zircons from rocks dredged in the Central Atlantic and their enrichment in REE and in U, Th, Hf, and P. According to them, the input of these trace elements into zircon grains was accompanied by disturbances in their U–Pb systems.

Table 10.3 Contents of rare earth and trace elements in zircons from rocks of the Berezovka massif, ppm.

Crystal	La	Ce	Pr	Nd	Sm	Eu	Gd	Tb	Dy	Ho	Er	Tm	Yb	Lu	Total REE	Rb	Sr	Y	Nb	Ba	Hf	Ta	Pb	Th	U
Sample 1597. Pyroxenite																									
1.1	0.056	3.09	0.047	1.27	2.75	0.298	14.1	5.39	63.3	18.9	94.3	21.0	207	29.4	461	0.051	0.218	551	2.271	0.045	7927	1.11	108	67.4	287
2.1	0.082	28.5	0.799	6.33	12.2	2.532	41.6	15.6	181.0	63.8	310.9	76.0	762	112	1613	0.258	0.694	1824	5.737	0.339	15580	2.56	47.4	338	330
3.1	0.030	5.51	0.087	1.61	2.30	1.184	11.1	4.33	53.6	19.3	98.5	22.2	232	33.8	486	0.155	0.357	492	1.241	0.134	5745	0.432	32.1	18.7	47.2
4.1	0.087	96.4	0.391	4.98	11.3	2.043	34.0	13.7	129.1	41.8	173.7	36.0	331	36.7	911	0.091	0.502	998	4.274	0.351	8652	1.85	15.3	113	73.4
5.1	0.681	24.0	0.449	4.57	5.12	0.522	25.2	8.26	106.6	36.1	171.6	34.8	340	46.4	805	0.302	1.106	964	4.582	0.429	10008	1.82	72.0	60.6	176
5.2	0.100	12.1	0.080	1.89	1.74	0.184	8.12	3.08	38.6	14.1	72.6	16.5	188	27.1	384	Н.О	0.286	356	9.152	0.036	5756	2.62	59.7	34.5	157
6.1	0.034	16.0	0.153	2.36	4.14	0.904	13.8	5.12	56.7	19.9	107.3	27.2	293	46.3	592	0.013	0.498	597	1.398	0.025	9488	0.458	43.1	60.3	101
7.1	0.041	10.5	0.143	1.51	2.80	0.996	8.47	3.28	37.1	15.1	76.7	19.7	223	37.4	437	0.155	0.483	478	1.768	0.172	7583	0.497	6.58	37.5	52.3
8.1	0.049	7.82	0.193	2.69	5.07	0.407	16.8	6.44	74.5	25.8	120.7	27.0	273	35.7	596	0.220	0.418	709	3.798	0.085	9183	0.937	35.5	43.9	125
9.1	0.053	15.2	0.087	1.03	3.65	0.637	11.0	4.23	53.8	21.0	109.8	25.7	295	39	580	0.156	0.587	638	1.316	0.109	8237	0.512	4.76	27.4	36.5
Sample 1607. Gabbro-pyroxenite																									
1.1	0.052	3.8	0.038	1.9	5.5	2.2	32	24	490	186	964	301	3598	550	6158	2.1	2.5	5028	4.2	0.23	15442	4.5	20	6.2	348
4.1	0.083	19.3	0.45	6.7	12.8	4.5	50	19.8	283	106	453	113	1253	176	2498	0.86	0.87	2749	5.8	0.48	9820	1.89	26	454	487
5.1	0.52	35	0.33	4	5	1.26	10.4	3.5	47	16.4	76	21	229	34	483	0.045	0.13	492	2.5	0.88	13194	1.16	242	185	288
6.1	0.091	2.8	0.088	0.33	2.7	0.84	12.1	6.2	116	49	279	91	1015	144	1719	0.51	0.18	1428	5.9	0.59	15832	5.7	22	26	236
7.1	0.13	1.16	0.15	2.7	5.5	0.52	13.4	2.5	15.5	2.3	5.2	1.29	7.1	1.05	59	0.86	0.09	59	1.11	10.8	14551	0.18	26	16.5	330
8.1	7	42	11.6	103	33	4.7	85	31	424	138	525	125	1136	139	2804	1.05	6.1	3582	16.6	4.7	13110	4	34	589	543
9.1	0.21	74	0.54	6.6	15.1	1.83	44	18.6	251	90	378	94	953	127	2054	0.52	0.08	2502	7.6	1.01	15713	2.1	12.2	391	285
10.1	1.16	9.8	1.42	1.86	4.7	0.6	9.5	4.9	86	40	220	77	912	136	1505	13	2.8	1195	8	14.6	15891	7.2	27	72	323
Sample 1610-2. Gabbro-pyroxenite																									
1.1	21	150	21	123	80	18.2	140	44	494	131	509	113	1093	137	3074	0.85	18.9	3599	22	11.6	13593	6.2	65	1358	1482
2.1	0.68	66	4.2	61	75	14.8	205	72	895	289	1354	355	3882	578	7852	3.3	3.4	8889	6.8	0.58	10379	0.88	21	577	586
3.1	1.05	18	0.91	8.8	10.7	3	29	10.3	161	60	301	91	1065	171	1931	0.71	0.96	1867	3.6	0.98	9667	0.96	10.1	112	347
4.1	0.47	9.5	0.65	5.3	6.7	1.88	16.5	7.6	108	41	201	67	845	143	1454	0.3	0.21	1305	2.3	0.48	10880	0.74	5.3	48	169
5.1	24	111	8.8	41	23	5.3	56	22	312	101	428	103	976	132	2343	1.21	6.4	2817	5.7	7.5	9565	2	14.3	574	285
6.1	1.2	24	1.64	13.4	14.9	4.7	45	16.2	211	78	398	110	1276	188	2382	0.83	1.22	2417	4.5	1.04	10015	3.3	21	218	730
7.1	0.71	14.1	0.55	4.2	5.7	1.38	16.4	6.2	86	28	128	30	319	43	683	0.048	0.7	810	3.3	1.01	8598	1.32	6.6	126	107
8.1	3.8	16.7	2.1	17.6	33	4.9	127	47	654	206	779	171	1530	211	3803	1.45	2.2	5076	5.4	1.23	9302	1.49	14.1	341	345
9.1	0.67	18.2	1.23	10.6	22	7.8	84	31	417	150	705	191	2176	392	4207	1.92	0.89	5125	4.3	2	11778	0.84	13.9	291	342
10.1	3.2	5.8	0.8	4.3	3.7	0.88	13.8	8	161	66	364	112	1394	207	2344	0.51	1.16	2026	3.6	0.3	15298	3	13.2	15.8	157
11.1	0.53	65	1.06	9.7	11.7	1.53	28	8.8	89	27	95	21	202	26	586	0.04	0.08	656	3.9	0.95	9862	1.46	8.6	109	59
12.1	0.26	27	0.85	10.6	18	4	60	25	329	119	582	180	1851	296	3503	0.91	2.1	3811	4.9	0.91	10105	0.83	12.5	213	382
13.1	219	850	177	958	612	121	868	187	1437	255	616	114	890	97	7401	1.54	1.72	6122	22	6.6	14565	7.1	67	Н.Д	611
14.1	88	633	93	625	503	129	730	223	1951	419	1229	267	2366	295	9551	2.8	70	8442	8.5	586	18446	9.1	165	Н.Д	Н.Д

(Continued)

Table 10.3 Continued.

Crystal	La	Ce	Pr	Nd	Sm	Eu	Gd	Tb	Dy	Ho	Er	Tm	Yb	Lu	Total REE	Rb	Sr	Y	Nb	Ba	Hf	Ta	Pb	Th	U
Sample 1606–1. Gabbro-pyroxenite																									
1.1	0.60	14.98	0.49	3.38	4.97	0.19	18.67	6.37	74.4	24.3	115	25.7	248	33	570	0.24	0.50	692	3.05	0.20	4456	1.04	16.0	79.0	143
2.1	0.31	19.32	0.97	12.53	15.43	1.53	40.21	12.19	129	41.1	177	35.1	315	40	840	0.26	0.40	1128	1.20	0.09	5985	0.63	14.9	156	123
3.1	0.05	7.00	0.29	4.27	6.92	0.47	21.38	6.98	77.0	23.5	104	22.6	210	28	512	0.09	0.80	570	4.36	0.04	6081	1.17	7.42	34.2	38.3
4.1	0.06	18.07	0.72	9.63	15.09	1.25	47.54	18.11	209	70.9	349	79.9	755	103	1678	0.75	0.48	1890	8.48	0.35	12359	2.57	40.1	85.0	139
5.1	0.02	11.63	0.15	2.42	4.90	0.67	12.97	5.23	59.8	21.8	105	23.9	243	36	527	0.25	0.48	617	1.51	0.08	8727	0.38	9.01	72.0	83.0
6.1	0.06	7.46	0.18	2.75	4.92	0.31	20.43	7.52	96.3	32.1	158	38.0	351	45	764	0.10	0.36	837	3.99	0.00	11666	1.54	24.3	40.7	109
7.1	0.01	23.38	0.28	3.11	7.43	1.88	25.99	9.18	98.6	32.1	158	36.4	347	46	789	0.16	0.48	874	2.23	0.11	10941	0.89	28.3	47.0	60.8
8.1	0.05	16.26	0.20	3.44	4.73	1.19	17.39	6.52	74.5	25.8	119	29.2	286	40	625	0.22	0.45	707	2.86	0.16	5241	0.98	8.83	68.4	76.8
9.1	0.02	8.54	0.12	1.40	3.19	1.09	19.37	8.82	120	51.8	311	80.2	975	168	1749	0.48	0.77	1544	1.88	0.11	7909	0.35	7.46	100	274
10.1	0.02	6.94	0.06	1.20	1.49	0.50	5.86	2.38	30.7	12.9	72.3	19.8	236	40	430	0.09	0.50	418	0.72	0.08	6522	0.23	11.3	54.2	104
Sample 1604. Melanocratic olivine gabbronorite																									
1.1	0.05	8.3	0.51	3.5	7.2	1.97	29	11.9	165	59	284	83	984	158	1795	0.31	0.62	1804	1.94	0.15	10566	0.56	6	83	200
2.1	2.1	121	0.93	4.9	5.5	1.83	27	11.1	169	64	300	83	901	140	1831	0.45	0.88	1908	9.9	0.78	11583	2.7	29	565	376
3.1	0.31	18.1	1.15	11.2	27	7.6	75	26	358	133	613	170	2078	344	3862	0.73	1.42	4083	3.3	0.35	11634	0.85	13.1	266	427
5.1	0.43	7.7	0.41	4.9	8.4	0.41	27	12.3	179	71	315	80	878	126	1711	0.69	0.75	1837	7.2	2	15391	4.5	387	493	1142
5.2	4.8	28	3.5	23	30	5.6	85	34	479	158	678	174	1804	241	3748	0.97	10.3	4415	14	9.2	17139	5.4	759	1129	Н.Д
6.1	0.18	29	0.25	2.6	8.3	1.6	26	11.9	207	85	449	142	1776	310	3049	0.88	0.82	2731	7.7	0.81	17757	2.2	19.2	194	488
7.1	28	150	26	176	133	30	194	43	321	56	148	24	177	17.7	1524	0.047	0.09	1508	7.3	2.6	17248	0.85	69	137	982
Sample 2622. Gabbro																									
1.1	0.006	11.4	0.13	2.6	6.5	0.27	34	15.1	191	75	342	78	797	104	1657	0.4	1.02	1961	4.1	0.14	13278	1.67	65	149	296
2.1	0.82	90	1.45	18.9	29	9.7	109	44	512	189	818	190	2017	280	4309	1.83	3.7	5346	9.4	3	16984	2.8	21	849	641
3.1	0.54	21	0.34	4	9.8	0.89	41	19.7	257	111	523	123	1341	180	2632	3.9	2.8	3017	7.6	3.8	17992	2.4	156	399	536
4.1	0.44	33	0.4	4.2	8.2	0.91	26	10.6	146	58	282	67	730	98	1465	0.49	1.61	1653	5.1	2	25493	2	449	777	777
5.1	0.11	28	0.29	5.3	11.7	1.33	45	18.4	245	102	496	116	1254	174	2497	0.88	1.46	1.46	2609	0.87	10492	1.46	5.2	300	360
6.1	0.49	32	0.32	4.5	6.3	1.6	20	7.2	97	39	187	47	583	97	1122	0.46	2.4	1068	3.8	4.2	17766	1.34	22	238	253
7.1	0.098	3.8	0.073	0.66	1.47	0.6	5.3	2.3	37	17.7	98	29	390	67	653	0.44	0.81	508	0.92	0.61	7167	0.4	1.16	11.8	49
8.1	0.017	3.6	0.11	1.67	3.4	0.2	7.2	1.58	11.4	2.5	6.8	1.17	9.9	0.82	50	0.044	0.017	66	0.43	0.064	3861	0.12	33	75	95
9.1	0.25	11.6	0.7	7.2	13.8	2.9	36	12.6	128	47	204	46	454	63	1027	0.23	0.34	1161	11.1	0.28	18137	3.8	10.4	14.4	29
Sample 1595. Gabbroid																									
1.1	0.2	25	1.58	28	57	4.3	159	57	730	234	875	198	1686	199	4254	1.53	1.58	5190	8.7	0.2	7267	2.5	38	177	147
2.1	31	54	15.4	76	54	10.7	71	20	216	55	227	63	593	90	1576	0.04	0.08	1555	8.7	5.4	15068	7.9	48	306	439
3.1	72	338	53	283	240	44	291	74	644	128	492	114	1117	156	4046	4.9	1.76	3385	13.1	8.7	18206	10	55	323	633
4.1	0.61	15.6	0.9	11.5	12.1	2.6	37	13.5	172	62	319	86	941	166	1840	0.29	0.7	1863	4.2	0.57	12058	4.1	23	166	701
5.1	0.11	3.2	0.041	0.45	3.6	0.86	12.5	7.7	150	61	312	94	1118	157	1920	0.49	1.04	1774	3.8	0.23	15950	2.9	40	166	326
6a.1	0.005	4.1	0.34	4.3	5.8	0.9	21	7.2	104	33	138	33	331	45	728	0.16	0.08	834	2.6	0.25	12474	0.87	5.1	17.2	31

Sample 1596-A1. Gabbroid

1.1	283	909	161	788	437	96	488	135	1176	245	699	151	1224	134	6926	1.71	25	36	16.6	14712	8.2	80	1416	1156
2.1	5	11.5	1.52	7.2	6.1	1.12	16.3	9.8	180	80	427	127	1326	253	2452	1.1	1.21	8	5.4	17307	2.9	46	32	224
3.1	0.02	8.7	0.056	0.27	1.93	0.42	4.1	1.35	24	7.8	41	12.1	141	22	265	0.07	0.61	3.9	0.64	12321	2.4	15.1	45	112
3.2	0.029	151	0.56	6.6	12.9	3.2	54	23	305	102	420	97	915	108	2198	0.59	1.07	18	98	12321	3.7	26	328	129
4.1	0.065	12.2	0.079	0.97	1.88	0.39	8.8	3.5	46	17.6	88	24	250	35	488	0.04	0.075	3.9	0.31	11142	2.2	226	98	475
4.2	0.15	15.1	0.19	1.12	2.5	0.63	8.9	3.3	57	23	117	35	385	56	705	0.04	0.54	10.4	1.24	15162	9.7	271	116	587

Sample 1596-6. Gabbroid

1.1	0.82	27	1.61	12.7	22	4.7	76	29	356	108	399	84	740	94	1955	0.51	3.6	36	9.9	13703	10.4	26	179	188
2.1	0.21	63	0.45	5.1	8.7	3.4	28	11.9	153	50	218	57	629	100	1328	0.27	0.21	5.8	0.6	10058	1.44	16.7	549	491
2.2	0.054	39	0.27	2.5	4.9	1.58	14	5.4	75	26	120	34	386	60	769	0.069	0.16	4.3	0.18	10032	1.67	11.9	256	364
3.1	0.013	1	0.084	1.62	6.6	0.56	9.5	1.85	10.7	1.65	3.7	0.71	3.6	0.59	42	0.04	0.068	0.69	0.13	13941	0.23	22	20	295
5.1	4.6	16.4	2.4	13.8	10.8	2.1	17.3	7.6	134	53	235	70	796	109	1472	0.42	0.52	4.5	1.01	14984	5.3	19.7	40	219
6.1	0.35	18.8	0.43	3.6	7.9	0.72	24	10.5	154	54	252	65	700	99	1390	1.04	0.52	6.8	1.47	8666	3.3	27	116	241

Sample 1596-4. Gabbroid

1.1	0.22	24	0.83	15.1	32	1.77	98	33	393	138	551	121	1122	143	2673	0.86	1.25	2.8	0.6	17522	1.49	16.7	513	523
2.1	1.47	26	0.85	10.1	14.4	4.3	34	12.5	166	63	315	80	878	148	1754	2.2	4.2	3.9	7.6	15696	0.87	3.7	252	178
3.1	1.12	17.4	0.67	5.4	12.9	0.99	43	18.7	259	105	508	121	1314	181	2588	0.97	0.94	5.9	0.2	15039	2.8	36	395	910
4.1	0.1	29	0.11	2.2	4.5	1.15	12	5.5	63	23	116	30	345	53	685	0.36	0.23	2.5	0.13	13505	1.23	8.7	246	252
5.1	0.19	24	0.79	9.9	19.4	1.98	56	20	225	86	404	93	966	129	2035	0.88	0.83	2.2	0.34	16270	0.79	6.8	602	498
6.1	0.097	28	0.42	4.7	8.5	1.97	25	9.3	125	46	218	51	580	78	1176	0.34	0.86	7.6	0.18	19096	3.3	2	87	81
7.1	0.11	36	0.18	4.9	8.9	2.1	32	13.7	201	81	433	108	1152	170	2243	0.97	0.94	4.7	0.68	19402	1.51	6.4	205	196
8.1	0.61	34	0.99	8.3	10.8	2.8	32	13.8	182	75	361	98	1179	166	2164	0.53	1.4	6	1.23	13111	2.7	42	326	638
9.1	0.074	1.99	0.25	4.2	13.3	0.32	53	24	290	105	457	99	954	121	2123	0.48	0.77	1.71	0.01	16583	1.32	209	138	575
10.1	8.2	47	1.81	8.8	9.9	1.23	34	14.3	189	79	415	107	1280	194	2389	0.81	3.8	4.3	0.3	15137	2	16.5	1067	994

Sample 1655. Gabbro-diorite

1.1	0.47	52	1.71	24	49	9.8	136	52	701	244	1138	297	3275	492	6472	2.2	1.76	7.1	1.1	10701	1.52	21	361	433
2.1	2.5	95	6.3	66	92	18.8	269	96	1192	426	1929	508	5449	756	10906	3.5	7.8	13.4	23	15976	4.7	37	585	1043
3.1	0.33	50	1.49	25	40	8.6	134	49	654	217	1029	271	2911	440	5830	1.79	3.2	7.1	0.38	10091	1.68	15.2	331	441
4.1	0.73	28	0.91	9.5	20	3.8	66	24	325	118	539	157	1742	266	3300	0.94	4.9	5.1	6.8	11892	3.4	16	201	503
5.1	0.63	21	1.79	20	23	4.9	51	18.7	265	90	409	111	1337	193	2546	0.58	1.84	4.9	1.3	10718	4.1	11.2	146	373
6.1	0.25	42	1.23	14.6	31	6.7	106	42	537	197	925	254	2755	432	5344	2.1	1.7	5.7	0.71	10896	1.09	15.3	297	367
7.1	0.85	86	1.73	25	52	7.3	170	62	957	348	1660	441	4715	693	9219	3.4	4.4	14.8	0.82	10159	2.9	36	937	1152
8.1	0.42	73	2.4	39	65	12.1	193	66	878	291	1265	340	3564	544	7333	2.6	2.5	7.2	0.66	9260	1.28	17.1	476	499
10.1	0.089	19.8	0.32	3.1	9	1.01	40	15.4	205	63	237	49	486	55	1184	0.26	<0.1	25	0.45	8581	5.5	21	73	128
11.1	0.3	48	1.5	16.9	35	6.4	111	40	556	202	969	250	2670	396	5302	1.42	2.2	7.1	1.38	10744	2.5	16.7	334	551

(Continued)

Table 10.3 Continued.

Crystal	La	Ce	Pr	Nd	Sm	Eu	Gd	Tb	Dy	Ho	Er	Tm	Yb	Lu	Total REE	Rb	Sr	Y	Nb	Ba	Hf	Ta	Pb	Th	U
Sample 1658. Gabbro-diorite																									
1.1	0.054	4.16	0.051	2.26	2.868	1.280	11.41	4.28	58.98	23.93	140.9	41.57	577	105	974	0.219	1.087	870	1.26	0.198	17088	0.571	3.24	23.61	111.6
2.1	0.021	6.81	0.376	4.00	6.839	2.044	27.52	10.37	119.8	46.92	249.9	70.41	798	135	1478	0.705	0.688	1536	1.45	0.001	7006	0.527	5.33	76.40	199.9
3.1	0.005	4.28	0.041	0.66	2.116	0.439	8.54	3.78	52.11	23.79	139.7	39.61	484	83	842	0.124	0.445	749	1.34	0.033	6424	0.293	3.45	30.89	135.3
4.1	0.177	12.07	0.151	2.38	3.666	1.157	21.92	11.29	170.3	75.40	410.3	103.5	1189	193	2195	0.885	0.612	2050	3.25	0.126	7993	0.721	10.25	144.7	392.6
5.1	0.054	4.66	0.229	3.53	5.695	1.584	20.19	8.52	100.3	37.17	200.8	51.15	633	104	1170	0.535	0.786	1100	0.94	0.082	6725	0.213	4.27	47.34	153.6
6.1	0.015	2.88	0.114	1.40	2.807	0.691	9.00	3.57	46.27	17.86	105.8	30.21	367	62	649	0.162	0.413	582	0.766	0.015	4869	0.193	3.00	24.68	103.9
7.1	0.053	7.59	0.060	1.75	2.956	0.845	17.95	8.38	117.5	50.82	301.3	79.75	941	155	1685	0.491	0.832	1567	1.49	0.277	7314	0.467	6.92	100.4	264.3
8.1	0.087	7.24	0.369	4.47	7.144	2.227	29.39	11.74	146.1	54.98	289.5	78.83	861	140	1633	0.590	0.910	1696	1.43	0.097	7601	0.412	6.11	81.33	237.9
9.1	0.086	8.17	0.391	4.57	8.630	2.992	35.13	11.52	143.1	53.63	295.5	79.55	1015	162	1821	0.519	0.643	1792	1.39	0.075	6789	0.342	6.58	94.08	240.2
10.1	0.038	5.50	0.249	3.28	6.496	2.018	27.05	9.44	113.3	41.19	221.4	58.62	718	119	1326	0.467	0.670	1334	1.23	0.021	4151	0.290	4.68	81.15	174.6
Sample 1652. Gabbro-diorite																									
1.1	0.14	24	0.81	12.8	24	4.7	68	26	332	140	770	207	2539	409	4557	1.39	2.3	4023	3.7	0.27	11413	1.02	11.5	204	464
2.1	8.9	84	10.7	103	103	29	210	70	683	244	1113	254	2897	414	6224	2.2	7.1	6798	5.5	3.3	7769	0.88	14.7	444	655
3.1	0.06	24	0.17	3.6	9.3	1.9	43	17.5	249	113	590	159	1883	295	3389	1.12	1.63	3263	3.5	0.11	6598	0.88	11.3	249	453
4.1	0.74	10.1	0.5	7.3	9	2.5	28	11.4	142	61	315	86	1113	185	1972	0.68	2.5	1836	20	0.62	22110	1.88	6.9	70	231
11.1	0.47	9.7	0.33	5.1	10.8	1.42	54	21	277	102	471	103	1081	152	2289	0.91	0.68	2749	4	0.26	17473	1.13	15.1	62	105
12.1	0.41	52	0.86	12.4	34	6.8	122	47	579	237	1258	323	3944	623	7239	2.6	5.4	7656	8.4	4.6	17050	1.71	25	554	871
13.1	0.29	46	1.32	28	54	12.1	187	64	759	287	1470	388	4469	715	8481	2.7	4.2	7935	5.7	0.41	17719	1.03	17.3	419	596
Sample 2612-2. Gabbro-diorite																									
1.1	0.2	20	0.6	8.4	14.5	2.3	55	20	236	85	359	75	720	88	1684	0.44	0.69	2182	1.63	0.11	8857	0.74	8.3	335	262
2.1	0.083	41	0.18	2.9	5.7	1.61	19	7.6	99	38	191	49	550	79	1084	0.34	0.4	1069	6.7	0.36	12911	1.95	3.9	227	237
3.1	0.5	15.3	2.9	53	93	1.87	322	120	1475	591	2623	541	5386	642	11867	4.1	5.4	11870	6.2	0.18	27184	2.4	66	940	371
4.1	0.064	13.1	0.15	2.8	7.2	1.74	28	12.5	167	66	328	80	915	133	1755	0.6	0.75	1813	4.3	0.05	15770	1.82	8.7	231	602
5.1	0.18	29	0.094	2.4	5.2	0.98	16.4	5.7	76	25	118	27	311	46	663	0.46	1.15	686	2.3	3	20145	0.77	7.1	103	120
6.1	0.17	40	0.94	12.3	17.5	3.9	52	19.1	208	73	291	63	600	82	1463	0.77	0.58	1782	2.3	0.23	13358	0.95	5	326	183
7.1	0.065	22	0.21	4.1	9.9	1.68	36	14.2	203	88	466	117	1405	218	2585	0.95	1.19	2599	2.9	0.31	16827	0.83	7	332	466
8.1	0.17	16.3	0.13	3.6	6.1	1.44	13.3	5.3	60	27	127	29	367	57	713	0.55	0.07	725	2.4	0.17	19111	0.92	10.5	99	127
9.1	0.1	13.2	0.75	4.7	2.9	0.67	7.9	3.1	35	14	64	14.5	169	23	353	0.28	2.6	371	0.72	0.41	7053	0.3	5.6	142	225
10.1	0.12	14.4	0.13	3.9	7.7	0.64	26	9.7	131	46	233	54	562	74	1163	0.64	0.47	1267	4.6	0.11	17199	1.99	5.9	236	366
Sample 160. Diorite																									
1.1	4.3	37	1.09	8.8	13.4	2.5	41	15.8	200	79	382	84	923	128	1920	20	23	1981	4.2	134	14931	0.91	5.4	166	141
2.1	0.16	26	0.19	3.7	8.1	1.16	29	11.5	148	54	243	58	597	85	1265	0.48	0.87	1530	2.6	0.24	21229	1.7	11.2	314	383
3.1	0.2	27	0.46	7.8	11.8	1.69	32	12.4	159	56	272	61	672	91	1404	0.44	0.45	1567	3	0.21	15759	1.8	13.3	337	300
4.1	0.79	18.2	0.4	5.3	9.1	3.4	36	13.6	166	63	288	62	660	95	1421	0.59	1.94	1665	1.21	1.85	4830	0.34	1.46	142	88
5.1	0.092	33	0.45	6.5	10.6	4.8	36	12.9	163	59	271	65	774	113	1549	0.52	0.58	1665	3.8	0.24	13013	1.31	28	332	340
6.1	1.25	41	0.93	13.7	23	3.1	85	32	410	149	719	162	1666	231	3537	1.38	1.63	4208	10	0.33	16645	3.4	11.1	680	649
7.1	0.045	7.5	0.013	0.75	1.99	0.36	6.6	2.7	40	18.7	112	35	420	62	708	0.21	0.37	494	4.1	0.082	9121	2.7	3.1	86	168
8.1	20	15.2	0.093	3.1	3.1	0.99	13	5.3	74	32	162	40	522	88	977	0.48	0.75	905	1.84	1.26	8529	0.62	2.8	148	231
9.1	0.11	14.2	0.14	1.41	2.7	0.67	7.6	3.4	40	14	67	16.8	181	30	379	0.11	0.34	433	1.4	0.17	7633	0.65	6.7	173	224

Sample 197. Diorite

1.1	0.03	2.6	0.03	0.7	1.3	0.48	5.6	2.5	35	15	96	28	321	54	562	0.06	0.41	531	0.7	0.001	4305	0.29	1.2	12	41
2.1	0.001	3.4	0.03	0.9	1.6	0.56	5.9	2.8	41	18	108	31	372	65	650	0.17	0.56	621	1	0.03	5227	0.36	1.4	15	53
3.1	0.12	4.7	0.05	1.4	2.3	0.65	9.5	4.4	60	27	162	46	538	90	946	0.2	0.76	930	1.3	0.34	6528	0.35	2.4	24	73
4.1	0.07	6	0.08	1.4	2.1	0.84	10.3	4.7	72	32	181	50	597	102	1059	0.26	1.4	1058	1.5	0.18	6322	0.36	2.3	29	77
5.1	0.1	6.7	0.1	2	3.9	0.94	12.1	5.6	86	39	223	67	835	129	1410	0.28	0.74	1268	1.6	0.13	8520	0.89	3.1	34	105
6.1a	0.04	5.3	0.08	1.6	3.5	1.11	13.5	4.7	63	28	169	51	640	108	1089	0.11	1.14	951	1.6	0.01	11228	0.41	2.5	20	85
6.1б	0.04	4.9	0.04	1.6	3	0.78	11.1	5.2	70	30	171	47	594	98	1037	0.34	1.04	909	1.2	0.15	7223	0.35	2.4	23	72
6.1в	0.04	5.9	0.06	2.1	1.9	1.03	9.6	4.6	67	30	169	51	667	103	1112	0.3	1.01	935	1.8	0.11	10897	0.5	2.6	19	87
6.1	0.04	5.4	0.0061	1.79	2.8	0.97	11.4	4.8	67	29	170	50	634	103	1080	0.32	1.06	932	1.52	0.13	9783	0.42	2.5	21	81
7.1	0.05	11.5	0.09	2.8	6.3	2.08	24.8	10.3	144	69	421	117	1433	232	2474	0.41	2.35	2251	3.8	0.39	18537	0.98	5	54	178
8.1	0.03	2.4	0.04	0.4	1.2	0.45	4.2	2	32	14	89	24	293	49	512	0.18	0.4	496	0.7	0.09	3297	0.18	0.9	12	34
9.1	0.02	2.7	0.005	1	1.7	0.58	6.5	3	39	16	98	29	373	63	634	0.14	0.87	599	0.9	0.08	7302	0.29	1.3	12	43
10.1	0.01	5.1	0.08	1.4	2.1	0.68	9.2	4.2	63	29	179	49	590	100	1033	0.06	0.69	940	1.1	0.07	6127	0.35	2	23	67

Sample 2612. Quartz diorite

1.1	0.09	11.8	0.13	1.78	4	1.54	14	5.1	68	31	164	48	588	100	1037	0.21	0.69	933	0.89	0.27	13411	0.15	1	21	54
2.1	46	117	10	38	13.3	1.71	19.5	5	54	19.7	88	21	248	38	719	14.5	73	547	3	19.6	13200	0.88	15.2	251	268
3.1	8.1	29	2	14.3	7.9	1.98	23	7.7	113	50	256	68	835	134	1550	0.32	9.7	1384	1.53	0.4	13390	0.52	1.73	59	112
4.1	6.3	46	1.51	8.3	7.3	0.78	22	9.4	128	55	286	74	872	124	1641	0.59	1.93	1445	8.3	0.19	11950	3.6	15.5	280	496
5.1	0.05	22	0.16	2.4	4.9	0.89	15.7	5.6	64	23	108	27	289	44	607	0.42	0.18	617	1.59	0.11	11594	0.53	3.3	93	110
6.1	9.1	49	2.5	11.4	9.3	2.6	36	15	203	84	457	118	1400	224	2621	1.06	10.9	2429	2.9	0.42	18678	55	3.2	155	215
7.1	0.033	6	0.1	1.1	2.2	0.91	7.3	3.5	45	20	107	31	362	64	650	0.21	0.22	639	0.82	0.08	9285	0.34	1.11	41	80
8.1	0.039	8.8	0.07	2.3	2.9	0.63	9.4	3.2	42	16.5	81	21	246	40	474	0.24	0.27	496	0.78	0.079	6142	0.31	4.5	77	71
9.1	0.13	16.7	0.064	0.92	2.4	0.62	10.2	4.6	69	32	174	50	635	104	1100	0.27	2.2	979	2.6	0.34	7605	0.59	2.4	83	173
10.1	0.11	19.1	0.34	5.3	9.3	1.61	37	15.5	196	79	405	100	1087	155	2110	0.72	1.17	2394	3.7	0.17	11599	1.51	26	488	561

Sample 156. Trachyandesite

1.1	0.047	1.9	0.28	2.5	11.9	0.41	58	26	425	147	607	155	1453	188	3076	0.81	0.57	3977	2.4	0.21	16018	1.42	108	98	512
1.2	0.35	1.77	0.48	4.5	6.9	0.57	37	17.8	275	92	357	88	820	111	1812	0.57	1.07	2531	1.76	0.76	14873	1.14	69	45	305
2.1	0.67	9	0.25	3.1	7.2	1.06	30	14.2	210	79	343	93	1036	146	1972	<0.05	1.98	2.275	4.4	3.3	12219	1.6	4	224	442
3.1	0.22	19.3	0.57	8.6	15.3	4.3	57	23	313	107	514	135	1391	217	2805	2.2	1.24	3216	3.5	7.1	10328	0.82	6.1	170	166
4.1	0.18	17.4	0.14	0.43	4	0.71	10.4	4.8	71	24	114	34	359	53	693	0.31	0.17	755	4	0.36	12338	2	4.6	164	234
5.1	0.13	3.7	0.17	2.8	4.6	0.26	13.6	6.3	99	36	185	52	592	89	1085	0.05	<0.1	1107	3.5	0.53	11804	1.61	184	74	468
5.2	0.03	6.2	0.2	2.2	5.6	0.13	22	9.2	128	49	211	58	608	84	1184	0.54	<0.1	1341	3.9	0.22	12769	2.2	202	87	463
5.3	0.44	6	0.25	3	3.8	0.34	12.5	5.2	88	33	168	46	529	83	979	0.43	<0.1	1031	3.1	0.49	13457	2.3	196	64	498
6.1	0.15	9.6	0.3	2.7	4	1.29	17.3	8.4	136	60	308	90	1055	197	1890	0.26	1.17	1622	3.5	0.37	13288	0.84	5.1	55	129

Note: Analyses were performed at the Analytical Center of the Sobolev Institute of Geology and Mineralogy (Novosibirsk) by LA ICP-MS with an Element mass-spectrometer (analyst S.V. Palesskii).

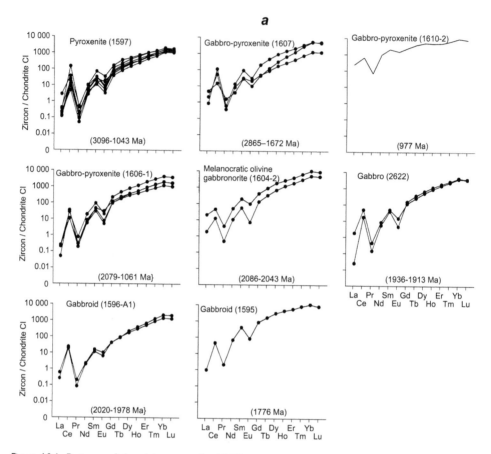

Figure 10.6   Patterns of chondrite-normalized REE contents in polychronous zircons from rocks of the Berezovka massif (after Table 10.3).
a – Ancient populations; *b* – young age populations; *c* – young age populations in which REE probably underwent epigenetic redistribution. Normalized to chondrite CI (after [Evensen *et al.*, 1978]).

Interestingly, the genetic nature of positive Ce anomalies and negative Eu anomalies, almost always present in the chondrite-normalized REE patterns typical of igneous zircons, can be accounted for by their ability to change valence depending on the redox conditions of the environment. According to experimental data, the partition coefficient of Ce^{4+} in the zircon–melt system is much higher (718) than that of Ce^{3+} (0.022) [Hinton and Upton, 1991]. It can be thus suggested that Ce^{4+} ions, predominant in oxidizing conditions, must better accumulate in zircon structure in contrast to Ce^{3+}, which will be expressed as intense positive Ce anomalies in the REE patterns. Moreover, oxidized Eu^{3+} is characterized by a lower partition coefficient in the zircon–melt system as compared with reduced Eu^{2+} [Burnham and Berry, 2012]. Therefore, under oxidizing conditions Eu^{3+} will worse incorporate into zircon structure than Eu^{2+},

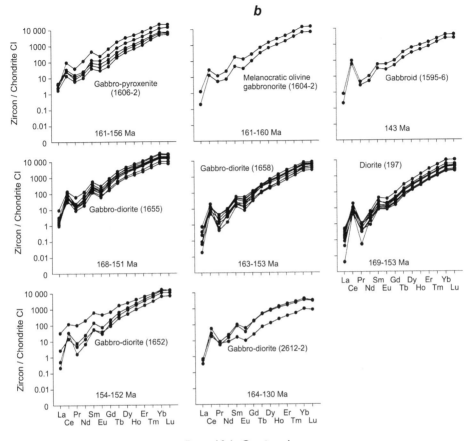

Figure 10.6 Continued.

which will yield negative Eu anomalies in the REE patterns. Hypothesis on the crystallization of zircons from terrestrial rocks under oxidizing conditions can be supported by the fact that positive Ce anomalies tend to be absent in the REE patterns of zircons from lunar rocks formed *a priori* under reducing conditions [Hinton and Meyer, 1991; Wopenka *et al.*, 1996].

In the discussions of factors that may have caused positive Ce anomalies in the REE patterns of zircons from rocks of the Urals mafic-ultramafic massifs, Fershtater [2013] also inferred that these anomalies were determined by higher oxidized Ce ions, inasmuch as Ce^{+4} will better incorporate in zircon structure in contrast to trivalent ions of other REE. However, Aranovich *et al.* [2013] had a different view on the nature of the Ce and Eu anomalies observed in the REE patterns of zircons. They attributed positive Ce anomalies to the close radii of Ce^{+4} and Zr^{+4} ions, whereas the presence of negative Eu anomalies in the zircon patterns was due to the mineral crystallization proceeding simultaneously with, or sometime later than, the coexisting plagioclase, holding significant Eu contents. The latter is evidenced by the presence of negative Eu anomalies in the patterns of zircon from rocks without plagioclase (e.g., dunites from

c

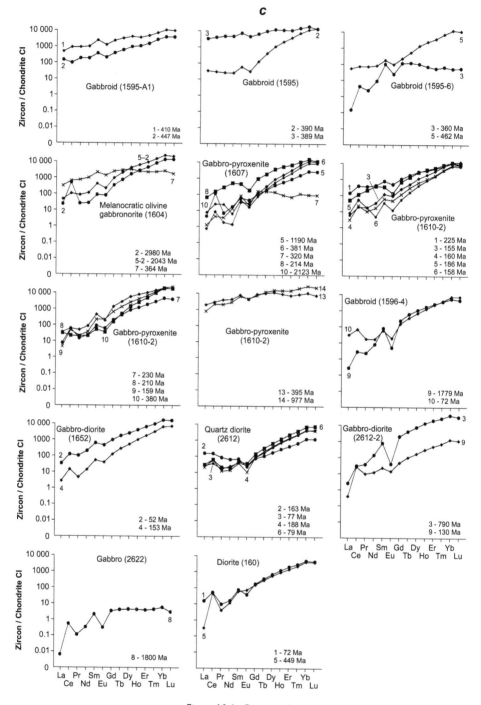

Figure 10.6 Continued.

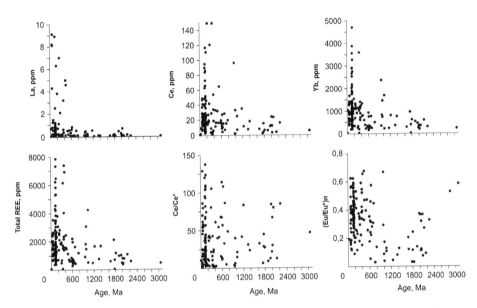

Figure 10.7 Relationship between the contents of La, Ce, Yb, and total REE, as well as $(Ce/Ce^*)_n$ and $(Eu/Eu^*)_n$, in zircons from rocks of the Berezovka massif and their isotopic ages (after Tables 10.2 and 10.3).

the Kos'va and Sakharinsk massifs in the Urals) [Fershtater *et al.*, 2009]. Ryabchikov [2012] provided some interesting insights into possible causes of occurrence of oxidizing conditions at crystallization of zircons proceeding in the ultramafic and gabbroid rocks, which was probably determined by the loss of hydrogen during the intense migration of hydrogen and hydroxyl into the surrounding environment owing to their diffusion coefficients found to be high in certain conditions.

Uranium, thorium, rubidium, yttrium, and strontium. Analyses of zircons by LA ICP-MS showed that U contents in zircons from the Berezovka massif rocks vary from 29 to 1482 ppm, with the values not greater than 700 ppm prevailing, while Th contents in these zircons range from 18 to 1416 ppm with the predominance of varieties showing not more than 600 ppm (Table 10.3). The contents of these two elements in zircons were also determined by SHRIMP II: U = 29–1381 ppm, Th = 11–1692 ppm (i.e., they are in approximately the same ranges as the contents measured by LA ICP-MS). The calculated U/Yb values vary from 0.06 to 1.9 in ancient zircons and from 0.10 to 1.33 in the mineral from age populations dated at ~190-140 Ma. Zircons in which either U or Th contents prevailed occurred in various age populations, and Th/U ratios were found, respectively, in the 0.1–4.8 interval with most of them concentrated in the 0.1–1.5 interval. U and Th contents in the total collection of the studied zircons and Th/U ratios tend to increase from oldest to younger populations (Fig. 10.8).

According to the LA ICP-MS data, Y contents in the zircons vary from 66 to 12,095 ppm, while they are localized in a narrower range (66–6122 ppm) in the grains

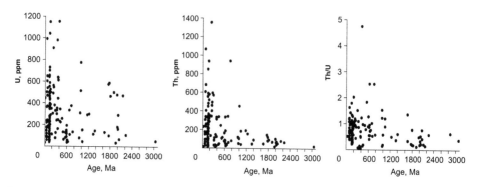

Figure 10.8 Relationship between U and Th contents, and Th/U in zircons from rocks of the Berezovka massif and their isotopic age (after Tables 10.2 and 10.3).

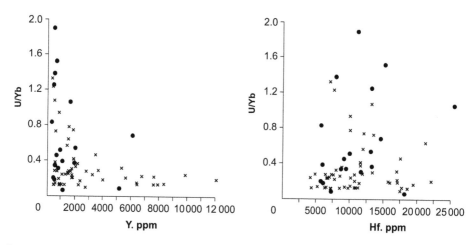

Figure 10.9 Relationship between U/Yb and the contents of Y and Hf in zircons from ultramafic and gabbroid rocks of the Berezovka massif and their isotopic age in the intervals of 3096–977 Ma (dots) and 188–103 Ma (crosses) (after Tables 10.2 and 10.3).

from the ancient population compared to the zircons from the population aged ~190–140 Ma (370–12095 ppm). The arrangement of representative points on the diagrams in Fig. 10.9 for the studied zircons suggests a negative correlation between Y contents and U/Yb ratios. The same correlation is presumed between the isotopic ages of zircons in the total collection and their Rb and Sr contents (Fig. 10.10).

Hafnium. According to the LA ICP-MS data, Hf contents vary from ~3300 to ~27200 ppm in the studied zircons, averaging ~12000 ppm. No differences between its contents in grains from different age populations have been revealed. Within the studied collection of zircons, there is no clear correlation between Hf contents and U/Yb ratios (Fig. 10.9). Given that the variations of the Lu/Hf ratios are limited in the diagrams, reflecting the variation in the values of parameters Lu/Hf and Sm/Nd, the

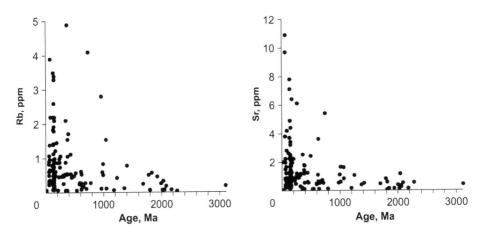

Figure 10.10 Relationship between Rb and Sr contents in zircons from rocks of the Berezovka massif and their isotopic age (after Tables 10.2 and 10.3).

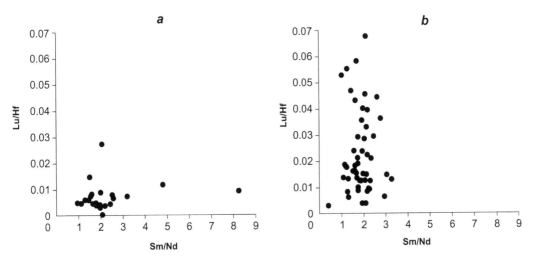

Figure 10.11 Relationship between Lu/Hf and Sm/Nd in zircons from rocks of the Berezovka massif. *a* – Populations with age ranges from 3096 to 991 Ma; *b* – population with the age range from 168 to 153 Ma (after Table 10.3).

compositional points for ancient zircons with ages in the range of 3096–991 Ma form a subhorizontal trend (Fig. 10.11, *a*). The compositional points for young zircons dated at 168–153 Ma show a subvertical trend due to high variations of the Lu/Hf ratios and minor variations of the Sm/Nd values (Fig. 10.11, *b*). Two zircon crystals with distinct crystal faces and well-defined oscillatory zoning (dated at 75–83 and 160 Ma, respectively) were scanned by LA ICP-MS along the cross sections to determine the contents of Hf, U, Ce, and Yb (Fig. 10.12). The resultant curves for element concentrations

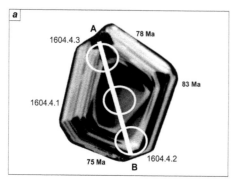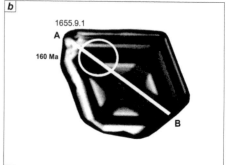

Figure 10.12   Microphotographs of zircon crystals with coated profiles A–B, along which the Hf, Yb, U, and Ce contents were measured by LA ICP-MS.
a – Grain 4.1 from sample 1604 (olivine gabbronorite), age 78–83 Ma; *b* – grain 9.1 from sample 1655 (gabbro-diorite), dated at 160 Ma. Microphotographs were taken in CL regime.

along the cross sections show a polymodal distrubution (Fig. 10.13). Comparison of the concentration curves and oscillatory zones on the CL images of grains (Fig. 10.12, *a*) showed that the light zones usually correspond to the maximum contents of Hf, Yb, Ce, and U, and the dark zones, to their minimum contents (Fig. 10.13). The same follows from the fact that the wide minima of Hf and other elements in the middle parts of the concentration curves for the grain from sample 1655 (Fig. 10.13) corresponds to the wide dark oscillatory zone shown in the center of the photomicrograph, almost not exhibiting cathodoluminescence (Fig. 10.12, *b*). This suggests that the light and dark oscillatory zones resulted from the regular alternations of zones with low/high contents of the above trace elements serving as a luminescence agent. The assumed relation between the zonal distribution of Hf, Yb, Ce, and U and the CL intensity of the oscillatory zones in the zircons explains the use of Hf, Ce, and some other elements as luminescent agents in equipment.

Taking into account the revealed relationship between the zonal distribution of some trace elements and the oscillatory zoning of the zircons, we assume that the extremely low CL intensity (down to full absence) and the lack of clear oscillatory zoning in most of the ancient grains are due to the diffusive redistribution of luminescence elements in the grain structure, which proceeded for a significant period of time, which largely depressed the variability of their contents and CL properties. The "homogenization" of luminescence element contents in ancient zircons and partial or total absence of oscillatory zoning in them and resorption of grains were probably favored by the thermal and chemical actions of later generated mafic melt and its fluids.

Rhenium and molybdenum contents were determined by LA ICP-MS in a few zircon grains [Lesnov and Palesskii, 2013], and the data on their contents in zircon are scarce. The contents of Re in nearly a half of the grains are above its LA ICP-MS

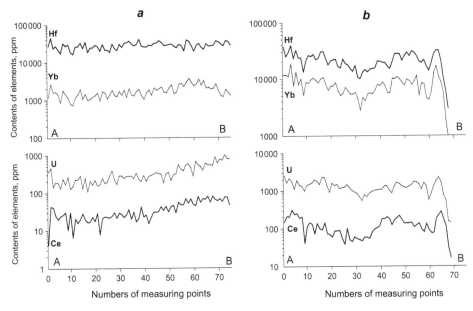

Figure 10.13 Distributions of Hf, Yb, U, and Ce contents in zircon grains along profiles A–B.
a – Grain 4.1, sample 1604, olivine gabbronorite, dated at 75–78 Ma (see Fig. 10.12, a);
b – grain 9.1, sample 1655, gabbro-diorite, dated at 160 Ma (see Fig. 10.12, b).

detection limit, 5–16 ppb; one grain contains 82 ppb Re (Table 10.4). High contents of Re were found in zircon grains from gabbro-diorite (151–168 Ma) and gabbro-pyroxenite (160–381 Ma). The arrangement of compositional points on the diagrams suggests a negative correlation between the Re content and the isotopic age of zircons, and, possibly, a positive correlation between the contents of Re and those of Mo, U, Yb, and Th (Fig. 10.14).

Multielement patterns of chlorite-normalized contents of trace elements in most of the studied zircon grains have a similar shape typical of their magmatic varieties (Fig. 10.15). The patterns of grains from the same rock sample mostly lie very close to each other in the diagrams. Most of the patterns show intense positive anomalies of Th, U, and Hf, weaker positive anomalies of Ce, and negative anomalies of Sr and La. In the shape and position of multielement patterns some zircon grains differ from others. The anomalous patterns point to the nonuniform LREE enrichment of the grains, to the low intensity (down to full absence) of positive Ce anomaly, and, sometimes, to lower HREE contents. These multielement patterns appear inherent in zircons from hybrid ultramafic rocks (gabbro-pyroxenites, samples 1607 and 1610-2), hybrid gabbroids (melanocratic olivine gabbronorites, sample 1604), and orthomagmatic gabbroids (gabbro, sample 2622). Probably, such geochemical anomalies in the multielement patterns were due to the epigenetic redistribution of trace elements, which proceeded with the LREE incorporation in variable contents and their accumulation in the intragrain microcracks in the form of nonstructural impurities.

Table 10.4 Contents of Re and Mo and isotopic age of zircons from the Berezovka massif.

Sample (grain)	Rock	$^{206}Pb/^{238}U$ age, Ma	Re, ppb	Mo, ppb	Samples (grains)	Rock	$^{206}Pb/^{238}U$ age, Ma	Re, ppb	Mo, ppb
1607 (4)	Gabbro-pyroxenite	257±4.9	<5	660	1596-6 (3)	Gabbroid	360±7.0	<5	750
1607 (5)	Gabbro-pyroxenite	1190±21	<5	520	1596-6 (5)	Gabbroid	462±9.2	<5	2500
1607 (6)	Gabbro-pyroxenite	356±6.9	<5	2900	1596-6 (6)	Gabbroid	394±7.7	9	500
1610-2 (1)	Gabbro-pyroxenite	225±2.2	82	1100	1595 (1)	Gabbroid	1058±24	<5	900
1610-2 (4)	Gabbro-pyroxenite	160±2.6	16	810	1595 (2)	Gabbroid	390±6.3	<5	1400
1610-2 (7)	Gabbro-pyroxenite	230±5.8	5	660	1595 (3)	Gabbroid	389±5.4	<5	1400
1610-2 (14)	Gabbro-pyroxenite	977±8.4	7	570	1595 (4)	Gabbroid	154±2.1	5	1200
1604 (1)	Melanocratic olivine gabbronorite	160±3.3	8	600	1595 (5)	Gabbroid	320±4.1	9	350
1604 (2)	Melanocratic olivine gabbronorite	298±5.7	<5	900	1595 (6)	Gabbroid	N.d.	<5	500
1604 (3)	Melanocratic olivine gabbronorite	161±3.3	<5	980	1655 (1)	Gabbro-diorite	167±3.6	9	850
1604 (4)	Melanocratic olivine gabbronorite	83±2.1	7	1200	1655 (2)	Gabbro-diorite	151±4.4	16	1300
1604 (5)	Melanocratic olivine gabbronorite	N.d.	<5	710	1655 (3)	Gabbro-diorite	160±3.2	14	1000
1604 (6)	Melanocratic olivine gabbronorite	161±3.8	7	870	1655 (4)	Gabbro-diorite	153±4.2	11	960
1604 (7)	Melanocratic olivine gabbronorite	364±6.8	<5	310	1655 (5)	Gabbro-diorite	164±3.5	11	710
1604 (8)	Melanocratic olivine gabbronorite	184±4.6	8	19000	1655 (6)	Gabbro-diorite	168±3.5	14	830
1596-A1 (1)	Gabbroid	410±3.5	7	1600	1655 (7)	Gabbro-diorite	164±3.2	13	940
1596-A1 (3)	Gabbroid	629±15	<5	780	1655 (8)	Gabbro-diorite	158±3.1	11	660
1596-A1 (4)	Gabbroid	2014±17	7	310	1655 (9)	Gabbro-diorite	160±3.1	11	480
1596-6 (1)	Gabbroid	633±12	7	830	1655 (10)	Gabbro-diorite	793±16	<5	970
1596-6 (2)	Gabbroid	143±2.8	<5	380	1655 (11)	Gabbro-diorite	152±3.2	14	720

Note: Analyses were performed at the Analytical Center of the Sobolev Institute of Geology and Mineralogy by LA ICP-MS with an Element mass spectrometer (from [Lesnov and Palesskii, 2013]).

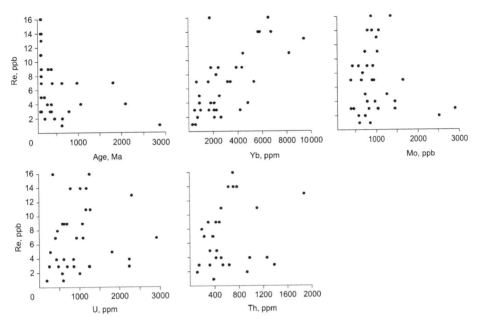

Figure 10.14 Relationship between Re contents and isotopic ages as well as between Re and Yb, Mo, U, and Th contents, in zircons from rocks of the Berezovka massif (after Tables 10.3 and 10.4).

10.6 SPECIFICS OF ZIRCON ISOTOPE DATING OF ROCKS OF MAFIC-ULTRAMAFIC MASSIFS

As was said above, the U–Pb zircon dating of rocks from igneous complexes is one of the highlights in modern petrology [Davis *et al.*, 2003]. Given the topicality of the research into this problem, including reconstructions of the conditions of formation of mafic-ultramafic massifs, we will make a brief review of some important publications giving results of study of polychronous formation of zircons, their host rocks, and igneous complexes.

In the early studies of the isotope dating of igneous rocks performed on zircons, Krasnobaev [1979] arrived at the conclusion that zircons from kimberlites ("kimberlitic zircons") may be older than their host rocks. In his opinion, there is no need to employ the mechanism of trapping ancient zircon grains from the host rocks to explain the revealed inconsistencies between the ages of kimberlites and that of the hosted zircons, since ancient zircon crystals from kimberlites were likely to be xenogenous (i.e., they may represent fragments of disintegrated peridotite xenoliths from the kimberlites).

Later, Kinny *et al.* [1989], having determined isotopic age of zircons from ultramafic xenoliths from the Botswana kimberlites (South Africa), divided the collection into two main populations: (1) zircons dated at 235 Ma corresponded to the time of formation of the host kimberlites; (2) zircons dated at ∼2800 Ma were regarded as

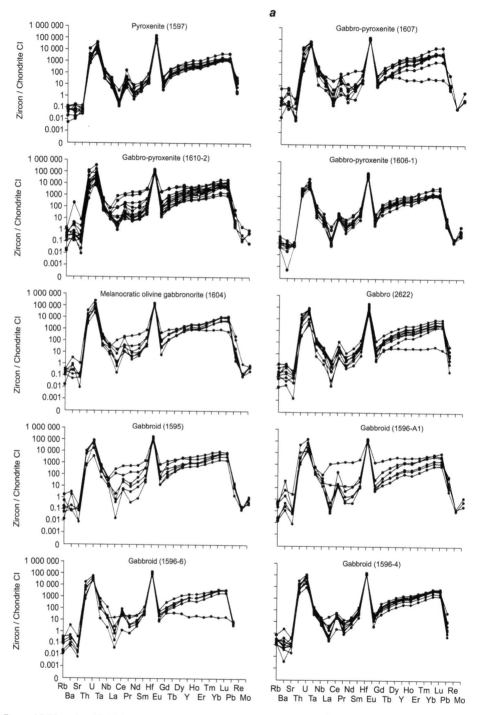

Figure 10.15 a, b – Multielement patterns of the chondrite-normalized impurity-element distribution in zircons from rocks of the Berezovka massif (after Table 10.3). REE data normalized to chondrite CI (after [Anders and Grevesse, 1989]).

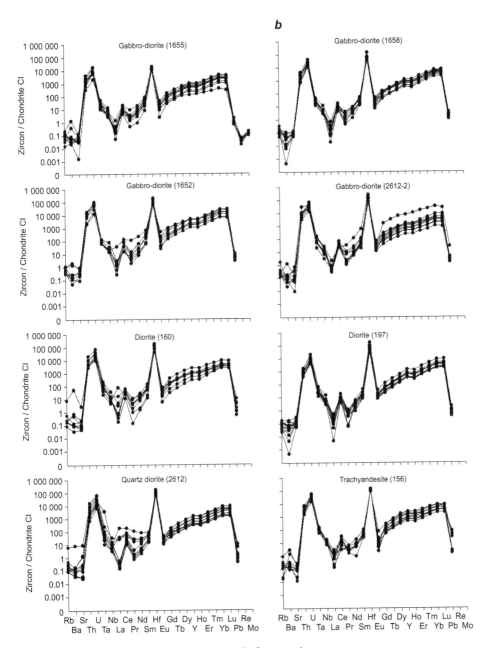

Figure 10.15 Continued.

xenogenous, i.e. those trapped in the kimberlites from the disintegrated xenoliths of the mantle ultramafic rocks.

Pilot *et al.* [1998] presented the first data on zircons of several populations in gabbroid rocks dredged from the Mid-Atlantic Ridge, with their age varying from relatively young (330 Ma) to very old (1623 Ma), which is considered to be abnormal for oceanic structures.

Bortnikov *et al.* [2008] obtained more U–Pb data on the age of zircons from the troctolites and gabbronorites dredged from the Mid-Atlantic Ridge (Sierra Leone Fracture Zone in the Markov depression), with the subsequent allocation of four age populations among them: 3126-3094, 2855-2689, 1843-1116, and 728-259 Ma. The authors explained the presence of Archean and Proterozoic zircons there (1) by trapping of the grains from mantle plume or (2) by their trapping by mafic melts penetrated by differently aged rocks of sheets of the ancient crust subducted into the upper mantle.

Skolotnev *et al.* [2009] performed a U–Pb study of the isotopic age of over 60 zircons from the ultramafic rocks, gabbroids, diorites, plagiogranites, and basaltoids dredged in the Central Atlantic. They divided the collection of dated zircons into seven age populations: 2715, 2470, 2250, 2070, 1850, 1650, and 420 Ma. More data on isotopic age of another batch of 50 zircon grains from lherzolites, gabbronorites, and gabbro dredged in the Mid-Atlantic Ridge are provided in another work by Skolotnev *et al.* [2010]. The grains were divided into five populations, comprising the interval of time spanning from 2712 to 100 Ma. The authors believed that the age of zircons of the youngest population corresponded to the time of formation of the host gabbroid rocks.

Savelieva *et al.* (Savelieva *et al.* [2006, 2007]), using U–Pb technique, discovered and dated zircons in chromitites of the Voikar–Syn'in mafic-ultramafic massif (Polar Urals) and divided them into three age populations: 2552, 622, and 585 Ma.

Malitch *et al.* [2009] estimated the U–Pb ages of zircons from dunites of the Nizhnii Tagil mafic-ultramafic massif (Urals) and recognized three age populations: 2656-2852, 1608, and 585 Ma. The authors believe that the age of zircons from the first two populations corresponds to the time of formation of the host dunites, in the conditions of the subcontinental mantle.

Badanina and Malich [2012] analyzed the collection of 24 zircons from dunites of the Konder massif (Aldan province) and recognized four populations by the U–Pb determinations: 2473, 1885, 176, and 143 Ma, with the age of the oldest one corresponding to the minimum age of the primary upper-mantle substratum.

K.S. Ivanov *et al.* [2012] dated zircons from olivine pyroxenites of the Klyuchevskoy mafic-ultramafic massif (Middle Urals) and divided them into two age populations: 1700 and 490–390 Ma. The authors concluded that the age of zircons of the ancient population corresponds to the minimum age of protoliths from the upper mantle generating mafic melts.

Kremenetskii and Gromalova [2013] summarized both published and their own data on isotopic age of zircon grains from samples of the ultramafic, gabbroid, and other rocks dredged in the Mid-Atlantic Ridge and in the Arctic oceanic structures. They suggested the following series of zircon age populations: 2700-2500, 2000-1500, 1200-1000, 600-400, 200-100, and 30-10 Ma, pointing out that the presence of ancient zircons in ultramafic and gabbroid rocks from structures of the ocean floor gives grounds for revision of the existing understanding of the origin of these rocks and that ancient zircons from ultramafic rocks and gabbroids on the oceanic floor are xenogenous. Besides, the authors inferred the existence of several sources of ancient zircons discovered in the bottom rocks of the ocean floor that might be (1) fragments of ancient oceanic crust, (2) fragments of preoceanic crust, (3) the "scattered" Precambrian substratum, (4) blocks of the lower continental crust, and (5) blocks of heterogeneous and tectonically layered lithospheric mantle.

Krasnobaev and Anfilogov [2014] presented a summary analysis of the data on isotopic age of zircons and dunites incorporated in many mafic-ultramafic massifs of the Urals. The authors believe that the age of ancient zircons corresponds to the time of formation of the host dunites, which originated in the mantle long before the start of formation of folded systems, like other restitic ultramafic rocks. Zircons from the mantle substratum which were not dissolved in melted basaltic magmas were inherited by dunites and other restitic ultramafic rocks.

Results of isotope studies of polychronous zircons from the mafic-ultramafic massifs, as well as evidence for the presence of ancient zircons in them, are given in many other publications (as cited in Table 10.5).

Along with the earlier published representative data on distributions of polychronous (including ancient) zircons in rocks from the mafic-ultramafic and some other similar massifs from various parts of the world, results of isotope dating of zircons from rocks of the fairly small Berezovka mafic-ultramafic massif provided well documented evidence of the gabbroid intrusion generated later than the spatially close ultramafic protrusion.

As was shown above, zircons from the Berezovka massif rocks strongly differ not only in isotopic age but also in grain size and morphology, CL intensity, kind of oscillatory zoning, and geochemical properties. The isotope dating allowed discrimination of several groups of age populations represented by (1) ancient zircons (~3090–990 Ma), (2) intermediate age populations (800-200 Ma), (3) young zircons (younger than ~200 Ma). Zircons from the ancient population discovered in the hybrid ultramafic rocks are regarded as a *relict phase* that was a product of later transformations of the upper-mantle ultramafic restites. Zircons close in age to the ancient ones but present in the hybrid and orthomagmatic gabbroids are assumed to be a *xenogenous phase* trapped by a mafic melt from the upper-mantle ultramafic source during its partial melting or during interaction between melt and ultramafic restites. Grains of the *relict* and *xenogenous* zircons belonging to intermediate populations are assumed to have been nonuniformly resorbed and "rejuvenated" under the thermal and chemical action of later generated mafic melts and their fluids. We assume that the impact of the above melts and fluids leveled and reduced the CL intensity of zircon grains, disturbed their primary oscillatory zoning, and led to nonuniform enrichment of the grains with LREE and the depletion with HREE.

As shown above, most grains of the studied collection of zircons from the Berezovka massif are represented by age populations ranging between 190 and 140 Ma, with the intense maxima of occurrence of their age corresponding to the interval of 170–150 Ma. Most of these zircon grains are present in orthomagmatic gabbroid rocks composing the intrusion breaking through the ultramafic protrusion and in hybrid gabbroids which crystallized from the great-depth mafic melt contaminated with the matter of the host rocks (amphibole gabbro, gabbro-diorites, and diorites). The majority of studied zircon grains have clear crystal faces with the concordant regular oscillatory zoning. It can be assumed with greater confidence that zircons from these populations

Table 10.5 The most important manifestations of polychronous zircons from rocks of mafic-ultramafic massifs and complexes.

Location	Massif (complexes, province)	Rocks	Age populations, Ma (number of determinations)	References
1	*2*	*3*	*4*	*5*
Africa. Botswana	Xenoliths in kimberlites	Mantle peridotites	2800; 235	Kinny et al. [1989]
Eastern Finland	Lahtojoki Tube	Xenocrysts from kimberlites	2665-1715 (21)	Peltonen and Manttari [2001]
Southern Italy	Northern Catena Costiera	Gabbro	296-262 (9); 247-227 (14)	Liberi et al. [2011]
South Korea	Baekdong massif	Harzburgites, dunites, metabasites	2522-1846 (10); 403-320 (6); 856 (1)	Oh et al. [2012]
China	Inner Mongolia (dikes)	Mafic rocks (dikes)	2563, 2461 (2); 329-251 (14); 135 (1)	Miao et al. [2008]
		Leucogabbro	2756-1859 (3); 597-249 (14)	
China	Liaoning province (dikes)	Dolerites	2510 (15); 2476 (8); 1908 (10); 1837 (8); 998 (10); 960.7 (14); 837.9 (8)	Liu et al. [2012]
Kurile Islands, Shikotan Island, Russia	Shikotan massif	Lherzolites, pyroxenites, gabbro-pyroxenites, gabbronorites, gabbro-diorites, diorites	2775-936 (55); 66.1-62.7 (2)	[Explanatory note ..., 2006]
Vitim Plateau, Russia	Xenoliths in basalts	Garnet-spinel peridotites	1694-1161 (8); 532 (1); 296-264 (13); 183.4-138.8 (10)	Saltykova et al. [2008]
Polar Urals, Russia	Voikar–Syn'in massif, Paity ore occurrence	Chromitites	2552 (1); 622 (1); 585.3 (7)	Savelieva et al. [2006, 2007]
Polar Urals, Russia	Voikar–Syn'in massif	Harzburgites, dunites, pyroxenites, chromitites	(*)2060-2050; (*)1900-1800; 600	Batanova et al. [2009]
Urals, Russia	Volkov massif	Medium-grained gabbro Olivine gabbro	1824 (1); 1698 (1); 603-336 (32); 1693 (1); 1498 (1)	Krasnobaev et al. [2007]
Urals, Russia	Volkov massif	Olivine gabbro	2700-340	Anikina et al. [2009]
Urals, Russia	Kytlym massif	Dunites, clinopyroxenites	2800-2600 (4); 1494 (1)	Knauf [2008]
Urals, Russia	Sakharinsk massif	Dunites	1687-1517; 378-374 (38)	Fershtater et al. [2009]

(Continued)

Table 10.5 Continued.

Location	Massif (complexes, province)	Rocks	Age populations, Ma (number of determinations)	References
1	*2*	*3*	*4*	*5*
Urals, Russia	East Khabarninsk massif	Dunites	2808-1911 (3); 407-402;	Fershtater et al. [2009]
		Gabbronorites	1554-1343 (4)	
Urals, Russia	Kos'va massif	Dunites	435-432	Fershtater et al. [2009]
Urals, Russia	Nizhnii Tagil massif	Dunites	2852-2656 (7); 1608 (1); 585 (1)	Malitch et al. [2009]
Urals, Russia	Nizhnii Tagil massif	Dunites, gabbroids	2852-2656; 1608; 585	Efimov [2010]
Urals, Russia	Klyuchevskoy massif	Olivine pyroxenites	~1700 (1); 490-390 (20)	Ivanov et al. [2012]
Urals, Russia	Sakharinsk massif	Dunites	1687-1517 (3); 378-320 (6)	Krasnobaev et al. [2009]
Urals, Russia	South Khabarninsk massif	Dunites	2808-1911 (2); 461-326 (9)	Krasnobaev et al. [2009]
Middle Urals, Russia	Sarana massif	Apodunite serpentinites	1794-1559 (13); 472-305 (14)	Krasnobaev et al. [2013]
Koryakia, Russia	Kuyul massif	Gabbro-pegmatite	181-136 (8)	Ledneva and Matukov [2009]
Koryakia, Russia	Gal'moenan massif	Dunites, clinopyroxenites	790 (1)	Knauf [2008]
Kamchatka. Peninsula, Cape Kamchatka, Russia	Olenegorsk massif, ophiolite association	Gabbro	2721; 2631; 2306; 1900; 1518; 1498; 64	Tsukanov and Skolotnev [2010]
Chukchi Peninsula, Russia	Ust'-Bel'sk massif	Amphibole gabbro	819-775; 361; 317-312	Ledneva et al. [2012]
Western Chukchi Peninsula, Russia	Aluchin massif	Gabbro	2698 (1); 1838 (1); 1794 (1); 283-273 (4); 91-89 (4)	Ganelin et al. [2012, 2013]
Aldan Shield, Russia	Konder massif	Dunite	1889-1860 (4); 1026-1009 (3); 399.9-383.5 (2); 129.6-123.2 (3)	Ronkin et al. [2013]
Aldan Shield, Russia	Konder massif	Dunites	2473; 1885; 176; 143 (total 24 grains)	Badanina and Malich [2012]
Mid-Atlantic Ridge	Kane fracture zone	Gabbro, olivine gabbro	1623; 330	Pilot et al. [1998]
Mid-Atlantic Ridge	Area of Sierra Leone fault, Markov depression	Troctolites, gabbronorites	3126-3094 (3); 2855-2689 (5); 1843-1116 (7); 728-259 (9)	Bortnikov et al. [2008]

(Continued)

Table 10.5 Continued.

Location	Massif (complexes, province)	Rocks	Age populations, Ma (number of determinations)	References
1	*2*	*3*	*4*	*5*
Mid-Atlantic Ridge	Central Atlantic	Ultramafic rocks, gabbroids, diorites, plagiogranites, basaltoids	2715 (4); 2470 (4); 2250 (4); 2070 (9); 1850 (37); 1650 (4); 420 (4)	Skolotnev et al. [2009]
Mid-Atlantic Ridge	Doldrums, Sierra Leone, and Vernadsky fracture zones	Lherzolites Gabbronorites Gabbronorites Ore gabbro Gabbro	2238-1137 (10); 2409-1856 (5); 2712-100 (24); 2359-1887 (7); 1409 (4)	Skolotnev et al. [2010]
Mid-Atlantic Ridge; Arctic Ocean	Mendeleev Rise	Ultramafic and mafic rocks	2700-2500; 2000-1500; 1200-1000; 600-400; 200-100; 30-10	Kremenetskii and Gromalova [2013]

Note: Age of zircon was determined by the U–Pb isotope method. (*) Age of the rocks was determined by the Re–Os isotope method.

crystallized exactly from the mafic melt which formed the bulk of the gabbroid intrusion. Therefore, they are regarded as *a syngenetic phase*. According to isotopic age determinations, most of the analyzed zircon grains indicate that the gabbroid intrusion breaking through the protrusion of ultramafic restites which jointly form the Berezovka massif is dated at 170-150 Ma, i.e, in the Middle–Late Jurassic.

Crystals dated at 99-65 Ma (Late Cretaceous) and 30–20 Ma (Oligocene–Miocene), which are relatively sparse in the rocks of the Berezovka massif, are also well-faceted and have moderate to intense CL and fine, regular oscillatory zoning. Probably, these zircons represent the *epigenetic phase* with respect to the host rocks and crystallized during the infiltration of heterogeneous fluids that were separated from mafic or felsic melts generated at the final stages of magmatic activity within the East Sakhalin Mts.

* * *

The results of the performed isotope-geochronological, optical, morphological, and geochemical studies of zircons from rocks of the Berezovka massif permitted the following conclusions:

1   Ages of zircons in the massif and in the particular rock samples vary from the Archean and Proterozoic to Jurassic, Late Cretaceous, and Miocene;

2   The entire studied zircon collection, numbering as many as 190 grains, is classified into four genetic types: relict, xenogenous, syngenetic, and epigenetic; zircons from each differ in age, size, morphology, oscillatory zoning, CL intensity, and distribution of impurity elements;

3   By morphology, the grains are divided into short-prismatic crystals with well-developed faces, edges, and pyramidal apices; long-prismatic crystals with well-developed faces, edges, and pyramidal apices; prismatic crystals with slightly resorbed faces and edges; prismatic crystals with strongly resorbed faces and edges; and intensely resorbed ovoid grains totally or almost totally lacking crystal faces;

4   CL intensity and oscillatory zoning of most zircon grains decrease from youngest to intermediate and oldest populations;

5   The oscillatory zoning of zircons is caused by a regular variability in the concentrations of luminescence agents, including Hf, Ce, Yb, and U;

6   A negative correlation is revealed between the isotopic age of zircons calculated from $^{206}Pb/^{238}U$, on the one hand, and the contents of La, Ce, Yb, total REE, $(Ce/Ce^*)_n$, and $(Eu/Eu^*)_n$, on the other;

7   Most of REE and multielement patterns of zircons are almost identical in shape common to magmatic zircons; the REE patterns of some zircon grains are of anomalous shape due to epigenetic redistribution of impurity elements, primarily LREE, and their nonuniform accumulation in the intragrain microcracks, in the form of nonstructural impurities;

8   A wide scatter of the isotopic ages of relict and xenogenous zircons is probably due to the nonuniform "rejuvenation" of the U–Pb isotope systems of their oldest varieties, which were present in the primary upper-mantle substratum;

9   The "rejuvenation" of ancient zircon grains was caused by the action of the mafic melt which later formed the gabbroid intrusion;

10   The isotopic age of most of the analyzed zircons is estimated at 170-150 Ma (Middle–Late Jurassic), which corresponds to the time of formation of the gabbroid intrusion of the Berezovka massif;

11   Results of isotope-geochronological and geochemical studies of zircons from rocks of the Berezovka massif agree with the earlier model of polygenetic formation of the massif, based on the geological and petrographic data;

12   Studies of isotope geochronology of zircons from rocks of the Berezovka massif have shown that the dating of ultramafic rocks and gabbroids and that of the massif composed of them, based on a small number of zircon grain collections, cannot yield reliable data.

Chapter 11

The chromitite occurrences in mafic-ultramafic massifs of Sakhalin Island

Occurrences and deposits of massive and disseminated chromitites, differing in size, location, body structure, and chemical composition of ores, are known in many mafic-ultramafic massifs of ophiolite associations. They are also present in massifs located in the folded structures of the Urals, Kazakhstan, West and East Sayan, Koryakia, Chukchi Region, Mongolia, Greece, Turkey, Cyprus, Albania, Bulgaria, New Caledonia, Cuba, etc. [Büchl *et al.*, 2004; Kocks *et al.*, 2007; Saveliev *et al.*, 2008; Zaccarini *et al.*, 2009]. Research in chromitite deposits and assessment of their platinum potential are still topical problems, particularly in Russia suffering from a scarcity of this type of mineral resources [Dodin *et al.*, 2003].

Mafic-ultramafic massifs of Sakhalin Island host several chromitite occurrences, including primary ones. The largest occurrences are localized within the Berezovka and South Schmidt massifs.

11.1 CHROMITITE OCCURRENCES IN THE BEREZOVKA AND SOUTH SCHMIDT MASSIFS

There are at least seven primary and diluvial-eluvial occurrences of massive and densely and scarcely disseminated chromitites within the Berezovka massif, most of which are localized in the upper reaches of the Geran' and Berezovka Rivers [Slodkevich and Lesnov, 1976; Lesnov and Agafonov, 1976; Danchenko, 2003; Explanatory note ..., 2009]. Chromitite bedrock exposures in the south of the massif are schlieren-like segregations and thin veins occurring among serpentinous dunites. Veined chromitite bodies are 2–100 cm in visible thickness and up to 10 m in length. Chromitites of the occurrence in the middle part of the massif form several subparallel lenticular bodies up to 0.2 m in thickness and 0.5–3.0 m in length. The content of Cr-spinel is 10–15% in densely disseminated chromitites and 90–95% in massive varieties, and the mineral grains measure 0.1–4 mm. The Cr-spinel interstices are usually filled with grains of serpentinous olivine. The crush zones crossing some veined chromitite bodies contain occasional silver microparticles ≤0.1 mm in size [Explanatory note ..., 2009]. The chromitites also host scarce grains of erlichmanite-laurite, which will be described in Chapter 12.

Larger chromitite occurrences were revealed in the South Schmidt massif. According to the information published at http://mpr.admsasakhalin.ru/uploads/files/PoleznIsk2.doc, these occurrences were found in 1908 and were intermittently studied till 1998. At present, two chromitite fields are known in the massif: Severo-Tominskoe (five occurrences), in the central part of the massif, and Yuzhno-Tominskoe (seven occurrences), in the southern part. The predicted resources of commercial chromite ores in the South Schmidt massif are estimated at 20 mln tons.

Chromitite clastics were discovered in a gold placer in the basin of the Derbysh Brook (a right tributary of the Langeri River) [Danchenko, 2003]. Their orebody is probably an ultramafic massif buried beneath detritus on the southern flank of the N–NW striking fault zone including the Berezovka, Shel'ting, and Komsomol'sk massifs. The above-linked site has information about discovered gold placers with up to 2.8 kg/m^3 Cr-spinel on the coast of Sakhalin Island, in the areas of Capes Obryvistyi, Shel'ting, and Terpeniya.

11.2 CHEMICAL COMPOSITION OF CHROMITITES

The first data on the chemical composition of chromitites of the Berezovka massif were presented by V.T. Sheiko *et al.* [Explanatory note …, 2009]. According to these data, the average content of Cr_2O_3 in these rocks was 48.6 wt.%. Later analyses of two samples from this massif revealed $Cr_2O_3 = 26.97$ wt.% in densely disseminated chromitites and $Cr_2O_3 = 43.93$ wt.% in massive chromitites (Table 11.1). They also

Table 11.1 Chemical composition of densely disseminated and massive chromitites from the Berezovka massif, wt.%.

	Densely disseminated chromitite	Massive chromitite
	Sample	
Component	1434-1	1434-2
SiO_2	21.20	4.30
TiO_2	0.06	0.25
Al_2O_3	5.27	14.28
Cr_2O_3	26.97	43.93
Fe_2O_3	3.96	7.12
FeO	9.54	14.31
MnO	0.03	0.08
MgO	25.98	14.97
CaO	0.12	0.12
P_2O_5	0.03	0.02
Na_2O	0.01	0.03
K_2O	0.05	0.05
Li_2O	Traces	Traces
V_2O_5	0.10	0.21
NiO	0.21	0.14
CoO	0.032	0.045
H_2O	0.7	0.3
LOI	5.5	N.f.
Total	99.76	100.16

Note: Analyses were carried out by a wet chemical technique at the Institute of Geology and Geophysics, Novosibirsk (analyst E.S. Guletskaya) N.f. – not found (after [Slodkevich and Lesnov, 1976]).

showed that the massive chromitites had higher contents of TiO_2, Al_2O_3, Fe_2O_3, FeO, MnO, Na_2O, V_2O_5, and CoO and lower contents of SiO_2, MgO, P_2O_5, and NiO than the densely disseminated chromitites. Later, an ICP-AES analysis of three more samples of massive chromitites of the Berezovka massif detected somewhat higher contents of Cr_2O_3 and somewhat lower contents of Al_2O_3 and MgO than those in the samples analyzed by other methods (Table 11.2).

The Sakhalin Prospecting and Exploration Expedition performed a large-scale sampling of chromitites at their deposit in the South Schmidt massif. The results showed $Cr_2O_3 = 28$–47 wt.% in the massive chromitite ores and $Cr_2O_3 = 11$–30 wt.% in the disseminated ores from the Severo-Tominskoe field. The massive ores of the Yuzhno-Tominskoe field contain $Cr_2O_3 = 48.9$–54.6 wt.%, and the disseminated ores contain $Cr_2O_3 = 11$–15 wt.%. Chromitites from the major ore shoot of the Yuzhno-Tominskoe field, containing, on the average, 49.3 wt.% Cr_2O_3, are assigned to high-grade metallic ores.

The chemical composition of massive and densely disseminated chromitites of the South Schmidt massif was estimated from the results of analysis of eight ore lump samples (Table 11.3). Two samples contained $Cr_2O_3 = 27.5$ and 27.8 wt.%, which is close to its content in the densely disseminated chromitites of the Berezovka massif. In the other six samples (massive chromitites) the Cr_2O_3 content varied from 33.7 to 49.6 wt.%, which is, on the average, somewhat higher than that in the massive chromitites of the Berezovka massif. The analyses also showed $Al_2O_3 = 4.47$–13.93 wt.% in the massive chromitites and $Al_2O_3 = 6.03$–33.03 wt.% in the densely disseminated ores

Table 11.2 Contents of major components and trace elements in massive chromitites from the Berezovka massif.

Major components and trace elements	Sample		
	Ev-2	Ev-3	1051B/1962
SiO_2, wt.%	2.83	2.40	1.80
TiO_2	0.070	0.074	0.215
Al_2O_3	10.1	9.5	10.6
Cr_2O_3	53.1	55.0	50.0
Fe_2O_3	18.0	15.5	20.3
MnO	0.154	0.141	0.156
MgO	13.5	12.1	10.3
CaO	0.049	0.050	0.030
Ba, ppm	8.8	5.7	3.0
Co	270	226	249
Cu	142	135	115
Li	123	<3	70
Ni	498	679	721
Sc	0.76	0.90	1.02
Sr	3.3	4.5	4.1
V	505	490	1169
Zn	239	263	371

Note: Analyses were carried out at the Analytical Center of the Institute of Geology and Mineralogy, Novosibirsk, by ICP-AES method using spectrometer IRIS Advantage (Intertex Corporation, USA) (analyst S.F. Nechepurenko).

Table 11.3 Chemical composition of densely disseminated and massive chromitites from the South Schmidt massif, wt.%.

Component	Densely disseminated chromitites		Massive chromitites					
	Sample							
	1475	1474	1475-1	1475-2	1475-3	12-22	102-28	35-56
SiO_2	21.50	5.35	7.60	6.90	10.40	7.82	3.5	2.61
TiO_2	0.20	0.13	0.38	0.25	0.28	N.d.	N.d.	N.d.
Al_2O_3	6.03	33.03	9.52	10.57	13.83	5.10	4.97	4.46
Cr_2O_3	27.48	27.75	48.30	49.60	38.79	33.7	48.3	48.9
Fe_2O_3	1.20	0.85	2.58	6.35	3.06	6.08	5.26	5.98
FeO	7.52	11.42	9.25	5.64	8.82	0.47	0.40	0.45
MnO	0.11	0.03	0.06	0.08	0.14	N.d.	N.d.	N.d.
MgO	28.03	19.83	18.98	18.97	20.85	11.00	6.10	4.30
CaO	N.f.	0.36	Traces	Traces	0.83	0.14	0.42	N.d.
P_2O_5	0.06	0.04	0.07	0.08	0.03	N.d.	N.d.	N.d.
Na_2O	0.01	0.03	0.03	0.03	0.03	N.d.	N.d.	N.d.
K_2O	0.03	0.05	0.06	0.04	0.05	N.d.	N.d.	N.d.
Li_2O	Traces	Traces	Traces	Traces	Traces	N.d.	N.d.	N.d.
V_2O_5	0.10	0.15	0.16	Traces	0.10	N.d.	N.d.	N.d.
NiO	0.34	0.21	0.21	0.21	0.24	N.d.	N.d.	N.d.
CoO	0.027	0.032	0.032	0.032	0.031	N.d.	N.d.	N.d.
H_2O	0.3	0.2	0.2	0.1	0.3	N.d.	N.d.	N.d.
LOI	7.2	1.1	2.3	1.7	2.9	N.d.	N.d.	N.d.
Total	100.14	100.56	99.73	100.55	100.68	N.d.	N.d.	N.d.

Note: Analyses were carried out at the Institute of Geology and Geophysics, Novosibirsk by a wet chemical technique (analyst E.S. Guletskaya). N.d. – no data. N.f. – not found (after [Lesnov and Agafonov, 1976]).

of the South Schmidt massif. Note that the massive chromitites from clastics found in the Derbysh Brook gold placer have $Cr_2O_3 = 28.9–41.0$ wt.% [Danchenko, 2003]. According to Danchenko's data, the massive chromitites from ore occurrences of the Sakhalin mafic-ultramafic massifs are generally of medium-Cr_2O_3 type.

* * *

There are several occurrences of massive and densely disseminated chromitites, differing in size, within the Berezovka and South Schmidt mafite-ultramafite massifs. According to preliminary assessments, the chromitite occurrences of the South Schmidt massif are considered medium-size deposits. The orebody of massive chromitites found as clastics in the Derbysh Brook gold placer is probably a mafic-ultramafic massif buried beneath the alluvium. Massive chromitites of the Berezovka massif have $Cr_2O_3 = 44–55$ wt.%, and disseminated chromitites have $Cr_2O_3 \approx 27$ wt.%. Massive and disseminated chromitites of the South Schmidt massif contain $Cr_2O_3 = 28.0–54.6$ wt.% and 11–27 wt.%, respectively. Chromitites from the major ore shoot of the Yuzhno-Tominskoe field in the South Schmidt massif, with average $Cr_2O_3 = 49.3$ wt.%, are assigned to high-grade metallic ores.

Chapter 12

Geochemistry of platinum group elements and rhenium in rocks and chromitites of mafic-ultramafic massifs of Sakhalin Island

The results of study of the platinum potential of mafic-ultramafic massifs and their complexes in different world regions, including Russia, were summarized by Dodin *et al.* [2001]. Later, the platinum potential of chromite deposits of different genetic types was considered, and a scheme of assumed Pt-metallogenic regionalization of Russia was proposed [Dodin *et al.*, 2003]. Sakhalin Island is also among the Pt-promising areas.

The platinum potential of Sakhalin has been poorly studied. There is only fragmentary information about this problem, in particular, summarized data on the abundance of PGE minerals in slime aureoles and gold placers of the island [Danchenko, 2003]. In some of its regions, grains of rutheniridosmine, ruthenosmiridium, osmiridium, iridosmine, Pt-iridosmine, laurite, and, more seldom, polyxene, ferroplatinum, and isoferroplatinum were found in slimes. Until the present time, research into geochemistry of PGE in rocks and chromitites of the Sakhalin mafic-ultramafic massifs has not been performed. This gap is partly closed with our first data on PGE and Re contents in some rock varieties and chromitites from these massifs.

12.1 GEOCHEMISTRY OF PGE AND RE IN ROCKS

The geochemistry of PGE and Re was studied by ICP-MS on a collection of ultramafite and gabbroid samples of the Berezovka and, to a lesser extent, Shel'ting, Komsomol'sk, and South Schmidt massifs. This collection included samples of harzburgites, lherzolites, plagiowehrlites, websterites, olivine websterites, orthopyroxenites, olivine gabbronorites, and anorthosites (Table 12.1). Since PGE minerals hardly pass into the analytical solutions, two techniques of different efficiency were used for acid digestion of weighted specimens: (1) digestion in a MARS-5 autoclave at <200°C with the following determination of PGE (except for Os) and Re contents [Palesskii *et al.*, 2009, 2010] and (2) digestion in a Carius tube at 260, 280, and 300°C with the following determination of PGE (including Os) and Re contents [Koz'menko *et al.*, 2011]. The analytical results showed that the second digestion technique ensured the more complete transition of PGE into the solution and thus the more accurate estimation of their contents in the samples.

Table 12.1 Contents of PGE and Re and their ratios in rocks from the Berezovka, Shel'ting, Komsomol'sk, and South Schmidt massifs, ppb.

Element and element ratio	Berezovka massif						
	Harzburgites			Lherzolites		Plagiowehrlites	
	Sample						
	1610-2	1610-2*	1610-3	145	1596	142	148-1a
Os	1.13	1.08	1.20	0.41	0.65	1.02	0.49
Ir	1.28	1.08	1.24	0.68	0.50	1.48	0.44
Ru	5.73	5.03	6.41	1.47	4.65	4.45	0.64
Rh	2.31	2.08	1.68	0.89	0.56	1.11	1.25
Pt	13.0	11.2	9.99	8.10	1.40	6.11	5.09
Pd	N.d.	25.5	7.19	0.96	0.79	12.3	3.44
Total (with Os)	23.45	45.97	27.71	12.51	8.55	26.47	11.40
Total (without Os)	N.d.	44.8	26.7	12.2	7.9	25.3	10.9
Re	0.15	0.11	0.17	0.04	0.09	0.02	0.01
Ir/Os	1.13	1.00	1.03	1.66	0.77	1.45	0.90
Pd/Ir	Н.д.	23.6	5.80	1.40	1.60	8.30	7.80
Pt/Ir	10.2	10.4	8.06	11.9	2.80	4.13	11.6
Ru/Ir	4.48	4.66	5.17	2.16	9.30	3.01	1.45
Ru/Rh	2.48	2.42	3.82	1.65	8.30	4.01	0.51
Pd/Pt	N.d.	2.28	0.72	0.12	0.56	2.01	2.28
Ir/Re	8.53	10.3	7.29	17.0	5.56	73.0	44.0
Pt/Re	87	102	59	203	15.6	306	509
Sample digestion technique	Carius tube						

Table 12.1 Continued.

Element or element ratio	Berezovka massif												
	Olivine websterites		Websterites		Olivine clinopyroxenites		Olivine gabbro	Gabbronorites			Gabbro		Anorthosite
	Sample												
	1593	1603-1	1605	143	1606-1	Skh-2	155-5	131	131a	1607	1612	Skh-1	148-1b
Os	N.d.	N.d.	N.d.	N.d.	N.d.	N.d.	0.14	N.d.	N.d.	N.d.	N.d.	N.d.	0.46
Ir	0.088	0.061	0.067	0.080	0.093	0.048	0.10	0.074	0.074	0.099	0.045	0.095	0.45
Ru	0.31	0.36	0.013	0.076	0.22	0.15	0.05	0.014	0.018	0.067	0.25	0.15	1.20
Rh	0.11	0.33	0.40	0.51	0.43	0.63	0.62	0.75	0.74	1.10	0.35	1.4	1.25
Pt	0.44	5.7	3.1	7.1	3.3	2.6	19.7	7.0	7.0	48	3.4	81	4.85
Pd	0.37	1.8	4.3	1.6	2.6	0.78	1.5	32	49	22	60	55	10.9
Total (with Os)	N.d.	N.d.	N.d.	N.d.	N.d.	N.d.	22.11	N.d.	N.d.	N.d.	N.d.	N.d.	19.11
Total (without Os)	N.d.	N.d.	N.d.	N.d.	N.d.	N.d.	22.1	N.d.	N.d.	N.d.	N.d.	N.d.	18.7
Re	0.072	0.16	0.13	0.063	0.088	0.048	0.02	0.14	0.14	0.14	0.15	0.097	0.04
Ir/Os	N.d.	N.d.	N.d.	N.d.	N.d.	N.d.	0.70	N.d.	N.d.	N.d.	N.d.	N.d.	0.98
Pd/Ir	4.20	29.5	64	20.1	28.0	16.3	15.0	432	662	222	1333	579	24.2
Pt/Ir	5.00	93	46.3	88.8	35.5	54.2	197	95	95	485	75.6	853	10.8
Ru/Ir	3.52	5.90	0.19	1.0	2.37	3.13	0.50	0.19	0.24	0.68	5.56	1.58	2.67
Ru/Rh	2.82	1.09	0.03	0.1	0.51	0.24	0.08	0.02	0.02	0.06	0.71	0.11	0.96
Pd/Pt	0.84	0.32	1.39	0.23	0.79	0.30	0.08	4.57	7.00	0.46	0.057	1.47	2.25
Ir/Re	1.22	0.38	0.52	1.27	1.16	1.00	5.00	0.53	0.53	0.71	0.30	0.98	11.3
Pt/Re	6.11	35.6	23.8	113	37.5	54.2	985	50.0	50.0	343	22.7	835	121
Sample digestion technique	MARS-5 autoclave						Carius tube	MARS-5 autoclave					Carius tube

(Continued)

Table 12.1 Continued.

Element or element ratio	Shel'ting massif			Komsomol'sk massif	South Schmidt massif			Reference samples			
	Enstatitite	Gabbronorites		Olivine gabbronorite	Lherzolites		Enstatitite	Geochemical standard		Primitive mantle	Chondrite
	Sample										
	174	182	183	191	161	163	162	"GP-13"	GP-13*	PM	CI
Os	N.d.	N.d.	N.d.	N.d.	2.32	5.13	N.d.	3.85	3.77	3.4	490
Ir	0.016	0.091	0.029	0.38	2.09	3.87	0.043	3.38	3.58	3.2	455
Ru	0.045	0.070	0.099	0.92	4.53	9.27	0.87	6.89	6.85	5.0	710
Rh	N.d.	N.d.	N.d.	N.d.	1.0	1.94	0.42	1.25	1.25	0.9	130
Pt	1.34	6.91	2.70	8.77	5.26	8.54	0.4	6.86	6.90	7.1	1010
Pd	0.14	0.29	0.39	13.7	7.05	9.75	0.65	6.71	6.40	3.9	550
Total (with Os)	N.d.	N.d.	N.d.	N.d.	22.25	38.50	N.d.	28.94	28.75	23.50	3345
Total (without Os)	N.d.	N.d.	N.d.	N.d.	20.2	33.7	2.38	25.4	25.3	20.4	N.d.
Re	0.01	0.04	0.07	0.41	0.24	0.33	0.043	0.31	0.32	0.28	40
Ir/Os	N.d.	N.d.	N.d.	N.d.	0.90	0.75	N.d.	0.88	0.95	0.94	0.93
Pd/Ir	8.60	3.20	13.3	35.6	3.40	2.50	15.1	2.00	1.80	1.20	1.20
Pt/Ir	84	76	93	23.1	2.52	2.21	9.3	2.00	1.93	2.22	2.22
Ru/Ir	2.81	0.77	3.41	2.42	2.17	2.40	20.2	2.00	1.91	1.56	1.56
Ru/Rh	N.d.	N.d.	N.d.	N.d.	4.53	4.78	2.07	5.51	5.48	5.56	5.46
Pd/Pt	0.10	0.04	0.14	1.56	1.34	1.14	1.63	0.88	0.95	0.94	0.93
Ir/Re	1.60	2.07	0.42	0.94	8.71	11.7	1.0	10.9	11.2	11.4	11.4
Pt/Re	134	157	39.1	21.7	21.9	25.4	9.3	22.1	21.6	25.4	25.3
Sample digestion technique	MARS-5 autoclave	MARS-5 autoclave			Carius tube		MARS-5 autoclave	Carius tube	N.d.	N.d.	N.d.

Note: Analyses were carried out by ICP-MS method at the Analytical Center of the Institute of Geology and Mineralogy, Novosibirsk (analysts I.V. Nikolaeva, S.V. Palesskii, and O.A. Koz'menko). 1610-2 and 1610-2* – analyses of two weighted specimens of the same sample; 1610-3 – apoharzburgite serpentinite. "GP-13" – analysis of the geochemical standard GP-13 carried out along with analyses of rocks; GP-13* – recommended composition of the geochemical standard, after Wittig et al. [2010]. PM – primitive mantle. CI chondrite (after [McDonough and Sun, 1995]. The contents of Ir in the samples digested in a MARS-5 autoclave might be slightly underestimated because of the incomplete digestion.

12.1.1   Peculiarities of PGE and Re distribution in rocks of different compositions

In the rock samples digested in a MARS-5 autoclave (without analysis for Os), the total content of PGE varied from 7.9 to 44.8 ppb. In the samples digested in a Carius tube (with analysis for all PGE), the total content of PGE varied from 8.63 to 46.0 ppb. Note that the total contents of PGE in restitic ultramafites of the Berezovka and South Schmidt massifs are generally close to those in the primitive mantle (23.5 ppb) [McDonough and Sun, 1995]. Below, we consider the distribution of each platinum group element in rocks of different petrographic types.

Osmium. The contents of this element in the studied samples vary from 0.14 to 5.13 ppb. The highest contents were found in lherzolites of the South Schmidt massif, and somewhat lower ones, in harzburgites, lherzolites, plagiowehrlites, olivine gabbronorites, and anorthosites of the Berezovka massif. For comparison, weakly depleted lherzolites in mantle xenoliths from alkali basalts contain 1.7–4.9 ppb Os; moderately depleted varieties, 2.9–4.4 ppb Os; and strongly depleted lherzolites, 1.45–6.1 ppb Os [Morgan, 1986]. Nearly the same content of Os (3.4 ppb) was established in the primitive mantle [McDonough and Sun, 1995].

Iridium. The content of this element in the studied samples is within 0.02–3.87 ppb. The maximum content, close to that in the primitive mantle (3.2 ppb), was found in lherzolites of the South Schmidt massif. The contents of Ir are somewhat lower in harzburgites and plagiowehrlites of the Berezovka massif and still lower in pyroxenites and gabbroids of the Berezovka, Shel'ting, and Komsomol'sk massifs. Weakly depleted lherzolites in mantle xenoliths from alkali basalts contain 2.9–6.3 ppb Ir; moderately depleted varieties, 3.5–4.8 ppb Ir; and strongly depleted lherzolites, 1.69–3.80 ppb Ir [Morgan, 1986].

Ruthenium. Elevated contents of this element were found in lherzolites of the South Schmidt massif (4.5 and 9.3 ppb) and in harzburgites (5.7 and 6.4 ppb) and lherzolites (4.7 ppb) of the Berezovka massif, and lower contents were established in pyroxenites (0.013–0.920 ppb) and some gabbroids of the Berezovka, Shel'ting, and Komsomol'sk massifs (0.014–1.200 ppb). Ultramafic rocks of the Sakhalin massifs are similar in Ru contents to the primitive mantle (5 ppb) [McDonough, Sun, 1995]. Restitic ultramafic rocks of the South Schmidt massif are somewhat richer in refractory PGE (Os, Ir, and Ru) than the same rocks of the Berezovka massif.

Rhodium. Its content decreases in the series from harzburgites of the Berezovka massif (1.7–2.3 ppb) to lherzolites of this and South Schmidt massifs (0.6–1.0 ppb) and then to plagiowehrlites (1.1–1.3 ppb) and olivine gabbronorites of the Berezovka massif (0.6 ppb). Ultramafic rocks of both massifs are somewhat richer in Rh than the primitive mantle (0.9 ppb) [McDonough and Sun, 1995].

Platinum. Elevated contents of this element were found in ultramafic rocks of the Berezovka (5.19–13.0 ppb) and South Schmidt (5.26–8.54 ppb) massifs and in olivine and olivine-free gabbroids of the Berezovka (3.4–81.0 ppb), Komsomol'sk (8.77 ppb), and Shel'ting (6.91 ppb) massifs. Lower contents of Pt were established in websterites and olivine clinopyroxenites of the Berezovka massif (0.44–2.6 ppb) and in enstatitites of the South Schmidt (0.40 ppb) and Shel'ting (1.34 ppb) massifs. In general, the contents of Pt in ultramafic rocks of the studied massifs are close to those in the primitive mantle (7.1 ppb) [McDonough and Sun, 1995].

Palladium. The content of this element in the studied rocks varies from 0.14 to 60 ppb. Gabbronorites and gabbro (22–60 ppb) of the Berezovka massif are the richest in Pd. Somewhat lower contents of Pd were found in harzburgites and plagiowehrlites of the same massif (3.44–12.3 ppb), in lherzolites of the South Schmidt massif (7.05–9.75 ppb), and in olivine gabbronorites of the Komsomol'sk massif (13.7 ppb). The lowest contents of Pd were established in websterites and clinopyroxenites of the Berezovka massif (0.37–1.8 ppb), in enstatitites and gabbronorites of the Shel'ting massif (0.14–0.39 ppb), and in enstatitites of the South Schmidt massif (0.65 ppb). Restitic ultramafic rocks of all these massifs are somewhat richer in Pd than the primitive mantle (3.9 ppb) [McDonough and Sun, 1995].

Rhenium. The content of this element increases in the series from plagiowehrlites, olivine gabbronorites, and anorthosites of the Berezovka massif (0.01 ppb) to harzburgites and lherzolites of the same massif, lherzolites of the South Schmidt massif (0.33 ppb), and, finally, olivine gabbro of the Komsomol'sk massif (0.41 ppb). Some regularities of Re distribution in different rocks of the mafic-ultramafic massifs were described earlier by Lesnov and Anoshin [2011].

Thus, the obtained data on PGE distribution in rocks of the Sakhalin mafic-ultramafic massifs evidence that refractory PGE (Os, Ir, and Ru) are concentrated mainly in restitic ultramafites of the South Schmidt massif and, to a lesser extent, in ultramafites of the Berezovka and other massifs. Easily fusible PGE (Rh, Pt, and Pd) are present in nearly equal contents in ultramafic rocks and gabbroids of the studied massifs.

12.1.2   Primitive mantle normalized PGE and Re contents patterns

The primitive mantle normalized PGE and Re patterns of rocks of the studied Sakhalin massifs are of different shapes; most of them are positively sloped (Fig. 12.1). The PGE patterns of harzburgites of the Berezovka massif show somewhat higher contents of these elements as compared with lherzolites (Fig. 12.1, 1, 2). The PGE patterns of plagiowehrlites of the same massif are similar in shape and the position of points to those of harzburgites and lherzolites. Both patterns of plagiowehrlites show an intense negative anomaly of Re, and one of them demonstrates a positive Pd anomaly (Fig. 12.1, 3). The steep positive slope of the patterns of websterites and olivine clinopyroxenites of the Berezovka massif indicates intense fractionation of PGE in these hybrid ultramafic rocks and a predominance of easily fusible PGE (Fig. 12.1, 5, 6). The steep positive slope of the patterns of gabbroids of the Berezovka, Shel'ting, and Komsomol'sk massifs testifies to a predominance of easily fusible PGE and a relative depletion in Re (Fig. 12.1). Lherzolites of the South Schmidt massif are similar in the shape and position of points of their PGE patterns to the primitive mantle. At the same time, their elements were not subjected to fractionation (Fig. 12.1, 14).

12.1.3   Parameters of PGE and Re distribution in rocks

Some specific geochemical features of PGE in rocks of the studied massifs are described on the basis of the element ratios and parameters (Table 12.1) depicted in corresponding patterns (Fig. 12.2). The Ir/Os parameter increases from 0.7 in olivine gabbro of the Berezovka massif (in which Ir dominates over Os) to 1.66 in lherzolites of the same

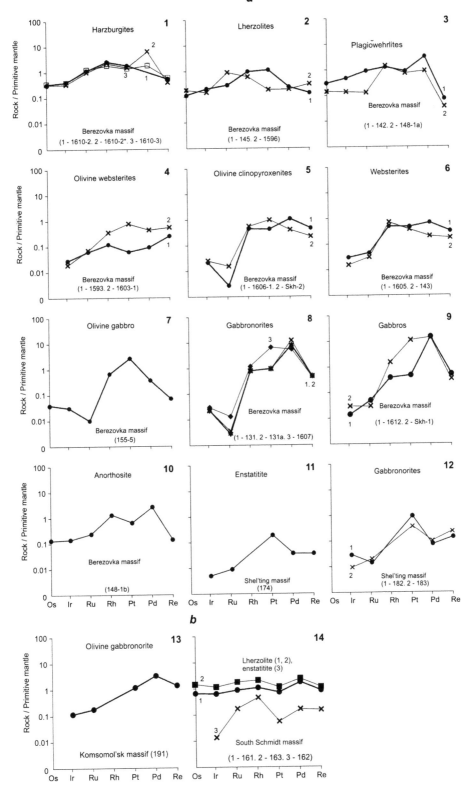

Figure 12.1 a, b. Primitive mantle normalized [McDonough and Sun, 1995] PGE and Re patterns of ultramafic and mafic rocks from: *a* – Berezovka and Shel'ting massifs; *b* – Komsomol'sk and South Schmidt massifs (data from Table 12.1).

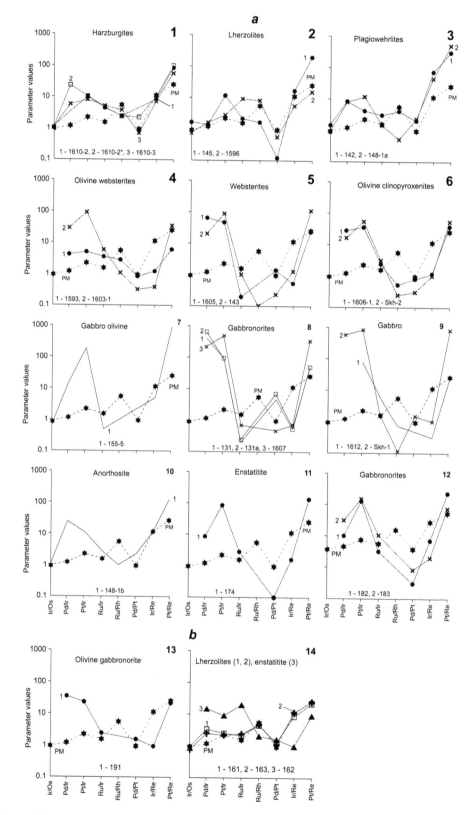

Figure 12.2 *a, b*. Variations in Ir/Os, Pd/Ir, Pt/Ir, Ru/Ir, Ru/Rh, Pd/Pt, Ir/Re, and Pt/Re in ultramafic and mafic rocks from the Berezovka (1–10), Shel'ting (11, 12), Komsomol'sk (13), and South Schmidt (14) massifs and in the primitive mantle (PM) (data from Table 12.1).

massif (in which Os dominates over Ir). The Pd/Ir parameter is generally accepted as an indicator of the degree of PGE fractionation. In the studied rocks it varies over a wide range of values: from 222–1333 in gabbroids (in which PGE are strongly fractionated) to 1.4–5.8 in lherzolites and harzburgites (in which PGE are much less fractionated). In continental and oceanic basalts Pd/Ir is usually higher than 100, whereas in peridotites it is close to 10 [Hulbert, 1997, and Barnes *et al.*, 1985 therein].

The studied rocks also show wide variations in Pt/Ir (2.21–853), Pt/Re (6–985), and Ir/Re (0.3–73). Much lower values and narrower variation ranges were established for Ru/Ir (0.19–20.2), Ru/Rh (0.2–8.3), and Pt/Pd (0.04–7.0). Their variation patterns show some difference in PGE and Re distribution in the studied rocks. The patterns for harzburgites, lherzolites, plagiowehrlites, and pyroxenites of the Berezovka massif and for enstatitites and gabbronorites of the Shel'ting massif are sinusoidal because of the Pd/Ir and Pt/Ir maxima and Ru/Rh and Pd/Pt minima (Fig. 12.2, 1–6).

12.1.4 Relationships between the contents of Ir and other PGE in the rocks

The constructed plots of relationships between the contents of Ir and other PGE show differences between ultramafic rocks and gabbroids of the Berezovka and South Schmidt massifs. On the Os–Ir plot, all composition points of rocks of both massifs, along with the point of the primitive mantle (PM), lie on a positively sloped trend (Fig. 12.3, 1). Probably, all these rocks contain Os and Ir in the same proportions as the primitive mantle. On the Ru–Ir plot, the composition points of restitic lherzolites of the South Schmidt massif are arranged along trend II, and the points of lherzolites and harzburgites of the Berezovka massif are localized on trend III. Both trends are remote from trend I with the point of PM (Fig. 12.3, 2).

On the Rh–Ir plot, the composition points also lie on two trends. The points of restitic lherzolites of the South Schmidt massif are localized on trend II, and the points of restitic and hybrid ultramafic rocks of the Berezovka massif, on trend III. Both trends are remote from trend I with the point of PM (Fig. 12.3, 3).

Difference between ultramafites of the South Schmidt and Berezovka massifs is also observed on the Pt–Ir plot (Fig. 12.3, 4). The points of lherzolites of the South Schmidt massif, point 4 of lherzolite of the Berezovka massif, and the point of PM lie on trend I, whereas the other points of ultramafic rocks of the Berezovka massif (having high Pt and low Ir contents) are localized on trend III. The Pd–Ir plot suggests no distinct relationship between these elements in the studied ultramafites and gabbroids, which might be due to the late redistribution of Pd (Fig. 12.3, 5).

On the Re–Ir plot (Fig. 12.3, 6), the composition points of rocks and PM lie on the same trend. As in the case of the Os–Ir relationship (Fig. 12.3, 1), this indicates that the rocks have the same proportions of Ir and Re as the primitive mantle. A slight difference between ultramafic rocks of the Berezovka and South Schmidt massifs is observed on the Os–Re plot (Fig. 12.3, 7). The points of lherzolites of the South Schmidt massif form positively sloped steep trend II, which intersects with trend I at the point of PM. The points of Os-depleted rocks of the Berezovka massif, on the contrary, form individual gentler trend III.

The Os–Re/Os plot (Fig. 12.3, 8) also shows a difference between the studied rocks. The composition points of lherzolites of the South Schmidt massif are arranged along negatively sloped steep trend I; the point of PM is located in the immediate vicinity of

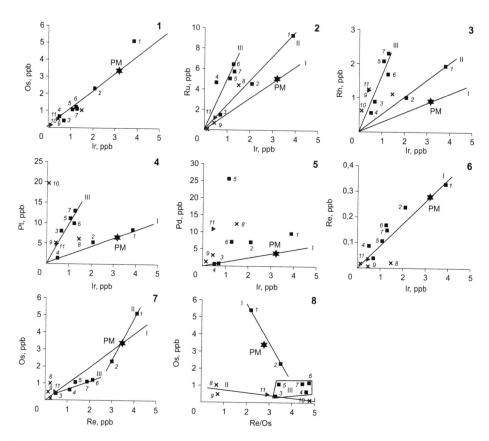

Figure 12.3 Relationships between the contents of PGE and Re in ultramafic and mafic rocks from the South Schmidt massif (1, 2) and Berezovka massif (3–11) (data from Table 12.1). 1–4 – lherzolites (samples 163, 161, 145, and 1596); 5–7 – harzburgites (samples 1610-2, 1610-2*, and 1610-3); 8 – plagiolherzolite (sample 142); 9 – plagiowehrlite (sample 148-1a); 10 – olivine websterite (sample 155-5); 11 – anorthosite (sample 148-1b). Analyses of all samples were performed with sample digestion in a Carius tube.

it. The points of hybrid ultramafic rocks and gabbroids of the Berezovka massif are localized along negatively sloped gentle trend II, and the points of harzburgites and lherzolites of the Berezovka massif lie within field III.

12.1.5   Relationships between the contents of PGE and REE in ultramafic rocks and gabbroids

Earlier studies of small collections of rock samples from mafic-ultramafic massifs showed an inverse relationship between the contents of PGE and REE with contrasting properties [Lesnov, 2009a]. We hypothetically related this to the differently directed fractionation of the above groups of elements during the partial melting of upper-mantle sources and the generation and crystallization of basaltic melts.

This "antagonism" between PGE and REE was confirmed by results of geochemical studies of rocks of the Sakhalin mafic-ultramafic massifs (Table 12.2). The plots

Table 12.2 Contents of PGE (ppb) and REE (ppm) in representative samples of ultramafic rocks and gabbroids from mafic-ultramafic massifs of Sakhalin Island.

Element	Sample 1610-2 / 1	1610-3 / 2	145 / 3	161 / 4	163 / 5	1596 / 6	142 / 7	1593 / 8	1603-1 / 9	143 / 10	1606-1 / 11
La	0.26	0.097	0.16	0.063	0.02	0.8	0.034	0.11	0.23	0.6	0.22
Ce	0.9	0.36	0.29	0.12	0.045	2.0	0.079	0.25	0.62	2.2	0.38
Pr	0.096	0.025	0.065	0.01	0.005	0.22	0.011	0.045	0.069	0.5	0.044
Nd	0.22	0.12	0.4	0.025	0.023	0.97	0.046	0.23	0.49	1.16	0.074
Sm	0.072	0.048	0.2	0.006	0.012	0.21	0.016	0.086	0.19	0.32	0.001
Eu	0.001	0.001	0.058	0.001	0.005	0.034	0.005	0.015	0.051	0.32	0.001
Gd	0.078	0.057	0.25	0.07	0.036	0.24	0.029	0.15	0.3	1.55	0.12
Tb	0.017	0.013	0.04	0.002	0.007	0.042	0.007	0.039	0.063	0.29	0.024
Dy	0.11	0.081	0.29	0.019	0.054	0.24	0.043	0.26	0.43	1.97	0.15
Ho	0.03	0.019	0.071	0.005	0.015	0.056	0.015	0.064	0.11	0.42	0.042
Er	0.091	0.059	0.18	0.016	0.059	0.21	0.045	0.22	0.36	1.29	0.13
Tm	0.012	0.011	0.036	0.004	0.013	0.034	0.009	0.033	0.056	0.21	0.024
Yb	0.077	0.069	0.23	0.033	0.088	0.24	0.063	0.23	0.36	1.3	0.15
Lu	0.008	0.007	0.034	0.008	0.015	0.041	0.01	0.038	0.054	0.2	0.021
Os	1.13	1.2	0.41	2.32	5.13	0.65	1.02	N.d.	N.d.	N.d.	N.d.
Ir	1.28	1.24	0.68	2.09	3.87	0.5	1.48	0.088	0.061	0.08	0.093
Ru	5.73	6.41	1.47	4.53	9.27	4.65	4.45	0.31	0.36	0.076	0.22
Rh	2.31	1.68	0.89	1.0	1.94	0.56	1.11	0.11	0.33	0.51	0.43
Pt	13.0	9.99	8.1	5.26	8.54	1.4	6.11	0.44	5.7	7.1	3.3
Pd	N.d.	7.19	0.96	7.05	9.75	0.79	12.3	0.37	1.8	1.6	2.6

Element	Sample 162 / 12	174 / 13	131 / 14	182 / 15	183 / 16	155-5 / 17	191 / 18	1607 / 19	1612 / 20	Skh-1 / 21	148-1b / 22
La	0.051	0.116	0.28	0.077	0.07	0.94	0.64	0.16	0.19	0.058	0.084
Ce	0.1	0.2	0.44	0.21	0.18	2.0	1.48	0.34	0.56	0.12	0.19
Pr	0.009	0.014	N.d.	0.039	0.025	0.32	0.23	0.026	0.11	0.017	0.023
Nd	0.045	0.061	0.48	0.2	0.18	1.57	1.06	0.13	0.22	0.08	0.092
Sm	0.021	0.031	0.16	0.095	0.098	0.54	0.34	0.036	0.088	0.033	0.039
Eu	0.007	0.007	0.075	0.051	0.057	0.19	0.13	0.001	0.02	0.001	0.047
Gd	0.026	0.033	N.d.	0.13	0.16	0.77	0.48	0.048	0.15	0.067	0.035
Tb	0.004	0.005	0.045	0.027	0.028	0.13	0.095	0.009	0.027	0.018	0.005
Dy	0.028	0.032	N.d.	0.17	0.2	0.86	0.67	0.055	0.19	0.085	0.021
Ho	0.01	0.006	N.d.	0.039	0.045	0.17	0.15	0.015	0.04	0.017	0.005
Er	0.038	0.022	N.d.	0.14	0.13	0.49	0.46	0.037	0.13	0.011	0.002
Tm	0.009	0.004	N.d.	0.024	0.027	0.074	0.069	0.007	0.018	0.067	0.01
Yb	0.059	0.03	0.245	0.16	0.17	0.47	0.45	0.037	0.13	0.067	0.01
Lu	0.011	0.05	0.03	0.027	0.021	0.061	0.069	N.d.	0.014	0.007	0.002
Os	N.d.	N.d.	N.d.	N.d.	N.d.	0.14	N.d.	N.d.	N.d.	N.d.	0.46
Ir	0.043	0.016	0.074	0.091	0.029	0.1	0.38	0.099	0.045	0.095	0.45
Ru	0.87	0.045	0.014	0.07	0.099	0.05	0.92	0.067	0.25	0.15	1.2
Rh	0.42	N.d.	0.75	N.d.	N.d.	0.62	N.d.	1.1	0.35	1.4	1.25
Pt	0.4	1.34	7.0	6.91	2.7	19.7	8.77	48.0	3.4	81	4.85
Pd	0.5	0.14	32.0	0.29	0.39	1.5	13.7	22.0	60.0	55.0	10.9

Note: Compiled after the data in Tables 6.3–6.5 and 12.1. Rocks: 1, 2 – harzburgites; 3–6 – lherzolites; 7 – plagiowehlite; 8, 9 – olivine websterites; 10 – websterite; 11 – olivine clinopyroxenite; 12, 13 – enstatites; 14–16, 19 – gabbronorites; 17 – olivine gabbro; 18 – olivine gabbronorite; 20, 21 – gabbro; 22 – anorthosite. N.d. – No data.

constructed from the table results show that in general, the PGE contents in these ultramafic rocks and gabbroids decrease as the REE contents increase (Fig. 12.4). The coefficients of PGE–REE pair correlation, calculated from the data in Table 12.2, vary from −0.08 to −0.46, not reaching the values statistically significant for the number of performed complex analyses. Nevertheless, the obtained analytical data on

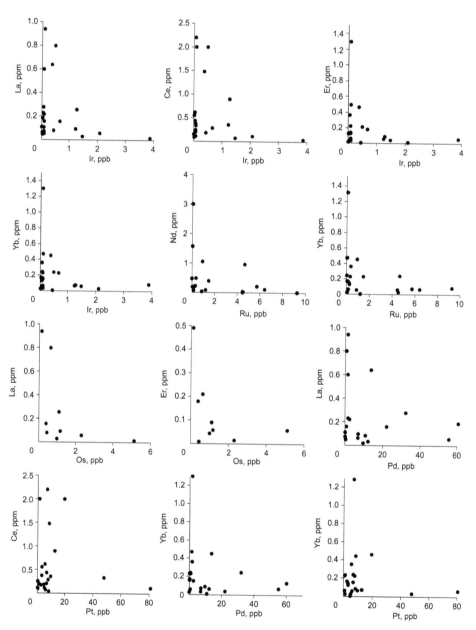

Figure 12.4 Relationships between the contents of PGE and REE in ultramafic rocks and gabbroids from the Berezovka, Shel'ting, Komsomol'sk, and South Schmidt massifs (data from Table 12.2).

the Sakhalin mafic-ultramafic rocks generally agree with the concept of an inverse relationship between PGE and REE contents in such rocks.

12.2   GEOCHEMISTRY OF PGE AND RE IN CHROMITITES

At the first stage of study of PGE geochemistry in two chromitite samples from the Berezovka massif, the contents of Pt and Pd were determined by the assay-spectral and radiochemical neutron activation methods [Slodkevich and Lesnov, 1976]. Later, the contents of Pt, Pd, Rh, Au, and Ag were measured by the atomic-absorption method in three additional samples (Table 12.3). In the following investigations, the contents of PGE and Re were determined in chromitites of the Berezovka massif by more perfect ICP-MS with acid digestion of weighted specimens in Carius tubes at 280 and 300°C for 24 h [Koz'menko et al., 2011; Lesnov et al., 2012]. The reproducibility of measurement results was controlled by analyses of duplicate specimens and by comparison of the determined contents of PGE in the geochemical standard GP-13 with their certified contents.

The performed analyses showed that the total contents of six PGE in chromitites of the Berezovka massif vary from 556 to 762 ppb (Table 12.4). These values are close to the estimated total contents of PGE (ppb) in chromitites of the Kempirsay (470–712), Rai-Iz (214–304), and Voikar-Syn'in (181–224) massifs in the Urals [Volchenko et al., 2011b] as well as the Shebenik (Albania) (90–729) [Kocks et al., 2007], Muğla (Turkey) (61–656) [Uysal et al., 2009], and Troodos (71–293) (Cyprus Island) [Büchl et al., 2004] massifs. The content of Os in chromitites of the Berezovka massif (31–106 ppb)

Table 12.3   Contents of Pt, Pd, Rh, Au, and Ag in massive chromitites from the Berezovka massif, ppm.

Sample	Pt	Pd	Rh	Au	Ag
Assay-spectral method					
1434-2	0.22	0.35	N.d.	N.d.	N.d.
1434-1*	0.10	0.16	N.d.	N.d.	N.d.
Radiochemical neutron activation method					
1434-2	1.44	0.09	N.d.	N.d.	N.d.
1434-1*	0.58	0.015	N.d.	N.d.	N.d.
Atomic-absorption method					
Ev-2	0.035	<0.005	0.03	0.0037	0.012
Ev-2a*	0.028	<0.005	0.023	N.d.	N.d.
Ev-3	0.12	≤0.005	0.044	0.019	0.014
Ev-3a*	0.13	≤0.005	0.043	0.024	0.018
1051b	0.66	0.015	0.046	0.012	0.017
1051b-1*	0.68	0.013	0.042	N.d.	N.d.

Note: Analyses by the radiochemical neutron activation method (analyst R.D. Mel'nikova) and atomic-absorption method (analyst V.G. Tsimbalist) were carried out at the Institute of Geology and Geophysics, Novosibirsk. Analyses by the assay-spectral method were performed at the Sibtsvetniiproekt Institute, Krasnoyarsk (analysts M. Lukicheva and N. Pantyukova). *Check analyses.

Table 12.4 Contents of PGE and Re and their ratios in massive chromitites of the Berezovka massif, ppb.

Element and element ratio	Temperature of sample digestion in Carius tubes						[Wittig et al., 2010]
	260°C	300°C	260°C	280°C	260°C	260°C	
	Sample						
	Ev-2	Ev-2a	Ev-3	Ev-3a	1051b	GP-13	GP-13*
Os	31.0	N.d.	37.3	N.d.	105.68	3.91	3.77
Ir	146.8	133.0	107.5	162.3	117.3	3.62	3.58
Ru	413.4	417.8	354.3	618.6	482.1	6.98	6.85
Rh	51.40	60.00	39.11	42.00	16.88	1.50	1.25
Pt	49.88	4.52	12.99	9.86	30.00	6.40	6.90
Pd	0.605	0.666	4.927	1.338	9.508	6.42	6.40
Re	0.077	0.097	0.075	0.097	0.116	0.223	0.32
Total with Os	693	N.d.	556	N.d.	762	29.1	29.1
Total without Os	662	616	519	834	656	25.2	25.3
Ir/Os	4.74	N.d.	2.88	N.d.	1.11	0.93	0.95
Pd/Ir	0.004	0.005	0.046	0.008	0.081	1.773	1.788
Pt/Ir	0.340	0.034	0.121	0.061	0.256	1.768	1.927
Ru/Ir	2.82	3.14	3.30	3.81	4.11	1.93	1.91
Ru/Rh	8.04	6.96	9.06	14.73	28.56	4.65	5.48
Pd/Pt	0.012	0.147	0.379	0.136	0.317	1.003	0.928
Ir/Re	1906	1371	1433	1673	1011	16	11
Pt/Re	648	47	173	102	259	29	22

Note: Analyses were carried out by ICP-MS with isotope dilution at the Analytical Center of the Institute of Geology and Mineralogy, Novosibirsk (analysts O.A. Koz'menko, I.V. Nikolaeva, and S.V. Palesskii). GP-13 – geochemical standard analyzed in parallel with chromitite samples; GP-13* – recommended element contents in the geochemical standard. N.d. – no data.

is the same as in chromitites of the Troodos massif (14–104 ppb). Thus, PGE form the following sequence according to their contents in chromitites of the Berezovka massif: Os < Ir < Ru > Rh >> Pt > Pd. The primitive mantle normalized PGE and Re patterns of these rocks are of convex shape, with maxima at Ru (Fig. 12.5). As mentioned above [Lesnov et al., 2012], the PGE patterns of chromitites of the Berezovka massif are similar to those of mantle podiform chromitites [Pagé et al., 2012]. The contents of Os, Ir, Ru, and Rh in them are an order of magnitude (or even more) higher than those of the primitive mantle, whereas the contents of Pt, Pd, and Re are much lower. The plots depicting Ir/Os, Ru/Rh, Pt/Re, Pd/Ir, Pt/Ir, and Pd/Pt in these chromitites are of nearly the same shape, lie close to each other, and have maxima at Ir/Os, Ru/Rh, and Pt/Re and minima at Pd/Ir, Pt/Ir, and Pd/Pt (Fig. 12.6). Note that in contrast to the plots for chromitites, the same plots for harzburgites and lherzolites of the Berezovka massif have minima at Pd/Pt and maxima at Pt/Re, Pt/Ir, and Pd/Ir (Fig. 12.2). This difference is due to the strong domination of Ir over Pt and Pd in chromitites and the domination of Pt and Pd over Ir in harzburgites and lherzolites. In general, chromitites of the Berezovka massif are similar in PGE contents and PGE ratios to chromitites of other mafic-ultramafic massifs of ophiolite associations.

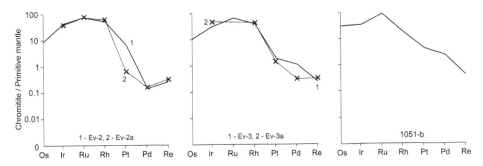

Figure 12.5 Primitive mantle normalized PGE and Re patterns of chromitites from the Berezovka massif (data from Table 12.4). The contents of elements in the primitive mantle are listed in Table 12.1.

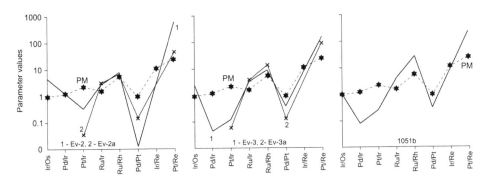

Figure 12.6 Variations in Ir/Os, Pd/Ir, Pt/Ir, Ru/Ir, Ru/Rh, Pd/Pt, Ir/Re, and Pt/Re in chromitites from the Berezovka massif (data from Table 12.4) and in the primitive mantle (PM) (data from Table 12.1).

12.3 ON MINERAL-CONCENTRATOR OF PGE IN CHROMITITES OF THE BEREZOVKA MASSIF

Chromitites from many occurrences and deposits usually contain unevenly distributed finely dispersed PGE mineral grains [Dodin *et al.*, 2003; Agafonov *et al.*, 2005; Volchenko *et al.*, 2011a]. The very small sizes of these grains, few to several hundreds of microns, make their discovery and identification difficult. Most of PGE mineral grains were first found during an analysis of heavy fractions of loose deposit and artificial concentrates of chromitites using scanning electron microscope (SEM).

On Sakhalin Island, PGE minerals were first found in studies of heavy fractions of concentrates from loose deposits during prospecting and survey works and during the mining of gold placers. These minerals were most often localized in concentrates sampled near mafic-ultramafic massifs [Danchenko, 2003]. According to this author, the majority of known concentrate aureoles of PGE minerals is localized on the Schmidt Peninsula, in the East Sakhalin Mountains, in the Pioner–Shel'ting and Susunai–Aniva regions, in the valleys of tributaries of the Langeri River (Derbysh

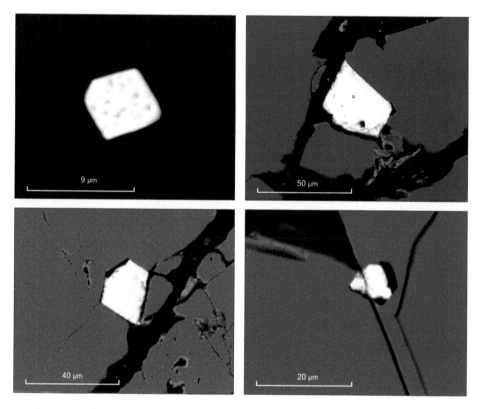

Figure 12.7 Microphotographs of laurite-erlichmanite grains (light gray) present as euhedral micro-inclusions in Cr-spinel grains (dark gray), cut by chlorite–serpentine veins (black), from chromitites of the Berezovka massif.
The position of a microprobe beam during PGE analysis is shown by a rectangle. The microphotographs were obtained with a LEO 1430VP scanning electron microscope in back-scattered electron mode (analyst A.T. Titov).

and Kukui brooks) and Rukosuev, Abramov, Ugol'naya Rivers, and at the Tesno, Uryu-Horokanai, Yubari, and Mukawa sites. The major PGE minerals found in slimes there were rutheniridosmine, ruthenosmiridium, osmiridium, iridosmine, platinian iridosmine, and laurite; polyxene, ferroplatinum, and isoferroplatinum were scarcer.

During a SEM study of large polished sections of chromitites from the Berezovka massif, we have first discovered occasional *in situ* grains of Ru sulfide [Lesnov *et al.*, 2010a]. The grains had polygonal cross sections and measured 10 to 40 μm. They were localized in chlorite–serpentine veinlets crossing Cr-spinel grains or near them (Fig. 12.7). A SEM analysis of the Cr-spinel grains with the use of energy-dispersive spectra showed that the mineral corresponds in chemical composition to the intermediate phase of the isomorphous series laurite (RuS_2) – erlichmanite (OsS_2), with Ru dominating over Os, and contains Ir and Rh impurities (Table 12.5). In addition, abundant submicron awaruite grains containing 71.4 wt.% Ni and 20.7 wt.% Fe

Table 12.5 Chemical composition of laurite-erlichmanite grains from chromitites of the Berezovka massif, wt.%.

Element and element ratio	Sample				
	Ev-2	Ev-3		1051b	
	Grain 1	Grain 1	Grain 2	Grain 1	Grain 2
Os	10.55	15.45	15.62	11.29	10.62
Ir	7.74	11.06	8.42	9.99	9.74
Ru	27.03	35.72	36.85	39.76	38.55
Rh	2.23	1.84	3.09	2.76	3.40
S	25.5	35.17	34.93	34.71	34.65
Cr	1.14	N.f.	N.f.	1.48	N.f.
Fe	0.62	0.69	N.d.	N.d.	N.d.
Os/Ir	1.36	1.40	1.86	1.13	1.09
Ru/Ir	3.49	3.23	4.38	3.98	3.96
Ru/Rh	12.12	19.41	11.93	14.41	11.34
Ir/Rh	3.47	6.01	2.72	3.62	2.86

Note: Analyses were carried out with a LEO 1430VP scanning electron microscope at the Analytical Center of the Institute of Geology and Mineralogy, Novosibirsk (analyst A.T. Titov). N.f. – not found, N.d. – no data.

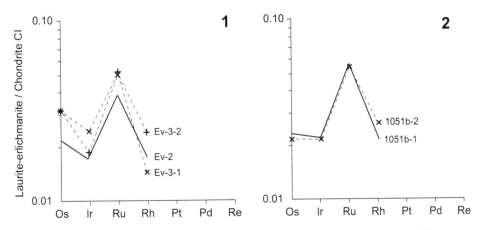

Figure 12.8 Chondrite Cl-normalized (after [Anders and Grevesse, 1989]) Os, Ir, Ru, and Rh patterns of laurite-erlichmanites from chromitites of the Berezovka massif (see Fig. 12.7). 1 – grains Ev-2 and Ev-3-1 (major analysis) and Ev-3-2 (check analysis); 2 – grains 1051b-1 (major analysis) and 1051b-2 (check analysis) (data from Table 12.5).

were found in the polished sections of chromitites. The presence of awaruite in these chromitites suggests their formation under reducing conditions.

The chondrite-normalized PGE patterns of laurite-erlichmanite grains are of the same shape and show a strong positive anomaly of Ru (Fig. 12.8). The chemical homogeneity of the mineral grains is also evidenced from the close location and the same

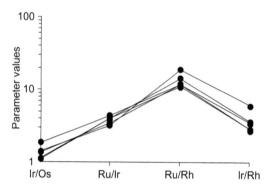

Figure 12.9 Variations in PGE ratios in laurite-erlichmanites from chromitites of the Berezovka massif (data from Table 12.5).

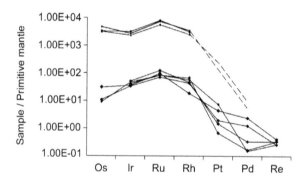

Figure 12.10 Primitive mantle normalized (after [McDonough and Sun, 1995]) patterns of PGE and Re in laurite-erlichmanites (top) (data from Table 12.5) and their host chromitites (bottom) from the Berezovka massif (data from Table 12.4). Dashed lines show the pattern segments constructed by extrapolation.

shape of the Ir/Os, Ru/Ir, Ru/Rh, and Ir/Rh curves (Fig. 12.9). The PGE contents of laurite-erlichmanite, reduced by 10,000 times, were normalized to the primitive mantle, together with the PGE contents of chromitites containing these mineral grains (Fig. 12.10). One can see that at the segments between Os and Rh, the PGE patterns of the mineral and the host rocks are of similar shape, i.e., the ratios of particular PGE in them are nearly equal. The same is evidenced from the arrangement of the composition points of laurite-erlichmanite and chromitites on the Os–Ru–Ir diagram (Fig. 12.11).

The above data show that laurite-erlichmanite is a strongly predominant or even the only mineral concentrating PGE in chromitites of the Berezovka massif. According to the available data, laurite and laurite-erlichmanite grains are often present in chromitites of various occurrences and deposits. For example, in chromitites of the Krasnogorsk mafic-ultramafic massif (Chukchi region), laurite-erlichmanite grains amount to ~28% of all grains of PGE minerals [Dmitrenko et al., 1987]. They are also abundant in chromitites of mafic-ultramafic massifs of Tuva and Mongolia [Agafonov

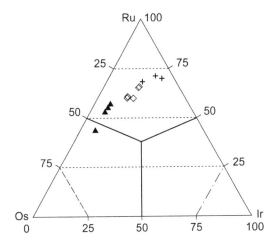

Figure 12.11 Relationship among the contents of Os, Ru, and Ir in laurites (rhombuses) and in the host chromitites (crosses) from the Berezovka massif (data from Table 12.5). Triangles mark laurite compositions (literature data). The upper quadrangular fragment of the diagram is the composition field of laurite, and the lower left quadrangular fragment is the composition field of erlichmanite.

et al., 2005], East Sayan [Kiseleva, 2014], Albania [Kocks *et al.*, 2007], Turkey [Uysal *et al.*, 2009], etc.

* * *

Applying different analytical methods, including ICP-MS, we have obtained the first data on PGE and Re contents in harzburgites, lherzolites, plagiowehrlites, websterites, olivine websterites, orthopyroxenites, olivine clinopyroxenites, olivine gabbronorites, and anorthosites composing the Berezovka, Shel'ting, Komsomol'sk, and South Schmidt massifs. The total contents of these elements in the rocks vary from 8.6 to 46.0 ppb. Elevated contents of Os, Ir, and Ru are observed in harzburgites and lherzolites, and lower ones, in wehrlites and olivine gabbronorites. The contents of easily fusible PGE (Rh, Pt, and Pd) in ultramafites and gabbroids of the studied massifs are nearly equal. The PGE patterns of lherzolites of the South Schmidt massif are similar in shape and the location on diagrams to the primitive mantle. They evidence that PGE in these rocks are almost not fractionated. The Ir/Os, Pd/Ir, Pt/Ir, and other PGE ratios were used as additional characteristics for the geochemical systematization of ultramafites and gabbroids of the studied massifs. The Pd/Ir ratio reflecting the degree of PGE fractionation shows the widest variations. The contents of all PGE and Re in chromitites of the Berezovka massif have been for the first time determined by ICP-MS. The total PGE contents in them vary from 693 to 834 ppb, which is close to or slightly higher than those in chromitites of other mafic-ultramafic massifs. Grains of PGE minerals (mainly those with prevailing Os, Ir, and Ru) have been found in some slime aureoles localized near mafic-ultramafic massifs of Sakhalin and in some gold placers. Laurite-erlichmanite grains have been first revealed by SEM in massive chromitites of the Berezovka massif. This is probably the major or the only mineral concentrating PGE in the above chromitites.

Chapter 13

The fundamentals of the concept of polygenetic formation of mafic-ultramafic complexes in ophiolite associations

Science is infallible, but scientists are often wrong …

A. France

… it is not necessary to stop the debate. One should always remember that a model which was popular at certain time can be inaccurate …

Yu.N. Avsyuk

On the global structure-tectonic schemes of occurrence of ophiolite associations, the mafic-ultramafic massifs in them form a system of differently directed belts of varying lengths (Fig. 13.1). Part of this system is hundreds of mafic-ultramafic massifs which are concentrated along the Asian shore of the Pacific from Chukchi Peninsula to Southeast Asia (Fig. 13.2). These massifs are structurally confined to the zones of long-lived deep faults which were the channels of repeated penetration of basaltoid melts and tectonic blocks of ultramafic restites from the upper mantle to the Earth's crust. It was established that the deep fault zones, or sutures, have mainly steep slopes, but, as a result of repeated folded-block deformations, on some sites these acquired relatively gentle slopes. Some researchers think that the mafic-ultramafic massifs confined to gently inclined fault zones took part in the formation of overlapped structures (overthrusts).

In the recent decades, the problems of structural position, geologic structure, and genesis of mafic-ultramafic massifs in the structure of ophiolite associations have been discussed from the viewpoint of the concept of plate tectonics [Coleman, 1979]. This concept is based on the idea that ophiolites are the ancient oceanic crust and that magmatic and terrigenous complexes in them, including mafic-ultramafic massifs, have a common stratified structure.

At the same time, results of earlier structural tectonic studies of a great number of mafic-ultramafic massifs occurring in ophiolite associations of the Chukchi Peninsula, Koryakia, Kamchatka Peninsula, northern Baikal region, Tuva, the Urals, Mongolia, Kyrgyzstan, and North Pamirs, as well as Sakhalin Island, evidence that, in a number of features, the structural position, morphology, inner structure, and material composition cannot be explained satisfactorily from the viewpoint of concept of plate tectonics and their petrological interpretation requires other approaches [Pinus et al.,

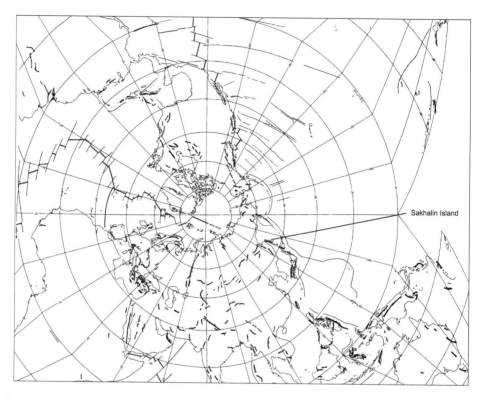

Sakhalin Island

Figure 13.1 Global scheme of location of ophiolite belts and their mafic-ultramafic massifs (after [Irwin, Coleman, 1972]).

1973, 1976, 1984; Lesnov *et al.*, 1973, 1982; Lesnov, 1976, 1979, 1980, 1981a–d, 1982, 1984, 1985, 1986a,b, 1988, 2007e, 2009c, 2011a; Slodkevich and Lesnov, 1976; Balykin *et al.*, 1991].

The observed contradictions with the basic statements of the concept of plate tectonics are mainly related to fundamental problems, such as space-time relationships between the bodies of ultramafic rocks and gabbroids, genesis of the so-called banded complexes, and lack of evidence for stratification features in the structure of mafic-ultramafic massifs and ophiolite associations in general. Some authors of the above-cited works revealed that in most cases gabbroid bodies do not "build over" the bodies of ultramafic rocks as it is postulated by the concept of plate tectonics, but occur along their hanging or lying tectonic contacts with the adjacent bodies of ultramafic rocks and rocks of enclosing series. It was also shown that the bodies of ultramafic rocks, which are mainly steeply inclined protrusions, had been intruded in a solid-plastic state into the Earth's crust along faults prior to the formation of gabbroid intrusions spatially close to these protrusions along the same faults. There is evidence that mafic melts which formed gabbroid intrusions under certain conditions actively interacted with the rocks of ultramafic protrusions, resulting in the formation of the so-called stratified complexes along the contacts between them. A "cumulative" mechanism of formation of these complexes has been developed in the concept of plate tectonics, but

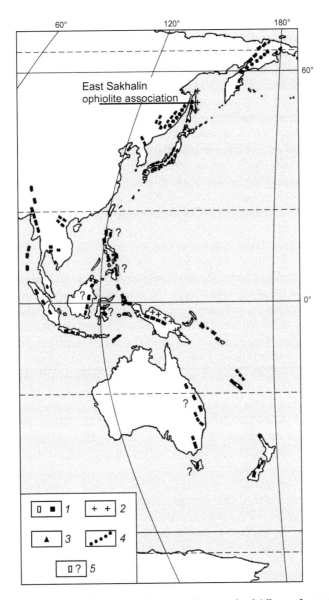

Figure 13.2 Scheme of location of belts of mafic-ultramafic massifs of different formations along western coasts of the Pacific (after [Zimin, 1973]).
1–4 – hyperbasic rock formations: *1* – dunite-harzburgite, *2* – of transitional composition, *3* – dunite-wehrlite-pyroxenite, *4* – olivinite-wehrlite; *5* – hyperbasic rocks from unidentified formation.

they can also be interpreted in a more consistent way as contact-reaction zones made up of hybrid ultramafic rocks and gabbroids.

Generalization of all these observations and established facts allowed us to suggest new approaches to the general problem of the genesis of complex mafic-ultramafic massifs in the composition of ophiolite associations. As a result, we have developed the bases of the concept of polygenous formation of mafic-ultramafic massifs that are part of ophiolite associations. To substantiate this concept, below we discuss the main structural features of some typical polygenous mafic-ultramafic massifs, beginning with those localized in the folded structures of the Chukchi Peninsula and Koryakia.

The Ust'-Bel'sk massif and adjacent *El'dynyr* massif, occurring on the Chukchi Peninsula, are exposed in the area of more than 970 km², of which 74% is the outcrop of ultramafic rocks, whereas the rest are the outcrops of gabbroids [Pinus *et al.*, 1973]. The Ust'-Bel'sk massif occurs among the Lower Cretaceous terrigenous-volcanogenic deposits and consists of one huge and numerous small bodies of ultramafic rocks represented by harzburgite, lherzolite, dunite, and their serpentinized varieties, as well as one large and several small bodies of gabbroids which intrude ultramafic rocks and consist of gabbronorites, gabbro, and amphibole gabbro. Along the endocontacts of gabbroid bodies with ultramafic rocks there occur taxitic rocks, including parallel-banded varieties of olivine gabbroids with significantly varying quantitative mineral composition. At places, these contain ultramafic xenoliths of lenticular and irregular shapes. The intrusive contact of gabbroids with ultramafic rocks was observed, in particular, in the headwaters of the Shirokii Brook, where in the direction from the contact of gabbroid intrusion toward the intruded body of ultramafic rocks, olivine-free orthomagmatic gabbro were replaced by their hybrid varieties with single olivine grains, the amount of which increased gradually, whereas the content of plagioclase, by contrast, decreased, resulting in a transition to melanocratic olivine gabbro. Then these melanocratic gabbroids were replaced by hybrid ultramafic rocks represented by plagiowehrlites, wehrlites, and plagioclase-bearing dunites with streak–banded distribution of plagioclase and, finally, plagioclase-free dunites. In addition, amphibole gabbro from this massif was found to contain two populations of zircons of different ages (819–775 and 361–312 Ma) [Ledneva *et al.*, 2012]. Therefore, the established unambiguous evidence suggests the later formation of gabbroid intrusion with respect to the protrusion of ultramafic rocks, which allowed us to regard the Ust'-Bel'sk massif as a polygenous mafic-ultramafic complex.

The Pekul'ney massif, also occurring on the Chukchi Peninsula, includes some close bodies of ultramafic rocks and gabrroids that outcrop in the area of about 200 km², of which about 40% are ultramafic rocks [Pinus *et al.*, 1973]. Bodies of ultramafic rocks border on enclosing Lower Cretaceous terrigenous-volcanogenic rocks along rather steeply inclined faults. In the contact zones of gabbroid intrusions with the bodies of ultramafic rocks of wide occurrence are hybrid ultramafic rocks and gabbroids represented by plagiodunites, wehrlites, plagiowehrlites, pyroxenites, troctolites, and olivine gabbro. On the basis of these features the massif is assigned to the category of polygenous mafic-ultramafic complexes.

The Aluchin massif, situated in the western part of the Chukchi Peninsula, is a meridionally elongate steep body composed of a protrusion of ultramafic rocks that form the southern part of the massif and a gabbroid intrusion forming its northern part [Pinus and Sterligova, 1973; Pinus *et al.*, 1973; Lesnov, 1988]. The massif occurs among the Middle Paleozoic, Jurassic, and Cretaceous terrigenous-volcanogenic rocks that have steep tectonic contacts with the protrusion of ultramafic rocks. Geomorphological observations suggest that in the modern time the protrusion of ultramafic rocks undergoes secular upward movements. The body of gabbroids has distinct intrusive contacts with the rocks of enclosing series and protrusion of ultramafic rocks. The contact zones of gabbroids with ultramafic rocks, the thickness of which varies from tens to hundreds of meters, are dominated by banded hybrid gabbroids, which contain streaky and lens-shaped xenoliths made up of wehrlites and pyroxenites that, in places, demonstrate plagioclase grains. Among zircons isolated from gabbro of this massif, three groups of populations with drastically different ages have been established: 2698, 1838, and 1794; 283–273; and 91–89 Ma [Ganelin *et al.*, 2012, 2013]. It is likely that zircons from the first group are xenogenic, trapped by mafic melts from the upper-mantle ultramafic source, and rejuvenated to some degree. Zircons from the population dated 283–273 Ma are, most likely, syngenetic with gabbroids and determine the time of formation of the intrusion composed of them. Finally, zircons from the population dated 91–89 Ma are, most probably, epigenic with respect to gabbroids and formed during the infiltration of the fluids which separated from basaltoid or granitoid melts at a later stage of magmatic activity in the regions of the western Chukchi Peninsula. Thus, if we take into account the available data, the Aluchin massif can be considered polygenous.

The Chirynay massif, localized in Koryakia, includes a great number of bodies of ultramafic rocks and gabbroids, which are exposed in the area of about 30 × 60 km [Pinus *et al.*, 1973]. The largest of them is, in fact, the Chirynai massif, which is made up of a protrusion of ultramafic rocks composed of harzburgites, dunites, and lherzolites outcropping in the area of about 54 km², and gabbroid intrusions that penetrate the protrusion in the north and in the south. Along the southern tectonic contact of the protrusion with enclosing series, inclined southward at an angle of 35°, ultramafic rocks are intensely serpentinized and foliated. The gabbroid intrusion localized near the southern contact of the ultramafic protrusion consists of olivine-free orthomagmatic gabbronorites and gabbro that, when approaching the contact with ultramafic rocks, are replaced by hybrid olivine-bearing varieties. Close to the contact with gabbroid bodies, ultramafic rocks are enriched in clinopyroxene, at places are cut by the veinlets of pyroxenites. In the western flank of the gabbroid intrusion, ultramafic xenoliths of varying sizes and shapes were found. On the basis of the above-mentioned features, this massif can be considered polygenous.

The Tamvatney massif is also localized in Koryakia and is exposed in the area of about 300 km² [Pinus *et al.*, 1973]. It consists of a large protrusion of ultramafic rocks represented by lherzolites and subordinate harzburgites, which occupies more than 90% of its area, and a number of small, predominantly lens-shaped gabbroid intrusions. The contact zones of gabbroid intrusions with the protrusion of ultramafic rocks are composed of alternating segregations of hybrid ultramafic rocks and gabbroids: wehrlites, plagiowehrlites, clinopyroxenites, taxitic olivine gabbro, and troctolites. In hybrid gabbroids one can observe lenticular and irregularly shaped xenoliths made up of wehrlites, clinopyroxenites, and their plagioclase-bearing varieties. Some of these xenoliths, reaching 10–20 m in cross section, are intersected by gabbroid veins. This massif possesses all characteristics that allow it to be considered a polygenous mafic-ultramafic body.

The Kuyul massif is another large mafic-ultramafic body in Koryakia. It is exposed in the area of about 360 km², 85% of which are the rocks of an ultramafic protrusion elongated in the northeastern direction and steeply inclined toward southeast. There also occur numerous small exposed intrusions of gabbroids penetrating ultramafic rocks [Pinus *et al.*, 1973; Lesnov *et al.*, 1980; Lesnov, 1988; Ledneva and Matukov, 2009]. Ultramafic rocks are dominated by significantly serpentinized harzburgites and lherzolites, which near the gabbroid intrusions, especially close to the largest of them, exposed in the basin of the Gankuvayam River, are replaced by wehrlites and plagiowehrlites. Toward the contact with gabbroid intrusion, wehrlites and plagiowehrlites are replaced by olivine gabbro, troctolites, and anorthosite, which mostly have a banded structure. At a distance from the contacts with ultramafic rocks, gabbroid bodies are composed of olivine-free gabbronorites and gabbro. Ultramafic rocks also contain veins of gabbro-pegmatites, in which the isotopic age of zircons is 181–136 Ma [Ledneva and Matukov, 2009]. Thus, it is reasonable to regard this massif as a polygenous mafic-ultramafic body.

The Cape Kamchatka massif is localized in the northern part of the Kamchatka Peninsula. It includes a protrusion of ultramafic rocks, called the Soldatsk massif, and a remote gabbroid intrusion, called the Olenegorsk massif [Velinskii, 1979; Lesnov, 1988]. The protrusion of ultramafic rocks is made up of harzburgites, lherzolites, and scarce dunites which in peripheral parts are intensely serpentinized and foliated. Ultramafic rocks are at places cut by veins of pyroxenites, occasionally passing into wehrlites, and veins of gabbro-pegmatites. Gabbroids from the Olenegorsk intrusion contain xenoliths composed of wehrlites close to which olivine-free gabbronorites and gabbro are replaced by olivine gabbro, troctolites, and anorthosites with a taxitic texture. The ages of isotope dating of most zircon grains from gabbro of the Olenegorsk intrusion vary in the range of 2721–1498 Ma, one of the grains being dated 64 Ma [Tsukanov and Skolotnev, 2010]. It is supposed that zircon grains from ancient populations are xenogenic and trapped by mafic melt from the upper-mantle ultramafic source during its partial melting, and the Cape Kamchatka massif is a polygenous mafic-ultramafic complex.

The Karaginsk massif, localized on a large island of the same name near the eastern shores of Kamchatka, is a northeastwardly stretched chain of close ultramafic bodies and elongate gabbroid intrusions that penetrate these bodies along the contacts with the framing rocks. The largest body of ultramafic rocks and contacting gabbroid intrusion are localized on the southwestern flank of this chain near Shapochka Mountain. The tectonic contacts of ultramafic protrusion with host rocks are inclined toward the northwest at angles of 45–60°. Ultramafic rocks are to a varying degree represented by serpentinized harzburgites, lherzolites, and dunites, which near the contacts with gabbroid intrusions are replaced by wehrlites, clinopyroxenites, olivine websterites, and orthopyroxenites. The same varieties of rocks make up xenoliths that occur in gabbroids. These data suggest that the gabbroid intrusion formed later than the protrusion of ultramafic rocks and that the Karaginsk massif is a polygenous mafic-ultramafic complex.

The Kroton massif, which occurs in the eastern part of the Kamchatka Peninsula within the Kumroch Ridge, is exposed in the area of about 380 km² and is the largest mafic-ultramafic body in this region [Sidorov, 2009]. It consists of a large protrusion of ultramafic rocks represented by serpentinized harzburgites and subordinate dunites and lherzolites, as well as intruding small gabbroid intrusions among which xenoliths of ultramafic rocks were found. Minor amounts of wehrlites and pyroxenites were also found in the massif. This massif can be also considered polygenous.

The Gal'moenan, Seinav, and similar mafic-ultramafic massifs occurring within Koryakia are typically assigned to a platinum-bearing gabbro-clinopyroxenite-dunite

(Urals–Alaskan) type [Vil'danova *et al.*, 2002]. The best studied one is the Gal'moenan massif, which borders on the placer deposit of platinum metals. The structure of the massif includes a large extended body of substantially dunite composition, a thick clinopyroxenite zone surrounding the massif along almost the entire perimeter, and a discontinuous chain of small gabbroid intrusions. The cited work provides a detailed description of the Gal'moenan massif, including the contact zones of dunites with the zone of clinopyroxenites and the contact zones of clinopyroxenites with gabbroid bodies. The authors, in particular, mention that at a distance of tens of meters from the contact with clinopyroxenites in dunites there occur individual segregations of clinopyroxene, the quantity and size of which, when moving from dunites, increase with transition to wehrlites, olivine clinopyroxenites, and olivine-free clinopyroxenites. In these zones, dunites are cut by the veins of clinopyroxenites ranging from several centimeters to several meters in thickness, which contain (called by the authors) "schlieren", "lenses", and "blocks" of dunites with signs of corrosion of olivine grains [Vil'danova *et al.*, 2002]. It is noteworthy that in the northern part of the massif such segregations of dunites occur in clinopyroxenites at a distance of up to 1 km from their contacts with dunites. In the same part of the massif one can observe zones that are up to several tens of meters thick, in which "bands" of dunites and clinopyroxenites alternate with smooth contacts ranging from a few millimeters to 10 cm in thickness. Similar alternation of the "bands" of dunites and clinopyroxenites was also observed in the Seinav massif. It was established that in the outer contact of clinopyroxenite zone, when approaching the contact with gabbroids, clinopyroxenites contained segregations of plagioclase, with the increasing amount of which clinopyroxenites were replaced by plagioclase clinopyroxenites and inequigranular banded gabbroids. The authors of the above-mentioned work report that the Gal'moenan massif is a zoned body with the features of apparent and "cryptic" zoning. In turn, Kozlov [2000] came to a conclusion that the main mechanism of formation of this massif is the directed fractionation crystallization of primary calc-alkali melts, which are similar in composition to plagioclase-bearing olivine pyroxenites.

However, on the basis of the data of other researchers and our own field observations within the Gal'moenan and Seinav massifs we think that the variously shaped segregations of dunites occurring among pyroxenes of the Gal'moenan massif are xenoliths preserved in later clinopyroxenites formed after them and that the Gal'moenan and Seinav massifs are polygenous. They consist of protrusions of restitic dunites, cut by gabbroid intrusions, and contact-reaction zones composed of hybrid clinopyroxenites, olivine clinopyroxenites, and wehrlites. The idea of the polygenous formation of the Gal'moenan massif is supported by the isotope dating of zircons from dunites which compose the massif [Knauf, 2008]. Among these zircons there are populations dated from 2591 to 1831 Ma. Knauf determined that the value of parameters $^{176}Hf/^{177}Hf$ in analyzed zircons changes in the range of 0.281110–0.281941. The reported facts evidence that dunites from the Gal'moenan massif are very ancient rocks formed within the upper mantle. In addition, approximately the same ancient age (2473–1885 Ma) was established for zircons from dunites of the Konder concentrically zoned clinopyroxenite-dunite massif [Badanina and Malich, 2012].

The Feklistov massif, localized on Feklistov Island, which is part of the Shantar Islands, and in the structure and composition it is similar to the Gal'moenan massif but much smaller. The massif has a concentrically zoned structure and is made up of dunites, olivine clinopyroxenites, and gabbroids [Lutskina, 1976]. The body of dunites, steeply inclined and subisometric in plan, is exposed in the center of the massif. It is surrounded by a wide zone composed of olivine clinopyroxenites and wehrlites. In the eastern part of the massif the zone of pyroxenites is separated from the host series of sandstones and conglomerates of the Devonian age by the discontinuous chain of gabbroid bodies, the largest of which is 1.5 km long and 0.2–0.3 km wide. Pyroxenites contain lens-shaped and plate-like dunite xenoliths with smooth contacts, and dunites demonstrate veins of pyroxenites. The available data suggest that this massif is polygenous and gabbroid bodies are fragments of a weakly eroded ring intrusion that intruded along the contacts of dunite protrusion with host rocks and that the contact-reaction zone composed of hybrid olivine clinopyroxenites and wehrlites formed under the influence of gabbroid melts [Lesnov, 1988]. Here again the Konder concentrically zoned clinopyroxenite-dunite massif is worth mentioning, dunite zircons from which are of a very ancient age (2473–1885 Ma) [Badanina and Malich, 2012].

The Shikotan massif is situated on Shikotan Island of the Kurile Ridge [Lesnov, 1988]. Its polygenous formation is supported by the fact that zircon grains (55 analyses) from lherzolites, which make up a small body occurring among predominant gabbroids, have isotopic age of 2775–936 Ma and single zircon grains from pyroxenites, gabbro-pyroxenites, and gabbronorites, among which the body of lherzolites occurs, are of much younger age (66–63 Ma) [Explanatory note ..., 2006].

The Dyukali massif, localized in the Khungari zone, is one of the largest mafic-ultramafic bodies, known within Sikhote Alin. Izokh [1965] and Romanovich [1973], who studied the massif, came to a conclusion that the gabbroid intrusion in the massif formed later that the spatially close body of ultramafic rocks, which is represented by serpentinized dunites, harzburgites, and lherzolites. The authors also report that near the contacts with gabbroid intrusion in ultramafic rocks there occur veins of pyroxenites and at the same sites of ultramafic rocks they observed irregularly distributed xenomorphic segregations of plagioclase. In the gabbroid intrusion itself they found xenoliths of ultramafic rocks of various sizes, which underwent feldspathization and pyroxenization up to transition to clinopyroxenites. At places near the contact with ultramafic rocks, olivine-free gabbroids that form the gabbroid intrusion are replaced by olivine-bearing varieties. E.P. Izokh pointed out that the relationships between ultramafic rocks and gabbroids observed in the Dyukali massif are also typical of the other mafic-ultramafic massifs of the Khungari zone, including the Gorbilya and Bogbasu massifs. It is especially worth noting that E.P. Izokh proposed to call the Dyukali and similar mafic-ultramafic massifs *polygenous.*

Complex mafic-ultramafic massifs are also widespread in the other regions of the Primorye territory: Kafen [Zimin, 1973], Ariadna and Sergeevka [Oktyabr'skii, 1971], Khanka [Shcheka *et al.*, 2001], and others. A great number of mafic-ultramafic massifs, which are part of ophiolite associations, occur on the islands of Japan. When studying the massifs of Horoman, Ogiwara, Oshima, Yakuno, Oeyama, etc., it was established that the bodies of gabbroids in them formed later than the spatially close bodies of ultaramafic rocks [Miyashiro, 1966; Ultrabasic ..., 1967; Hirano, 1977; Ishiwatari, 1985, 1991; Kurokawa, 1985].

The evidence for the fact that the bodies of ultramafic rocks and gabbroids of the ophiolite mafic-ultramafic massifs formed at different times and for the discrepancy between their structure and postulates of the concept of plate tectonics was found within the Urals folded area. Schteinberg *et al.* [1986] criticized the ideas of some colleagues who considered the problems of the structure and genesis of ophiolite associations and mafic-ultramafic massifs of this region from the viewpoint of the concept of plate tectonics. It was pointed out by the example of many similar massifs that gabbroid bodies do not occur stratigraphically higher than the bodies of ultramafic rocks but intrude them, have intrusive relationships with them, and metamorphose them. However, gabbroid intrusions not only occur higher than the bodies of ultrabasic rocks but also "underlie" them. The authors emphasized that in the Ural region, as in the whole world, platinum-bearing dunite-clinopyroxenite complexes are similar to those that occur along the borders of the bodies of Alpine-type harzburgites with younger bodies of gabbroids and, therefore, the apoharzburgite, reaction-metasomatic origin of these complexes can be considered proven.

The ideas of Schteinberg *et al.* agree with the later data of Bazhin [2013], who established by the example of some mafic-ultramafic massifs of the Southern and Middle Urals that gabbroid intrusions formed later than spatially associated bodies of ultramafic rocks. In turn, Selivanov [2011] came to a conclusion that the dunite-wehrlite-clinopyroxenite complex, which is part of the Voikar–Syn'in mafic-ultramafic massif (Polar Urals) and is localized along the contacts of mantle ultramafic rocks and gabbroid intrusion, formed as a result of interaction between them.

The views that the bodies of ultramafic rocks and gabbroids in the mafic-ultramafic massifs of the Urals formed at different times are also supported by the results of isotope dating of zircons from constituent rocks. It was established that the age of zircons from dunites of the Nizhnii Tagil massif is 2852–2656 Ma, whereas the age of zircons from gabbroids of this massif is 585 Ma [Efimov, 2010]. Study of zircons from dunites from the Voikar-Syn'in [Batanova *et al.*, 2009], Sakharinsk and East Khabarninsk [Fershtater *et al.*, 2009; Krasnobaev *et al.*, 2009], and Sarana [Krasnobaev *et al.*, 2013] massifs showed their very ancient age (~1500 Ma and older).

Pertsev *et al.* [2009] analyzed the problems of petrology and space-time relationships between the bodies of ultramafic rocks and gabbroids, which are part of the ophiolite association of the Mid-Atlantic Ridge. They have established that restitic harzburgites in this association were injected by later mafic melts, which intruded the rocks through the systems of subparallel joints and, under the influence of these melts and their fluids, the petrographic, mineralogic, and geochemical properties of ultramafic restites changed to a varying degree. It is noteworthy that the results of these studies showed that, during the influence of mafic melts and their fluids on more ancient ultramafic rocks, clinopyroxenes from these rocks were enriched with Ti, Na, Zr, and LREE.

Mafic-ultramafic massifs of a varying size, inner structure, and material composition are widespread in the composition of ophiolite associations localized in the Caledonian and Hercynian folded structures in Mongolia, Tuva, and the Baikal region [Pinus *et al.*, 1958, 1984]. Results of studies evidence later injection of gabbroid intrusions than the spatially close protrusions of ultramafic rocks, which allows these massifs to be considered polygenous. We have obtained convincing evidence of the polygenous formation of massifs, such as Naran (Western Mongolia) [Lesnov, 1982], Shishkhid-gol (Northern Mongolia) [Lesnov *et al.*, 1977], Idzhim [Lesnov *et al.*, 2005b] and Ergak [Lesnov *et al.*, 2008] (Northeastern Tuva), and Chaya [Lesnov, 1972, 2007e] (Northern Baikal region). This can be, in particular, proven by the

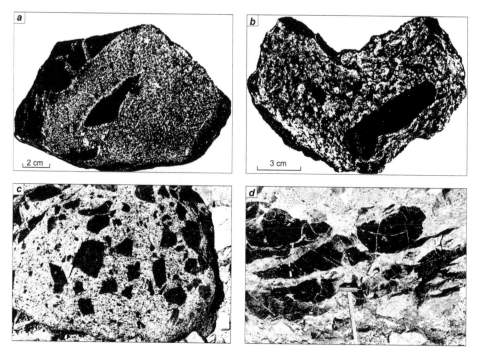

Figure 13.3 Lens-shaped and angular xenoliths of serpentinized peridotites, contained in the veins of orthomagmatic (*a*) and hybrid (*b*) gabbroids and pyroxenites (*c, d*) (Naran massif, Western Mongolia).

photographs of the rocks from some massifs, illustrating the relationships between ultramafic rocks and gabbroids (Figs. 13.3, 13.4, 13.5, *a*, *b*).

The arguments that are discussed below and taken as the basis of the proposed concept of polygenic formation of mafic-ultramafic massifs include the data obtained during the detailed study of the geologic structure and material composition of the well-known Lanzo massif, which is part of the ophiolite association in the Western Alps. [Piccardo *et al.*, 2005]. On the basis of these data the authors make a conclusion that parallel-banded structure and geochemical inhomogeneity in harzburgites, dunites, plagioperidotites, and pyroxenites from this massif are the result of infiltration of mafic melts and their active interaction with the preliminarily foliated ultramafic restites (see Fig. 13.5, *c, d*). This interaction resulted in the nonuniform distribution of porphyroblast segregations of olivine, orthopyroxene, clinopyroxene, and plagioclase in ultramafic restites. Moreover, Italian researchers showed that the nonuniform distribution of newly formed clinopyroxene crystals in ultramafic rocks from this massif is responsible for the nonuniform distribution of REE in these rocks.

An illustrative example of the polygenous mafic-ultramafic massif is, in our opinion, the Berezovka massif, described in detail in this monograph. The concepts of the later formation of gabbroid intrusion compared to the ultramafic protrusion in this massif are based on the following facts: (1) gabbroid intrusion contains xenoliths of ultramafic rocks; (2) ultramafic protrusion is cut by gabbroid vein bodies, at places

Figure 13.4 Relationships between ultramafic rocks and gabbroids in the Idzhim massif, Northern Tuva. *a* – Xenoliths of serpentinized peridotites of angular and oval shapes in pyroxenites; *b* – branching veins of pyroxenites in serpentinized peridotites; *c* – hybrid ultramafic rocks (wehrlites) (gray) with a parallel-banded structure, resulting from alternation of veinlets of hybrid gabbroids (light gray) and flat xenoliths of serpentinites (black); *d* – branching veins and veinlets of anorthosite (light gray) in plagioclase-bearing wehrlites (black). All images are reduced by 2.5 times.

passing into the veins of pyroxenites and wehrlites; (3) along the borders of the ultramafic protrusion and gabbroid intrusion there is a thick contact-reaction zone made up of hybrid ultramafic rocks and gabbroids, which resulted from the reaction of mafic melts with ultramafic rocks; (4) important evidence of the later formation of the gabbroid intrusion compared to the protrusion of ultramafic rocks is the representative isotope data on the polychronous nature of contained zircons, which were obtained for the first time. Zircons include rather young varieties, which are syngenetic with gabbroids, and much more ancient (relict and xenogenic) varieties contained in hybrid ultramafic rocks and gabbroids. As it was shown, until now the populations of ancient and much younger zircons have been found in different petrographic types of rocks from mafic-ultramafic massifs that are part of many ophiolite associations on the continents and those localized within the mid-ocean ridges.

When summarizing the above-mentioned data from comprehensive studies of mafic-ultramafic complexes and the results of investigations of many petrologists, we proposed a generalized geological section of polygenous mafic-ultramafic massifs, which gives the idea of the main features of their structure (Fig. 13.6). Let us

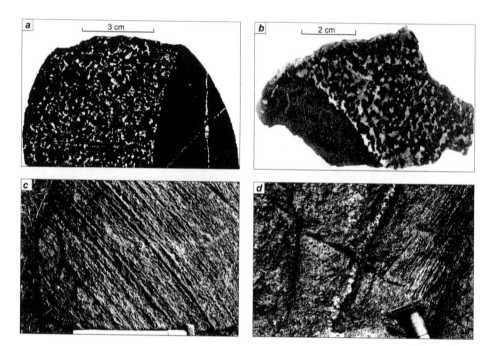

Figure 13.5 Evidence of polygenous formation of mafic-ultramafic massifs.
a – Contact of xenolith of serpentinized peridotite (black) with plagioclase-enriched plagiowehrlite (spotty gray), which contains small xenoliths of serpentinized peridotite (Chaya massif) (after [Lesnov, 1988]); *b* – contact of xenolith of serpentinized dunite with coarse-grained plagiowehrlite (spotty gray) (Lukinda massif) (after [Lesnov, 1988]); *c* – parallel-banded taxitic structure of hybrid ultramafic rocks, resulting from their enrichment in pyroxene along the cracks of foliation (Lanzo massif, Western Alps (after [Piccardo *et al.*, 2005])); *d* – the same as in the previous photograph but with a thin veinlet of leucocratic gabbro.

discuss the most important features of the geologic structure and the reconstruction of the probable conditions of formation of structure–material components which formed polygenous mafic-ultramafic massifs: (a) protrusions of restitic ultramafic rocks, (b) gabbroid intrusions, and (c) contact-reaction zones.

Protrusions of restitic ultramafic rocks, the outcropping area of which ranges from fractions to hundreds of square kilometers, are mainly lens-shaped and occur as belts of various lengths and their branches, tracing the zones of long-lived deep faults. The protrusions have predominantly steep, at places rather gentle tectonic contacts with enclosing volcanogenic-terrigenous and metamorphic rocks. Close to the faults surrounding the protrusions, ultramafic rocks are mainly represented by their serpentinized varieties, which in most cases underwent plastic and brittle deformation, foliation, and brecciation of various intensities. In some cases, in the zones of such contacts ultramafic rocks preserved thin and coarse platy cleavage (Figs. 13.7, 13.8) [Lesnov, 1981a], which was in the majority of cases destroyed during later brittle deformations. Most likely, during the later infiltration of mafic melts and their fluids

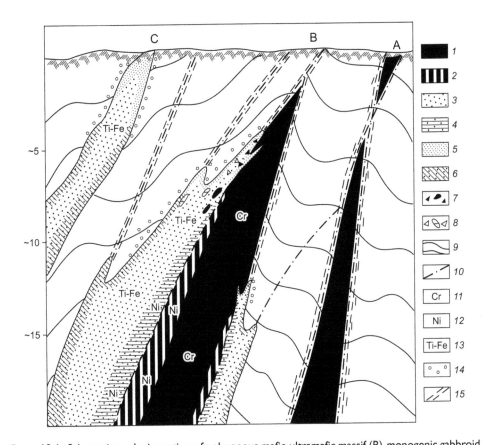

Figure 13.6 Schematic geologic section of polygenous mafic-ultramafic massif (B), monogenic gabbroid intrusion (C), and monogenic ultramafic massif (protrusion) (A) (after [Lesnov, 1986a]). *1* – restitic (orthomagmatic) ultramafic rocks (lherzolites, harzburgites, dunites, and their serpentinized varieties); *2* – hybrid (paramagmatic) ultramafic rocks (wehrlites, websterites, enstatitites, bronzitites, clinopyroxenites and their olivine- and plagioclase-bearing varieties, plagiolherzolites, plagioharzburgites, plagiodunites, metamorphic olivine, olivine–serpentine, amphibole–olivine, harzburgite, and lherzolite parageneses); *3* – orthomagmatic gabbroids (gabbro, gabbronorites, norites, their hornblende varieties); *4* – hybrid (paramagmatic) gabbroids from the contact zones of gabbroid intrusions with ultramafic protrusions (olivine gabbro, olivine gabbronorites, troctolites, their leuco- and melanocratic varieties, anorthosites); *5* – orthomagmatic gabbroids from the quenched zones of shallow-depth gabbroid intrusions (ophitic, often fine-grained gabbro, gabbro-diabases, diabases); *6* – hybrid (paramagmatic) gabbroids from the contact zones of gabbroid intrusions with host volcanic and terrigenous rocks and their metamorphic derivates (meso- and leucocratic amphibole, quartz- and biotite-bearing gabbro, gabbro-diorites, diorites); *7* – ultramafic xenoliths; *8* – host rock xenoliths; *9* – volcanic and terrigenous rocks and their metamorphic derivates; *10* – faults; *11–13* – sites of mineralization: *11* – chromitites with platinum-metal micromineralization; *12* – sulfide copper–nickel ores with platinum-metal micromineralization; *13* – iron–titanium ores; *14* – hornfelses; *15* – brecciated and foliated zones (not shown in ultramafic rocks).

Figure 13.7 Thin-platy parting in serpentinites (Alagula massif, western Mongolia), fragmented (*a*) and nonfragmented (*b*) microlithons.

Figure 13.8 Coarse-platy parting in dunites (Kytlym massif, Urals) (after [Efimov and Efimova, 1967]).

along the cracks of platy jointing, hybrid ultramafic rocks and gabbroids with parallel-banded and other banded structures, which are part of contact-reaction zones, formed in ultramafic rocks (Fig. 13.9).

Within many ultramafic protrusions one can observe irregularly alternating zones of streaky, lens, or irregular shape, made up of lherzolites, harzburgites, or dunites, without sharp borders. Many of ultramafic protrusions are spatially close to the intruding gabbroid intrusions, forming polygenous mafic-ultramafic massifs, but the type of

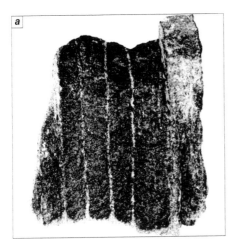

Figure 13.9 Samples of parallel-banded structures (after [Lesnov, 1988]).
a – Parallel-oriented veinlets of clinopyroxenite (gray) in serpentinite (dark gray), formed during infiltration of mafic melt into the system of cracks of platy parting (Nurali massif, Urals, magnification ∼1.2); *b* – subparallel-oriented flat xenoliths of serpentinized wehrlite (black) containing rare porphyroblasts of plagioclase, alternating with subparallel veins of olivine-bearing anorthosite (light gray) (Troodos massif, Cyprus).

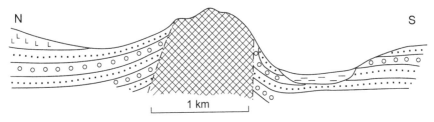

Figure 13.10 Schematic geologic section of the Nogontsav ultramafic massif (shown by a rectangular grid), occurring as a subvertical protrusion in the Cretaceous molasse deposits (southern Mongolia) (after [Lesnov, 2007d]).

ultramafic protrusions which is also widespread is represented by independent bodies of these rocks, which occur at a distance from gabbroid intrusions.

By the example of the Nogontsav mafic-ultramafic massif, which is localized in the Hercynian structures of southern Mongolia, it was established that ultramafic protrusions after their intrusion into the upper levels of the Earth's crust underwent permanent upward movements. It is shown in the schematic section of this massif that the ultramafic protrusion penetrated as a subvertical tectonic block into the overlapping terrigenous deposits. Directly close to the contacts with the protrusion, the bedding of deposits becomes steeply inclined on both sides of it (Fig. 13.10). It is also worth noting that straight toward the west and east of the Nogontsav massif there occur several small volcanic cones composed of Quaternary basalts. This evidences that the deep fault zone in which the Nogontsav massif occurs was active until now [Lesnov, 2007d]. The evidence for the permanent ascending displacements of ultramafic protrusions is also the fact that most of them are exposed in the positive forms of the relief. As far back as 1964, F.R. Boyd and I.D. MacGregor, after summarizing the available data on the structure and conditions of occurrence of ultramafic bodies, drew a conclusion, proven later, that Alpine-type peridotites are the products of the uppermost mantle intruded in solid state [Boyd and MacGregor, 1968].

Gabbroid intrusions, spatially close to the protrusions of ultramafic rocks, varying in size and morphology, occur mostly along hanging tectonic contacts of protrusions with framing rocks; at places they are localized along the lying contacts or inside the protrusions. The intrusions can be less extended than the adjacent protrusions of ultramafic rocks, comparable with them or more extended. In addition to those gabbroid intrusions that occur at the immediate contact with the protrusions of ultramafic rocks and form polygenous massifs with them, mafic-ultramafic belts typically include (independent) gabbroid intrusions at a distance from ultramafic protrusions. Shallow-depth (hypabyssal) intrusions are commonly made up of nearly exclusively orthomagmatic olivine-free gabbronorites and gabbro. In mid-depth (mesoabyssal) gabbroid intrusions, when approaching the protrusions of ultramafic rocks, olivine-free gabbroids are generally replaced by their hybrid olivine-bearing varieties that have a varying quantitative mineral composition and taxitic, including parallel-banded, structures. In many mafic-ultramafic massifs the composition of gabbroid intrusions includes ultramafic xenoliths of varying sizes and shape, which is one of main criteria of their later intrusion compared to the protrusions of ultramafic rocks. The direct contacts of gabbroid intrusions with the protrusions of ultramafic rocks are often disturbed by faults and overlapped by loose sediments. The endocontact zones of abyssal and mesoabyssal gabbroid intrusions along their borders with enclosing series are typically made up of amphibole- and quartz-bearing gabbro, gabbro-diorites, diorites, and quartz diorites, which crystallized from mafic melts contaminated with the substance of host rocks and, thus, are hybrid rocks. In the endocontacts of hypabyssal gabbroid intrusions along the borders with host rocks of predominant occurrence are fine-grained olivine-free gabbro and gabbro-diabases, whereas the host rocks contacting with them have signs of amphibolization.

The contact-reaction zones in most polygenous mafic-ultramafic massifs are localized along the hanging, in places along lying contacts of ultramafic protrusions with penetrating gabbroid intrusions formed under abyssal and mesoabyssal conditions. The thickness of these zones ranges from a few tens of meters to a few kilometers. Hybrid ultramafic rocks and gabbroids making up these zones have a more nonuniform quantitative mineral composition than the orthomagmatic rocks from the ultramafic protrusions and gabbroid intrusions.

The structure of contact-reaction zones includes petrographic varieties of hybrid ultramafic rocks, such as plagioharzburgites, plagiolherzolites, wehrlites, plagiowehrlites, websterites, clinopyroxenites and their olivine- and plagioclase-bearing varieties as well as hybrid gabbroids, such as melano-, meso-, and leucocratic olivine gabbronorites and gabbro, troctolites, and, rarely, anorthosites. All these varieties of rocks are characterized by a diversity of structures and textures, but commonly they have parallel-banded structures. The borders between alternating segregations of ultramafic rocks and gabbroids varying in the quantitative mineral composition change from sharp to gradual, "blurred". Hybrid ultramafic rocks from the contact-reaction zones often contain porphyroblastic segregations of olivine, orthopyroxene, clinopy-roxene, and plagioclase and subparallel or differently directed veins and veinlets of pyroxenites, olivine gabbroids, and anorthosites. It was established that the content of anorthite component in plagioclases from some plagioclase-bearing ultramafic rocks is lower than that in the mineral from gabbroids associated with ultramafic rocks. Hybrid gabbroids from the contact-reaction zones of many massifs contain xenoliths of ultramafic rocks, which have an angular, lens, oval, or irregular shape. The contacts of xenoliths with enclosing gabbroids also change from sharp to gradual, "blurred" (see Figs. 3.5, 3.6, 3.10–3.13, 13.4, 3, 13.5, 3, 13.9).

As it was mentioned above, parallel-banded structures of gabbroids and ultramafic rocks, which make up the transition zones between the bodies of ultramafic rocks and gabbroids, have long been interpreted as the main evidence that complex mafic-ultramafic massifs and all the rocks composing them are the result of intrachamber gravitative crystallization differentiation of mafic melts, which are responsible for the formation of different banded rocks called cumulates which inherit both visible and "cryptic" layering. However, as it was shown in Chapter 8, analyses of plagioclase from the parallel-banded wehrlite-gabbroid rock, a sample of which was picked at the Berezovka massif, confirmed that plagioclase grains from the alternating "bands" of this rock with different quantitative mineral compositions have a nearly stable chemical composition (i.e., "cryptic" layering is absent in this banded rock). Thus, we can refer to the data by Kushiro [1983], who showed that the presence of xenoliths of ultramafic rocks in basalts or gabbro evidences that the parental melts of these rocks could not undergo fractional crystallization and gravitative separation of crystals, at least, after ultramafic xenoliths occurred in these melts.

The contact-reaction zones of many mafic-ultramafic massifs typically contain clinopyroxene-enriched varieties of hybrid ultramafic rocks, such as wehrlites, clinopy-roxenites, websterites, and their plagioclase-bearing varieties. To establish the reasons that could be responsible for this petrologic composition of contact-reaction zones, we may turn to the data of McBirney [1983]. Using the results of physical experiments, he showed that in those experiments, in which olivine grains were added into the melt dur-ing crystallization of clinopyroxenes from mafic melt, the quantity of clinopyroxene crystallized from such contaminated melt increased. These observations suggest that clinopyroxene-rich hybrid ultramafic rocks, which make up contact-reaction zones, formed as a result of crystallization of mafic melts that had been to a varying degree contaminated with ultramafic restites. The data of physical experiments by McBir-ney and our field and petrographical observations agree with the results of numerical experiments of Velinskii *et al.* [1999]. Having conducted thermodynamic calculations of influence of mafic melt on ultramafic rocks at temperatures from 1200 to 25°C and pressure 1 kbar, the authors concluded that this interaction must lead to contamina-tion of mafic melt with ultramafic material and further crystallization of rocks, such as melanocratic olivine gabbronorites, with initial ultramafic rocks transforming into plagioclase-bearing peridotites, olivine websterites, and olivine clinopyroxenites.

One more indication of the heterogeneity of ultramafic rocks and gabbroids, the bodies of which form mafic-ultramafic massifs, is the commonly observed discrete-ness of chemical compositions of this group of rocks. This discreteness is manifested, first of all, in the bimodal or polymodal structure of histograms for the frequency of occurrence of main chemical components, constructed on the basis of sampled chemical analyses in an individual massif, and in the isolated character of the fields of figurative points on the two-dimensional diagrams of relationships between the contents of some oxides or those constructed in the coordinates of parameters CaO/Al_2O_3 – $FeO_{tot}/(FeO_{tot} + MgO)$ and others. As it was already pointed out, contents of chemical components in hybrid ultramafic rocks and gabbroids are distributed less uniformly than in restitic ultramafic rocks and orthomagmatic gabbroids. Belousov [1982] made a conclusion that revealing of the discreteness of chemical compositions of a particular set of rocks is a rather reliable feature of their genetic autonomy.

Let us discuss the reconstruction of the supposed multistage "scenario" of geodynamic and petrogenetic processes that formed those mafic-ultramafic massifs which are proposed to be determined as polygenous. This reconstruction is based on the collection of data from the previous chapters and the author's earlier published works in which different aspects of the structural position of these massifs, their internal structure, petrography, petrochemistry, geochemistry, mineralogy, isotope geochronology, and minerageny were discussed.

The first stage of "scenario" under study began in different segments of the Earth nonsimultaneously. As a result of heat penetration from the deeper levels of the asthenosphere and decompression in the root zones of long-lived faults at a depth of ~50–150 km, there formed chambers of partial melting of the upper-mantle source, which is assumed to have the chemical composition of pyrolite and corresponded to one part of basalt and three parts of dunite [Ringwood, 1972]. As the main mineral phases, the upper-mantle source contained olivines, orthopyroxenes, clinopyroxenes, pyrope garnet, accessory Cr-spinels, zircons, and some other minerals. In the petrographic composition the upper-mantle source could be similar to garnet and spinel lherzolites.

On partial melting of the upper-mantle substance, the first to pass into the melt were clinopyroxenes and garnet, and basalt melts of tholeiite composition and complementary refractory residue (ultramafic restites) were generated. After accumulation of rather large volumes, basaltic melts began to filter through the fault zones upward to the above-lying levels of asthenosphere and lithosphere. After reaching the crust surface, basaltoid melts erupted on the bed of ocean basins and within oceanic islands, forming thick series of initial volcanics of basic composition. At the same time, greater volumes of melts, before reaching the surface, cooled at the intermediate levels of feeding channels and formed gabbroid intrusions of varying depths.

It is likely that during the partial melting of the upper-mantle sources and extraction of basaltic melts, the latter trapped some quantity of zircon grains as xenogenic phase. The zircon grains were syngenetic with the ultramafic substance of these sources and, hence, had the same ancient isotopic age as the ultramafic substance. This assumption can be made, in particular from findings of zircon grains in basaltoids. Tables 10.1 and 10.2 show data on the resorbed grains of very ancient zircons (2170–1188 Ma) found in the samples from trachyandesibasalts of the Bogataya Formation of probably Cretaceous age, which are exposed on the framing of the Berezovka massif. Another example is the resorbed zircon grains from dolerites of the complex of parallel dikes near Mt. Azov (Middle Urals), dated 1984–1555 Ma [Ivanov, Berzin, 2013]. The authors report that ancient zircons were extracted by mafic melts during partial melting of the upper mantle and, thus, may evidence its age. It is worth noting that from the information published in *Nature Geoscience* the beach sands of Mauritius Island, formed during the washout of young basalts, contain zircon grains that are dated 1971 to 840–660 Ma. These data suggest that the xenogenic grains of ancient zircons can be found both in ancient and in young basaltoids from other provinces.

Ultramafic restites, which formed from partial melting of the upper-mantle substrate, are mainly variously depleted varieties of lherzolites, clinopyroxene-free and clinopyroxene-bearing harzburgites, and dunites. In addition to widely spread accessory Cr-spinel and microparticles of PGE, they sometimes contain relict pyrope and zircon grains. At the same time, it is reasonable to assume that a very ancient isotopic age of some zircon grains from ultramafic restites corresponds to the minimum age of their upper-mantle sources.

During formation, ultramafic restites were nonuniformly depleted with easily fusible chemical elements, including REE, and simultaneously enriched with refractory chemical elements, mainly Mg, Cr, and refractory PGE. Evidence for this fractionation of elements on partial melting of the upper-mantle sources is the inverse relationship, revealed in the samples of rocks from some mafic-ultramafic massifs between REE and PGE contents.

Irregular alternation of small bodies composed of lherzolites, harzburgites, or dunites and having no distinguished border, which is observed in many ultramafic protrusions, and wide variations in the values of Cr# parameter in contained accessory Cr-spinels suggest that the degree of partial melting of the upper-mantle sources during formation of ultramafic restites even within their rather small volumes was nonuniform. Most likely, at this stage of evolution of the upper-mantle sources, bodies of massive and impregnated chromitites of varying scale, morphology, and composition formed in ultramafic restites.

At the second stage, ultramafic restites resulting from the partial melting of upper-mantle sources, which were initially strongly heated plastic masses, cooled during a long period and, losing their plasticity, became solid and rather brittle rocks. Then, during the cycles of stretching and compression, which periodically replaced each other, blocks of ultramafic restites underwent multiple translational movements through deep faults upward to the zones of lower pressure, some of their volumes as protrusions reaching the bed of ocean basins and day surface. During the upward displacements, the protrusions of ultramafic rocks transformed into elongate (in plan) and upwardly pinching-out blocks with their long axis oriented along the strike of traced zones of deep faults, forming belts and their branches of different lengths. The protrusions of ultramafic restites during their upward displacements underwent multiple plastic and brittle deformations, which led to the formation of zones of foliation and platy parting of different thicknesses at the periphery. The latter zone was later destroyed by brittle deformations with formation of tectonic breccias (mélange). The dynamometamorphic transformations of restites are responsible for their intense serpentinization in the marginal parts of protrusions under the effect of descending currents of ocean waters. Because of the decreased strength of serpentinized ultramafic rocks that make up the edges of protrusions, during the next upward displacement of the latter, these zones underwent "scraping", owing to which the protrusions, moving to higher levels, became smaller in volume. Many ultramafic protrusions, after reaching the highest levels of the crust, still underwent upward displacements and washout. It is known that in most cases the protrusions of ultramafic rocks are exposed within the positive forms of the relief, which is additional evidence for their advanced movement relative to the blocks of wall rocks in the modern epoch of relief formation.

At **the third stage** of the upper-mantle evolution, which came after a long break in its activity, in the root zones of long-lived deep faults there began another cycle of partial melting of its substrate, both already depleted and unaffected by this process. It is worth noting that the resulting basaltoid melts trapped, as xenogenic phase, zircons grains from the upper-mantle source, the age of which was equal to that of the upper-mantle substance. Later, the basaltoid melts started to filter upward through the refurbished channels of deep faults and, after reaching the upper levels of the Earth's crust and starting to crystallize, formed gabbroid intrusions of different sizes, both spatially close to the earlier intruded ultramafic protrusions, giving rise to *polygenous* mafic-ultramafic massifs and monogenic gabbroid bodies remote from the protrusions. These mafic melts were injected mainly along more permeable hanging tectonic contacts of ultramafic protrusions with framing rocks, less commonly, rising along less permeable lying tectonic contacts of protrusions with framing rocks and along the dislocations with a break in continuity, crosscutting the ultramafic protrusions.

The fourth stage began prior to the completion of the previous one, i.e., during the injection of mafic melts along the tectonic contacts of ultramafic protrusions. In that period, during cooling and crystallization of mafic melts and their fluids, they interacted, with a varying degree of intensity, with differently disintegrated rocks of ultramafic protrusions and adjacent enclosing series. Depending on the depth of formation of gabbroid intrusions and consequent time of melt crystallization, as well as on the degree of disintegration of wall rocks, the interaction of melts with them is responsible for the formation of contact-reaction zones of varying thickness, structure, and chemical composition, occurring along the borders of gabbroid intrusions with the protrusions of ultramafic and framing rocks. In the zones of contacts with ultramafic protrusions there formed hybrid ultramafic rocks and gabbroids with nonuniform chemical and quantitative mineral compositions, structure, and texture. The time of crystallization of mafic melts at different depths, their fluid saturation, and the degree of disintegration of bordering sites of ultramafic protrusions and host rocks, to a considerable degree, determined the scales of magma-metasomatic transformations of wall rocks, the intensity of contamination of mafic melts with these rocks, and the thicknesses of contact-reaction zones. It is noteworthy that transformations of ultramafic restites under the influence of mafic melts and their sulfur-enriched fluids were commonly followed by redistribution and local concentration of ore components, such as Ni and PGE.

Long-term interaction of mafic melts and their fluids with dynamometamorphosed ultramafic rocks in abyssal and mesoabyssal conditions was the most important reason for the origin of indistinct, "blurred" contacts and gradual transitions between gabbroid intrusions and their apophyses, on the one hand, and contained xenoliths of ultramafic and host rocks, on the other. During the infiltration of mafic melts and their fluids into ultramafic rocks through the systems of subparallel cracks of platy parting and foliation, there took place interchange of chemical components, contamination of melts with the substance of ultramafic and host rocks, and, as a result, formation of various hybrid ultramafic rocks and gabbroids with parallel-banded and other taxitic structures.

Intensification of such interaction and interchange of components was mostly due to the high level of dynamometamorphic disintegration of ultramafic rocks. Under certain conditions, hybrid gabbroids preserved ultramafic xenoliths of different sizes and shapes, represented by their hybrid species. During the formation of gabbroid intrusions in hypabyssal conditions and, respectively, at elevated crystallization rates of mafic melts, interaction of the latter with ultramafic rocks and rocks of enclosing series was less durable and active. This is the reason of much smaller-scale magma – metasomatic transformations of wall rocks. Owing to this, xenoliths of ultramafic and host rocks are of angular shape and have sharp contacts [Lesnov, 2009f].

Thus, within the framework of the discussed multistage reconstruction of formation of mafic-ultramafic massifs, which are part of ophiolite associations, it is assumed that the main process that is responsible for the formation of the whole diversity of petrographic varieties of ultramafic and mafic rocks making up these massifs was *differently effective magma-metasomatic interaction and integration of mafic melts with the rocks of intruded ultramafic protrusions and host terrigenous-volcanogenic and metamorphic rocks*, rather than intrachamber gravitative crystallization differentiation of mafic melts with the formation of "cumulative" rocks. Processes of such integration were typically followed by larger- or smaller-scale contamination of mafic melts with the substance of ultramafic restites and rocks of enclosing series with formation of hybrid gabbroids. A similar magma-metasomatic interaction could be the most intense in abyssal conditions on long-term cooling and crystallization of mafic melts in the presence of fluid components.

* * *

It is proposed to distinguish the mafic-ultramafic massifs being part of ophiolite associations, in which later injection of gabbroids relative to spatially close protrusions of ultramafic restites was established, as a special category of *polygenous mafic-ultramafic massifs*. The main features that evidence the later formation of gabbroid bodies than the bodies of ultramafic rocks are (1) presence of ultramafic xenoliths in gabbroid intrusions that are spatially close to ultramafic protrusions, (2) presence of veined gabbroid bodies and associated pyroxenite veins in ultramafic protrusions, (3) presence of contact-reaction zones along the borders of ultramafic protrusions and gabbroid intrusions, (4) lack of "cryptic layering" in banded gabbroids and ultramafic rocks which make up these contact zones, (5) steep up to subvertical contacts of ultramafic protrusions and penetrating gabbroid intrusions with the rocks of enclosing series, (6) discreteness of chemical composition of rocks that make up the ultramafic protrusions and gabbroid intrusions, and (7) presence of ancient relict and xenogenic grains in ultramafic rocks and hybrid gabbroids and much younger syngenetic zircon grains in gabbroids.

In the proposed concept of polygenous formation of mafic-ultramafic massifs one of the basic points is that in mesoabyssal and especially abyssal conditions the upper-mantle mafic melts could actively interact with the rocks of more ancient ultramafic protrusions. As a result of this interaction, in the contacts of gabbroid intrusions with ultramafic protrusions there formed contact-reaction zones of varying thickness and heterogeneous petrographic composition, which were made up of hybrid ultramafic rocks and gabbroids. In the formation of these zones the leading process was magma-metasomatic integration of the substance of ultramafic restite rocks with injected mafic melts.

The proposed concept of polygenous formation of mafic-ultramafic massifs allows consistent interpretation of many established facts and regularities concerning the internal structure and space-time relationships between the bodies of ultramafic rocks and gabbroids, as well as interpretation of their petrochemical, petrographic, geochemical, and mineralogic properties, and isotope-geochronological parameters. However, this concept, undoubtedly, requires a more strict substantiation involving additional geological, structural, geochemical, and isotope data and results of physical and numerical experiments. Without this kind of new information, this model does not allow complete reconstruction of the entire sequence of geodynamic events in the formation of such massifs and determination of the main cause-effect relationships between the processes that led to their formation.

Conclusions

Differently aged mafic-ultramafic massifs of diverse structure and petrographic composition are widespread in the Earth's main structures, from midocean ridges and folded areas to ancient shields. The study of such massifs, including those belonging to ophiolite associations, by structure-geological, geophysical, petrographic, petrochemical, geochemical, mineralogical, isotope-geochronological, and experimental methods and petrogenetic modeling of their formation is a priority in modern petrology. Along with the solved problems in this field of geological knowledge, there remain many disputable questions concerning the geologic structure of mafic-ultramafic massifs, spatiotemporal relations between ultramafic and gabbroid bodies, their composition at the macro- and microlevels, and physicochemical interpretation of the processes which produced the diversity of the constituent rocks.

In the last decades, petrological studies at many science centers and reconstructions of the processes which produced the wide range of mafic-ultramafic massifs and their rocks have conventionally been carried out based on the classical theory of gravitative crystallization fractionation of mafic melts and on plate tectonics, postulating the stratified structure of mafic-ultramafic massifs of ophiolite associations.

On the other hand, the new petrological data obtained over the past decades contradict the above theories, including the ideas of stratified structure of ophiolite associations and mafic-ultramafic massifs composing them; nearly simultaneous formation of ultramafic and gabbroid bodies; formation of banded rock series as the most important component of these massifs, resulting from intrachamber gravitative crystallization differentiation of basaltoid melts, and other views on the origin of such plutonic bodies.

According to regional studies, numerous belts of mafic-ultramafic massifs of different lengths are localized among fold–block structures of different ages along the Pacific coast of the Asian continent in the N–S direction from Chukchi Peninsula to Koryakia, through Kamchatka Peninsula, Sakhalin Island, Primorye, and islands of Japan, and then to Southeast Asia. These belts, consisting of many adjacent branches, mark global zones of long-lived deep-seated faults and feathering faults. One of these branches is formed by the mafic-ultramafic massifs confined to the structure of the fault zone along the eastern shore of Sakhalin Island. These massifs are assigned to the East Sakhalin ophiolite association. The major ones are the differently studied Berezovka, Shel'ting, Komsomol'sk, and South Schmidt massifs. They differ in structural position, shape, size, internal structure, and the occurrence of ultramafic and gabbroid rocks, which outcrop at the present-day erosion levels.

We determined the Berezovka and other studied mafic-ultramafic massifs of the East Sakhalin ophiolite association as *polygenic* plutonic bodies, based on results of field observations and petrographic, petrochemical, geochemical, mineralogical, and isotope-geochronological data. Numerous mafic-ultramafic bodies belonging to ophiolite associations of Northeastern Russia, Mongolia, and other regions are assigned to this type of massifs.

Four main genetically distinct structure–material components are generally distinguished in polygenic mafic-ultramafic massifs: (1) restitic ultramafic protrusion; (2) orthomagmatic gabbroid intrusion cutting the ultramafic protrusion; (3) contact-reaction zone localized along the boundaries of the gabbroid intrusion and ultramafic protrusion; it is composed of hybrid ultramafic and gabbroid rocks resulting from interaction between mafic melts and rocks of the ultramafic protrusion and from their contamination with ultramafic material; and (4) contact-reaction zone of hybrid gabbroids formed by interaction between mafic melts and enclosing rocks and by contamination with their substance.

Ultramafic restites occur as lherzolites, harzburgites, dunites, and their serpentinized varieties in different proportions. Orthomagmatic gabbroids include olivine-free gabbronorites, gabbro, and rarer norites. Hybrid ultramafic rocks are plagioharzburgites, plagiolherzolites, and wehrlites; websterites, orthopyroxenites, and clinopyroxenites and their olivine- and plagioclase-bearing varieties; as well as rare olivinites and amphibole peridotites. Hybrid gabbroids from contact-reaction zones localized along the boundaries of ultramafic protrusions and gabbroid intrusions are melano-, meso-, and leucocratic olivine gabbro and gabbronorites as well as rarer troctolites and anorthosites. Hybrid gabbroids at contacts with enclosing rocks are amphibole- and quartz-bearing gabbro, gabbro-diorites, diorites, and quartz diorites. In total, >20 petrographic varieties of ultramafic and gabbroid rocks were described in the studied massifs of Sakhalin Island. In general, restitic ultramafic rocks, like orthomagmatic gabbroids, are more homogeneous in texture, structure, and quantitative mineral and chemical compositions than hybrid ultramafic rocks and gabbroids. The main minerals in most of the varieties of rocks making up polygenic massifs occur in different proportions: olivine, ortho- and clinopyroxene, plagioclase, amphibole, and accessory minerals (Cr-spinel, magnetite, and rarer titanomagnetite, zircon, apatite, titanite, awaruite, iron sulfides, and microparticles of PGE minerals).

Petrochemical data on the studied massifs testify to the more or less discrete chemical compositions of ultramafic rocks and gabbroids. Restitic ultramafic rocks from the Berezovka massif are characterized by low values of the Mg/Si parameter, elevated values of the Ca/Al parameter, and their wider variations compared to those for ultramafic rocks of the Shel'ting and South Schmidt massifs. The low values of the Ca/Al parameter in ultramafic rocks from the Shel'ting and South Schmidt massifs make them similar to pyrolite. As seen from the values of the above parameter, these ultramafic rocks were more depleted during the partial melting of the upper-mantle source.

Ultramafic rocks and gabbroids from the studied mafic-ultramafic massifs are depleted in REE, which are mainly concentrated as an isomorphous (structural) impurity in clinopyroxenes and amphiboles. A much smaller contribution to the general balance of REE in these rocks is made by orthopyroxenes, plagioclases, and accessory minerals (apatites, zircons, and titanites). Also, as a result of epigenetic processes, variable contents of these elements were concentrated in the rocks and their minerals as a nonstructural impurity in inter- and intragranular microcracks. Gabbroid and ultramafic rocks from the studied massifs are characterized by low Rb and Sr contents and low values of the $^{87}Rb/^{86}Sr$ and $^{87}Sr/^{86}Sr$ parameters, which points to the limited contribution of crustal rocks to contamination of mafic melts.

The Mg-number in olivines increases in passing from olivine gabbro and olivine orthopyroxenites to plagioclase lherzolites. In orthopyroxenes this parameter increases as we pass from olivine orthopyroxenites to websterites. Orthopyroxenes from gabbronorites are similar in composition to bronzite. The Mg# values in clinopyroxenes from lherzolites and websterites of the South Schmidt massif are higher than those in clinopyroxene from plagiolherzolites, plagiowehrlites, and olivine and olivine-free gabbroids of the Berezovka massif. Plagioclases from rocks of this massif vary in composition from bytownite in gabbronorites to anorthite in other gabbroids. According to incomplete chemical analyses of plagioclase grains localized along the profile oriented across the strike of banding in wehrlite–gabbroid rock from the contact-reaction zone of the Berezovka massif, the composition of these plagioclases remains almost unchanged; this testifies to the lack of "cryptic layering". Parallel-banded rocks, widespread in the contact-reaction zones of polygenic mafic-ultramafic massifs, are produced by the infiltration of mafic melts and their fluids through platy-parting systems in ultramafic rocks and interaction with them.

Isotope analyses of zircons from the Berezovka massif rocks have shown that they belong to different age populations. Note that zircons from different populations differ in grain size and morphology as well as in optical and geochemical characteristics. Occasional zircons in the total set are from the oldest population ($\sim$3100–990 Ma). The overwhelming majority of these grains makes a population aged $\sim$190–140 Ma. Zircon grains from the oldest population, which are strongly resorbed and almost lack crystal faces and cathodoluminescence, occur predominantly in hybrid ultramafic rocks (pyroxenites and gabbro-pyroxenites). Zircons of this type are regarded as a relict phase originally contained in the upper-mantle source; therefore, their age must show the minimum ages of this source. The predominant zircon grains in the set, from a population aged $\sim$170–150 Ma, mainly occur in orthomagmatic gabbronorites and gabbro (less often, in hybrid amphibole gabbro, gabbro-diorites, and diorites). All these grains are characterized by clear crystal faces, moderate to intense cathodoluminescence, and fine rhythmic oscillatory zoning parallel to the crystal faces. This type of zircon grains is viewed as a syngenetic phase which crystallized from the mafic melt that formed the gabbroid intrusion; so, their age corresponds to the time of formation of this intrusion. Zircon grains aged from $\sim$790 to $\sim$200 Ma are observed both in hybrid ultramafic rocks and in gabbroids. Also, many of them have a resorbed surface and different (generally weak) cathodoluminescence intensity. Zircon grains of this type are considered xenogenic. It is presumed that they were initially contained in the upper-mantle source and yielded corresponding ancient age. Later a part of these grains was preserved in hybrid ultramafic rocks, whereas the other was trapped by basaltoid melts, within which the zircon grains were transported to the upper crust. Maybe, they were subjected to the thermal and chemical action of mafic melt; this caused disturbances of different intensity in their trace-element composition and U–Pb isotope systems and, therefore, their more or less significant "rejuvenation". In general, the results of the performed isotope-geochronological studies of zircons from rocks of the Berezovka massif agree with the polygenic model of its formation.

The first data on the distribution of PGE in rocks from mafic-ultramafic massifs of Sakhalin Island have shown that the total contents of Os, Ir, Ru, Rh, Pt, and Pd in them vary from 8.6 to 46.0 ppb. Harzburgites and lherzolites have elevated Os, Ir, and Ru contents, whereas nearly equal contents of Rh, Pt, and Pd are observed in harzburgites, lherzolites, wehrlites, pyroxenites, and gabbroids. In massive chromitites from the Berezovka massif the total PGE contents were 693–834 ppb, which is close to those in chromitites from some other chromite-bearing mafic-ultramafic massifs. During a SEM study of massive chromitites, PGE sulfide belonging to the laurite–erlichmanite isomorphous series was first found in situ for that massif. This mineral might be the main, if not the only, PGE concentrator in the Berezovka massif chromitites.

The total set of data obtained on the structural position of mafic-ultramafic massifs on Sakhalin Island, their geologic structure, petrography, petrochemistry, geochemistry, mineralogy, and zircon isotopic age has provided a better basis for the proposed theory of polygenic formation of such massifs. This process is divided into several tentative stages: (1) heating and partial melting of upper-mantle sources, which produced ultramafic restites and complementary mafic melts; ascending along deep-fault zones, they crystallized in feeding channels, forming intrusions, and partly erupted on the ocean floor and on land, with the formation of volcanic series; (2) emplacement of solid–plastic blocks (protrusions) of ultramafic restites into the Earth's crust along the same but "renewed" deep-fault zones; (3) the next cycle of activity of deep-fault zones and related partial melting of the upper-mantle source, generating mafic melts; (4) ascent of these mafic melts along fault zones to the upper crust and their crystallization, during which gabbroid intrusions formed along the tectonic contacts of ultramafic protrusions with the framing rocks and at a distance from them; and (5) magma-metasomatic interaction of different intensity between the emplaced mafic melts and ultramafic restites, producing the compositionally heterogeneous hybrid ultramafic rocks and gabbroids of contact-reaction zones.

***

The performed studies of petrology of mafic-ultramafic massifs of the East Sakhalin ophiolite association provide a somewhat better basis for the theory of polygenic formation of such igneous bodies. The author hopes that, in the nearest future, petrologists will deal more with the problems that were discussed and, if possible, solved in the monograph and better substantiate (or, on the contrary, question) the presented petrogenetic reconstructions. After all, as Aristotle said, *"it will contribute towards one's object, who wishes to acquire a facility in the gaining of knowledge, to doubt judiciously"*.

17 January 2015

References

Agafonov L.V., Lkhamsuren Zh., Kuzhuget K.S., and Oidup Ch.K. (2005) Platinum mineralization of ultramafic–mafic massifs of Mongolia and Tuva. Ulaanbaatar. 224 p. (in Rus.).

Anders E. and Grevesse N. (1989) Abundances of the elements: Meteoritic and solar. *Geochim. Cosmochim. Acta, 53* (1), 197–214.

Anikina E.V., Krasnobaev A.A., Alekseev A.V., Busharina S.V., and Lepekhina E.N. (2009) Results of U–Pb dating of zircons from ore-bearing gabbro of the Baron gold–palladium ore occurrence (Volkov massif). *Ultrabasic–basic complexes of folded regions and related ore deposits.* Volume 1. Yekaterinburg: Publishing House of the Institute of Geology and Geochemistry, Uralian Branch of the Russian Academy of Sciences, 50–54 (in Rus.).

Aranovich L.Ya, Zinger T.F., Bortnikov N.S., Sharkov E.V., and Antonov A.V. (2013) Zircon in gabbroids from the axial zone of the Mid-Atlantic Ridge, Markov Deep, 6 N: correlation of geochemical features with petrogenetic processes. *Petrology, 21* (1), 4–19 (in Rus.).

Badanina I.Yu. and Malich K.N. (2012) Polychronous age of zircons in dunites from the Konder massif (Aldan Province, Russia). *Geochronometric isotopic systems, methods of their study, and the chronology of geologic processes.* Moscow: Publishing House of IGEM RAS, 49–52 (in Rus.).

Balashov Yu.A. (1976) Geochemistry of rare-earth elements. Moscow: Publishing House Nauka, 268 p. (in Rus.).

Balykin P.A., Polyakov G.V., Bognibov V.I., and Petrova T.E. (1986) Proterozoic mafic–ultramafic formations of the Baikal–Stanovoi area. Novosibirsk: Publishing House Nauka, 206 p. (in Rus.).

Balykin P.A., Krivenko A.P., Konnikov E.G., Lesnov F.P., Lepetyukha V.V., Litvinova T.I., Pushkarev E.V., and Fershtater G.B. (1991) Petrology of postharzburgite intrusions of the Kempirsay–Khabarninsk ophiolite association (Southern Urals). Sverdlovsk: Publishing House of the Institute of Geology and Geochemistry, Uralian Branch of the USSR Academy of Sciences. 160 p. (in Rus.).

Baryshev V.B., Kolmogorov Yu.P., Kulipanov G.N., and Skrynskii A.N. (1986) X-ray fluorescence elemental analysis using synchrotron radiation. *Journal of Analytical Chemistry, 41* (3), 389–401 (in Rus.).

Batanova V.G., Brigman G.E., Savelieva G.N., and Sobolev A.V. (2009) Using the Re–Os isotope system for dating mantle processes by the example of ophiolite complexes. *Ultrabasic-basic complexes of folded regions and related ore deposits.* Volume 1. Yekaterinburg: Publishing House of the Institute of Geology and Geochemistry, Uralian Branch of the Russian Academy of Sciences, 77–80 (in Rus.).

Bazhin E.A. (2013) Geodynamic conditions of formation of ultrabasic rocks in the zone of convergence of the Southern and Middle Urals. *Lithospheric structure and geodynamics.* Irkutsk: Publishing House of the Institute of the Earth's Crust, Siberian Branch of the Russian Academy of Sciences, 11–13 (in Rus.).

Bazylev B.A. (2003) Petrology and geochemistry of oceanic and Alpine-type spinel peridotites in connection with the problem of evolution of mantle material. *Abstract of dissertation.* Moscow: Publishing House of GEOKHI RAS. 36 p. (in Rus.).

Bazylev B.A., Popević A., Karamata S., Kononkova N.N., Simakin S.G., Olujić J., Vujnović L., and Memović E. (2009) Mantle peridotites from the Dinaridic ophiolite belt and the Vardar zone western belt, central Balkan: A petrological comparison. *Lithos, 108 (1–4),* 37–71.

Bekhtol'd A.F. and Semenov D.F. (1978) New data on the composition and structure of the Shel'ting gabbro-peridotite pluton (Sakhalin Island). *Doklady of the USSR Academy of Sciences, 243* (2), 445–448 (in Rus.).

Bekhtol'd A.F. and Semenov D.F. (1990) Metabasites and ultrabasites of the Susunai Ridge (Sakhalin Island). *Russian Journal of Pacific Geology,* (1), 121–126 (in Rus.).

Belousov A.F. (1982) Petrological interpretation of petrochemistry of igneous rocks. *Petrochemistry. Aspects of petrology and metallogeny.* Novosibirsk: Publishing House of the Institute of Geology and Geophysics, Siberian Branch of the USSR Academy of Sciences, 3–30 (in Rus.).

Bogatikov O.A., Petrov O.V., and Morozov A.F. (Eds.) (2009) Petrographic code of Russia. St. Petersburg: Publishing House of VSEGEI. 200 p. (in Rus.).

Bogdanov N.A. and Khain V.E. (Eds.) (2000) Explanatory note to the tectonic map of the Okhotsk Sea region. Moscow: Publishing House ILOVM, Russian Academy of Sciences. 193 p. (in Rus.).

Bortnikov N.S., Sharkov E.V., Bogatikov O.A., Zinger T.F., Lepekhina E.N., Antonov A.V., and Sergeev S.A. (2008) Finds of young and ancient zircons in gabbroids of the Markov Deep, Mid-Atlantic Ridge, 5°54′–5°02.2′ N (Results of SHRIMP-II U–Pb Dating): Implication for deep geodynamics of modern oceans. *Doklady Earth Sciences, 421* (2), 240–248 (in Rus.).

Boyd F.P. and MacGregor I.D. (1968) Ultrabasic rocks. *Petrological constitution of the upper mantle.* Moscow: Publishing House Mir, 278–282 (in Rus.).

Büchl A., Brügmann G., and Batanova V.G. (2004) Formation of podiform chromitite deposits: implications from PGE abundances and Os isotopic compositions of chromites from the Troodos complex, Cyprus. *Chemical Geology, 208* (1–4), 217–232.

Burnham A.D. and Berry A.J. (2012) An experimental study of trace element partitioning between zircon and melt as a function of oxygen fugacity. *Geochim. Cosmochim. Acta, 95,* 196–212.

Chekhov A.D. (2000) Tectonic evolution of Northeast Asia (marginal sea model). Moscow: Publishing House Nauchnyi Mir, 204 p. (in Rus.).

Coleman R.G. (1979) Ophiolites. Moscow: Publishing House Mir. 262 p. (in Rus.).

Cox K.G., Bell J.D., and Pankhurst P.J. (1982) The interpretation of igneous rocks. Moscow: Publishing House Nedra, 414 p. (in Rus.).

Danchenko V.Ya. (2003) Geologic position and substance-genetic types of ores of rare and precious metals in the South Okhotsk region on the Pacific framing. Yuzhno-Sakhalinsk: Publishing House of the Institute of Marine Geology and Geophysics, Far Eastern Branch of the Russian Academy of Sciences, 227 p. (in Rus.).

Davis D.W., Williams I.S., and Krogh T.E. (2003) Historical development of zircon geochronology. Zircon. *Reviews in Mineralogy and Geochemistry, 53 (1)* (Eds. Hanchar, J.M., Hoskin, P.W.O.), 145–182.

Dilek Y. and Polat A. (2008) Suprasubduction zone ophiolites and Archean tectonics. *Geology,* *36* (5), 431–432.

Dilek Y. and Robinson P.T. (Eds.) (2003) Ophiolites in Earth History. *Geological Society of London, Special Publication 218,* 723 p.

Dmitrenko G.G., Mochalov A.G., Palandzhan S.A., and Akinin V.V. (1987) Accessory minerals of platinum group elements in Alpine-type ultramafic rocks of the Koryak Highland. *Russian Journal of Pacific Geology,* (4), 66–76 (in Rus.).

Dobretsov N.L., Moldavantsev Yu.E., Kazak A.P., Ponomareva L.G., Savelieva G.N., and Saveliev A.A. (1977) Petrology and metamorphism of ancient ophiolites (for example, the Polar Urals and East Sayan). Novosibirsk: Publishing House Nauka. 222 p. (in Rus.).

Dodin D.A., Chernyshov N.M., and Cherednikova O.I. (2001) Metallogeny of platinum group elements in large regions of Russia. Moscow: Publishing House Geoinformmark, 302 p. (in Rus.).

Dodin D.A., Landa E.A., and Lazarenkov V.G. (2003) Platinum and titanomagnetite-bearing chromite deposits. *Platinum-bearing deposits of the world.* Volume 2. Moscow: Publishing House Geoinformtsentr, 410 p. (in Rus.).

Efimov A.A. (2010) The results of a century of studying the Platinum Belt of the Urals. *Litosfera* (5), 134–153 (in Rus.).

Efimov A.A. and Efimova L.P. (1967) The Kytlym platinum ore-bearing massif. Moscow: Nedra, 356 p. (in Rus.).

Evensen N.M., Hamilton P.J., and O'Nions R.K. (1978) Rare-earth abundances in chondritic meteorites. *Geochim. Cosmochim. Acta,* 42 (8), 1199–1212.

Explanatory note to the State Geological Map of the Russian Federation, scale 1 : 200,000. 2nd ed. Kurile Series. Sheet K-55-III (Malokuril'skoe) (Rybak-Franko *et al.*). St. Petersburg: Publishing House VSEGEI Map Factory. 2006. 157 p. (in Rus.).

Explanatory note to the State Geological Map of the Russian Federation, scale 1 : 200,000. 2nd ed. Sakhalin Series. Sheet M-54-XXIV (Pervomaisk) (Gal'versen *et al.*). St. Petersburg: Publishing House VSEGEI Map Factory. 2009. 157 p. (in Rus.).

Faure G. (1989) Principles of isotope geology. Moscow: Publishing House Mir, 590 p. (in Rus.).

Fershtater G.B. (2013) Paleozoic intrusive magmatism of the Middle and Southern Urals. Yekaterinburg: Publishing House of the Uralian Branch of the Russian Academy of Sciences, 368 p. (in Rus.).

Fershtater G.B., Krasnobaev A.A., Bea F., Montero P., Levin V.Ya., and Kholodnov V.V. (2009) Isotopic-geochemical features and age of zircons in dunites of the platinum-bearing type Uralian massifs: Petrogenetic implications. *Petrology, 17* (5), 539–558 (in Rus.).

Filonenko V.P. (2002) The structure of the Earth's crust of the South Okhotsk depression. *Modern problems of geology. Materials of youth conference. Yanshin readings, 2.* Moscow: Publishing House Nauchnyi Mir, 65–68 (in Rus.).

Ganelin A.V., Sokolov S.D., Leier P., and Simonov V.A. (2012) The first isotope dating of plutonic rocks of the Aluchin ophiolite complex (Western Chukotka Peninsula) and their geodynamic consequences. *Ultrabasic–basic complexes of folded regions and their mineragency.* Ulan-Ude: Publishing House of the Geological Institute of the Siberian Branch of the Russian Academy of Sciences, 37–40 (in Rus.).

Glazunov O.M. (1979) Contamination and ore content of gabbroid magma in the geological aspect. *Contact processes and mineralization in gabbro-peridotite intrusions.* Moscow: Publishing House Nauka, 108–124 (in Rus.).

Glebovitsky V.A., Nikitina L.P., Vrevskii A.B., Pushkarev Yu.D., Babushkin M.S., and Goncharov A.G. (2009) Nature of the chemical heterogeneity of the continental lithospheric mantle. *Geochemistry International,* 47 (9), 910–936 (in Rus.).

Godard M., Jousselin D., and Bodinier J.-L. (2000) Relationships between geochemistry and structure beneath a palaeo-spreading centre: a study of the mantle section in the Oman ophiolite. *Earth Planet Sci Lett., 180* (1–2), 133–148.

Grannik V.M. (2008) Geology and geodynamics of the southern part of the Sea of Okhotsk region in the Mesozoic and Cenozoic. Vladivostok: Publishing House Dal'nauka, 297 p. (in Rus.).

Green D.H. and Ringwood A.E. (1968) The genesis of basaltic magmas. *Petrology of the upper mantle.* Moscow: Publishing House Mir, 132–227 (in Rus.).

Green T.H. (1994) Experimental studies of trace-element partitioning applicable in igneous petrogenesis — Sedona 16 years later. *Chem. Geol. 117* (1–4), 1–36.

Hawthorne F.C., Oberti R., Harlow G.E., Maresch W.V., Martin R.F., Schumacher J.C., and Welch M.D. (2012) Nomenclature of the amphibole supergroup. *American Mineralogist, 97,* 2032–2048.

Herzberg C. (2004) Geodynamic information in peridotite petrology. *Journal of Petrology, 45* (12), 2507–2530.

Herzberg C. and O'Hara M.J. (2002) Plume-associated ultramafic magmas of Phanerozoic age. *Journal of Petrology, 43* (10), 1857–1883.

Hinton R.W. and Meyer C. (1991) Ion probe analysis of zircon and yttrobetafite in a lunar granite. *Lunar Planet. Sci., 22,* 575–578.

Hinton R.W. and Upton G.J. (1991) The chemistry of zircon: variation within and between large crystals from syenite and alkali basalt xenoliths. *Geochim. Cosmochim. Acta, 55* (11), 3287–3302.

Hirano H. (1977) Genesis of harzburgite- and wehrlite-series ultramafic rocks in the Maizuru belt, Japan. *The Journal of the Geological Society of Japan, 83* (11), 707–718.

Holland T.J.B. and Powell R. (1998) An internally consistent thermodynamic dataset for phases of petrological interest. *J. Metamorph. Geol., 16* (3), 309–343.

Hoskin P.W.O. and Schaltegger U. (2003) The composition of zircon and igneous and metamorphic petrogenesis. *Zircon. Reviews in mineralogy and geochemistry, 53.* MSA, Geochem. Soc., 27–62.

Hulbert L.J. (1997) Geology and metallogeny of the Kluane mafic-ultramafic belt, Yukon territory, Canada: Eastern Wrangellia – a new Ni-Cu-PGE metallogenic terrane. *Geological Survey of Canada. Bulletin 506,* 265.

Irvin W.P. and Coleman R.G. (1972) Preliminary map showing global distribution of Alpine-type ultramafic rocks and blueschists.

Ishiwatari A. (1985) Igneous petrogenesis of the Yakuno ophiolite (Japan) in the context of the diversity of ophiolite. *Contrib. Mineral. Petrol., 89* (2), 155–167.

Ishiwatari A. (1991) Ophiolites in the Japanese islands: Typical segment of the circum-Pacific multiple ophiolite belts. *Episodes, 14,* 274–279.

Ishiwatari A., Sokolov S.D., and Vysotskiy S.V. (2003) Petrological diversity and origin of ophiolites in Japan and Far East Russia with emphasis on depleted harzburgite. *Ophiolites in Earth History. Geol. Soc. London. Spec. Publ.* (218), 597–617.

Itoh S., Terashima S., Imai N., *et al.* (1993) 1992 compilation of analytical data for rare- earth elements, scandium, yttrium, zirconium and hafnium in twenty-six GSJ reference samples. *Geostand. Geoanal. Res. 17* (1), 5–79.

Ivanov K.S. and Berzin S.V. (2013) The first data on the U–Pb age of zircons from the relict dolerite zone of backarc spreading of Mt. Azov (Middle Urals). *Litosfera* (2), 92–104 (in Rus.).

Ivanov K.S., Krasnobaev A.A., and Smirnov V.N. (2012) Zircon geochronology of the Klyuchevskoi gabbro–ultramafic massif and the problem of the age of the Mohorovicic paleoboundary in the Central Urals. *Doklady Earth Sciences, 442* (4), 516–520 (in Rus.).

Ivanov O.K. (1997) Concentrically zoned pyroxenite–dunite massifs of the Urals. Yekaterinburg: Publishing House of the Uralian State University. 487 p. (in Rus.).

Izokh E.P. (1965) Ultrabasic–gabbro–granite formational series and formation of high-alumina granites. Novosibirsk: Publishing House of the USSR Academy of Sciences. 140 p. (in Rus.).

Kaulina T.V. (2010) Generation and transformation of zircon in the polymetamorphic complexes. Apatity: Publishing House of the Kola Science Center of the Russian Academy of Sciences. 145 p. (in Rus.).

Khain E.V. and Remizov D.N. (2009) Six types of ultramafic–mafic complexes of folded structures, or the problem of existence of asthenospheric windows beneath continental margins which have undergone ophiolite obduction. *Ultrabasic–basic complexes of folded regions and related mineral deposits.* Yekaterinburg: Publishing House of the Institute of Geology and Geochemistry, Uralian Branch of the Russian Academy of Sciences, 237–242 (in Rus.).

Khanchuk A.I. (Ed.) (2006) Geodynamics, magmatism, and metallogeny of Eastern Russia. Books 1, 2. Vladivostok: Publishing House Dal'nauka, 202–204 (in Rus.).

Kinny P.D., Compston J., Bristow J.W., and Williams I.S. (1989). Archaean mantle xenocrysts in a Permian kimberlite: two generations of kimberlitic zircon in Jwaneng DK2, southern Botswana. *Kimberlites and related rocks, Volume 2* (Eds. Ross J. *et al.*). *Geol. Soc. Austr. Spec. Publ.*, *14*. Blackwell, Melbourne, 833–842.

Kiseleva O.N. (2014) Chromitites and PGE mineralization in ophiolites from the southeastern East Sayan (Ospin–Kitoi and Kharanur massifs. *Abstract of dissertation.* Novosibirsk: Publishing House of the Institute of Geology and Mineralogy, Siberian Branch of the Russian Academy of Sciences, 16 p. (in Rus.).

Kiseleva V.Yu., Lesnov F.P., Avdeev D.V., and Dokukina G.A. (2010) Distribution and isotope ratios of Rb and Sr in the rocks from the Berezovka mafic–ultramafic massif, Sakhalin Island (the first data). *Metallogeny of the ancient and modern oceans–2010. Ore-bearing rift and island-arc structures. Proceedings of the 16th Youth Scientific School.* Miass: Publishing House of the Institute of Mineralogy, Uralian Branch of the Russian Academy of Sciences, 239–242 (in Rus.).

Kislov E.V., Konnikov E.G., Posokhov V.F., and Shalagin V.L. (1989) Isotope evidence for crustal contamination in the Ioko-Dovyren pluton. *Soviet Geology and Geophysics, 30 (9)*, 140–144 (in Rus.).

Knauf O. (2008) The age of dunite-clinopyroxenite core of Kytlym and Galmoenan zonal Ural-type massifs by U-Pb data of zircons. *33rd International Geological Congress. Abstracts.* Oslo, 105–107.

Kocks H., Melcher F., Meisel T., and Burgath K.-P. (2007) Diverse contributing sources to chromitite petrogenesis in the Shebenik Ophiplitic Complex, Albania: evidence from new PGE- and Os-isotope data. *Mineralogy and Petrology, 91 (3)*, 139–170.

Kovtunovich Yu.M., Rozhdestvenskii V.S., and Semenov D.F. (1971) Igneous complexes of Sakhalin Island and their metal content. *Igneous complexes of the Far East.* Vladivostok, 229–236 (in Rus.).

Kozlov A.P. (2000) The Gal'moenan mafic–ultramafic massif, Koryakia: geology, petrology, and ore potential. *Abstract of dissertation.* PhD. Moscow: Publishing House of Moscow State University. 16 p. (in Rus.).

Koz'menko O.A., Palesskii S.V., Nikolaeva I.V., Tomas V.G., and Anoshin G.N. (2011) Improving methods of chemical preparation of geological samples in the Carius tubes for determination of platinum group elements and rhenium. *Analysis and control. 15* (4). 378–385 (in Rus.).

Krasnobaev A.A. (1979) Mineralogical and geochemical characteristics of zircons from kimberlites and questions of their genesis. *Proceedings of the Academy of Sciences of the USSR. Geological Series* (8), 85–96 (in Rus.).

Krasnobaev A.A. and Anfilogov V.N. (2014) Zircons: Implication for dunite genesis. *Doklady Earth Sciences*, 456 (3), 310–313 (in Rus.).

Krasnobaev A.A., Bea F., Fershtater G.B., and Montero P. (2007) The polychronous nature of zircons in gabbroids of the Ural Platinum Belt and the issue of the Precambrian in the Tagil Synclinorium. *Doklady Earth Sciences*, 413 (6), 785–790 (in Rus.).

Krasnobaev A.A., Fershtater G.B., and Busharina S.V. (2009) Character of zircon distribution in dunites of the South Urals (Sakharin and East Khabarnin Massifs). *Doklady Earth Sciences*. 426 (4), 523–527 (in Rus.).

Krasnobaev A.A., Rusin A.I., Busharina S.V., and Antonov A.V. (2013) Zirconology of dunite from the Sarany chromite-bearing ultramafic complex (Middle Urals). *Doklady Earth Sciences*, 451 (1), 81–86 (in Rus.).

Kremenetskii A.A. and Gromalova N.A. (2013) Nature of ancient zircons from rocks of the Mid-Atlantic Ridge and the Mendeleev Rise in the Arctic Ocean. *Basic Research* (10, part 3), 594–600 (in Rus.).

Kropotkin P.N. and Shakhvarstova K.A. (1965) The geologic structure of the Pacific mobile belt. Moscow: Publishing House Nauka. 365 p. (in Rus.).

Kurokawa K. (1985) Petrology of the Oeyama ophiolitic complex in the Inner zone of Southwest Japan. *Sci. Rep. Niigata Univ. Ser. E (Geology and Mineralogy)*, 6, 37–113.

Kushiro I. (1983) Fractional crystallization of basaltic magma. *Evolution of the igneous rocks: Fiftieth anniversary perspectives*. Moscow: Publishing House Mir, 172–202 (in Rus.).

Kutolin V.A. (1972) Problems of petrochemistry, and petrology of basalts. Novosibirsk: Publishing House Nauka. 208 p. (in Rus.).

Kuznetsov Yu.A. (1964) The main types of magmatic formations. Moscow: Publishing House Nedra. 388 p. (in Rus.).

Latypov R.M. (2009) The predominant mechanism of mafic–ultramafic intrusions: gravity settling or directional solidification? *Ultrabasic–basic complexes of folded regions and related deposits. Volume 2*. Yekaterinburg: Publishing House of the Institute of Geology and Geochemistry, Uralian Branch of the Russian Academy of Sciences, 11–13 (in Rus.).

Lavrenchuk A.V. and Latypov R.M. (2009) The Sudbury intrusive complex, Canada: the origin from one or two layers of magma? *Ultrabasic–basic complexes of folded regions and related deposits. Volume 2*. Yekaterinburg: Publishing House of the Institute of Geology and Geochemistry, Uralian Branch of the Russian Academy of Sciences, 5–8 (in Rus.).

Ledneva G.V. and Matukov D.I. (2009) Timing of crystallization of plutonic rocks from the Kuyul ophiolite terrane (Koryak highland): U–Pb microprobe (SHRIMP) zircon dating. *Doklady Earth Sciences*, 424 (1), 71–75 (in Rus.).

Ledneva G.V., Bazylev B.A., Kononkova N.N., and Ishawatari A. (2009) Ultramafic and mafic rocks of the Pekul'nei complex (Chukchi Peninsula): high-pressure island-arc cumulates. *Ultrabasic–basic complexes of folded regions and related deposits. Volume 2*. Yekaterinburg: Publishing House of the Institute of Geology and Geochemistry, Uralian Branch of the Russian Academy of Sciences, 17–20 (in Rus.).

Ledneva G.V., Bazylev B.A., Lebedev V.V., Kononkova N.N., and Ishiwatari A. (2012) U–Pb zircon age of gabbroids of the Ust'-Belaya mafic–ultramafic massif (Chukotka) and its interpretation. *Geochemistry International*, 50 (1), 48–59 (in Rus.).

Lesnov F.P. (1972) Geology and petrology of the Chaya gabbro-peridotite–dunite nickel-bearing pluton (northern Baikal region). Novosibirsk: Publishing House Nauka. 228 p. (in Rus.).

Lesnov F.P. (1976) On the structural and textural criteria of impact of gabbroids on ultrabasic rocks in basic–ultrabasic plutons of folded regions. *Proceedings on genetic and experimental mineralogy. Volume 10*. Novosibirsk: Publishing House Nauka. 75–80 (in Rus.).

Lesnov F.P. (1979) On basic–ultrabasic associations of Mongolia. *Geology and magmatism of Mongolia.* Moscow: Publishing House Nauka. 10–12 (in Rus.).

Lesnov F.P. (1980) Ultrabasic xenoliths in gabbroids and questions of genesis of polygenic basic–ultrabasic plutons. *Mantle xenoliths and problems of ultramafic magmas.* Novosibirsk: Publishing House of the Institute of Geology and Geophysics, Siberian Branch of the USSR Academy of Sciences, 153–154 (in Rus.).

Lesnov F.P. (1981a) Platy cleavage in ultrabasic rocks and the problem of genesis of banded structures in rocks of polygenic basic–ultrabasic plutons. *Questions of genetic petrology.* Novosibirsk: Publishing House Nauka, 203–213 (in Rus.).

Lesnov F.P. (1981b) Polygenic basic–ultrabasic plutons as a special type of magmatic formations of folded regions. *Petrology of the lithosphere and ore content.* Abstracts of the VI All-Union petrographic meeting. Leningrad: Publishing House of VSEGEI, 206–207 (in Rus.).

Lesnov F.P. (1981c) Structural and genetic relationship between ultrabasic and gabbroid rocks in the ophiolite belts of Mongolia. *Questions of magmatism and metallogeny of the MPR.* Novosibirsk: Publishing House of the Institute of Geology and Geophysics, Siberian Branch of the USSR Academy of Sciences, 62–71 (in Rus.).

Lesnov F.P. (1981d) The polygenic basic–ultrabasic plutons of folded regions. *The evolution of ophiolite complexes.* Sverdlovsk: Publishing House of the Institute of Geology and Geochemistry, Uralian Science Center of the USSR Academy of Sciences, 6–7 (in Rus.).

Lesnov F.P. (1982) The Naran polygenic basic–ultrabasic pluton (western Mongolia). *Ultrabasic associations of folded regions. Issue 1. Geology, petrography, petrochemistry, and geochemistry.* Novosibirsk: Publishing House of the Institute of Geology and Geophysics, Siberian Branch of the USSR Academy of Sciences, 58–95 (in Rus.).

Lesnov F.P. (1984) Petrology of polygenic basic–ultrabasic plutons of folded regions. *Proceedings of the USSR Academy of Sciences. Geological Series* (2), 71–78 (in Rus.).

Lesnov F.P. (1985) Polygenic basic–ultrabasic plutons of folded areas as a chemically discrete association of igneous rocks. *Formational division, genesis, and metallogeny of ultrabasic rocks.* Sverdlovsk: Publishing House of the Institute of Geology and Geochemistry, Uralian Science Center of the USSR Academy of Sciences, 14–16 (in Rus.).

Lesnov F.P. (1986a) Petrochemistry of polygenic basic–ultrabasic plutons of folded regions. Novosibirsk: Publishing House Nauka. 136 p. (in Rus.).

Lesnov F.P. (1986b) To the relationship between ultrabasic and gabbroid rocks in the ophiolite associations. *Ophiolites on the eastern margin of Asia.* Khabarovsk: Publishing House of the Institute of Tectonics and Geophysics, Far Eastern Science Center of the USSR Academy of Sciences, 88–90 (in Rus.).

Lesnov F.P. (1988) Petrology of polygenic basic–ultrabasic plutons of folded regions. *Absrtact of dissertation.* Novosibirsk: Publishing House of the Institute of Geology and Geophysics, Siberian Branch of the USSR Academy of Sciences, 24 p. (in Rus.).

Lesnov F.P. (1991) The plagioclases of polygenic basic–ultrabasic plutons. Novosibirsk: Publishing House Nauka. 112 p. (in Rus.).

Lesnov F.P. (2000a) Status of study of geochemistry of rare-earth elements in rock-forming olivines, clinopyroxenes, and orthopyroxenes. *Petrography on the threshold of the 21st century: results and prospects. Volume 3. Proceedings of the 2nd Russian petrographic meeting.* Syktyvkar: Publishing House Geoprint, 67–69 (in Rus.).

Lesnov F.P. (2000b) On the trends of change of the distribution coefficients of rare-earth elements in the system clinopyroxene–melt. *Petrography on the threshold of the 21st century: results and prospects. Volume 3. Proceedings of the 2nd Russian petrographic meeting.* Syktyvkar: Publishing House Geoprint, 64–66 (in Rus.).

Lesnov F.P. (2000c) Geochemistry of rare-earth elements in plagioclases from the Berezovka mafic–ultramafic massif (Sakhalin Island). Petrology and metallogeny of basic–ultrabasic complexes of the Kamchatka Peninsula. Petropavlovsk-Kamchatskii: Publishing House of the Institute of Volcanology, 51–52 (in Rus.).

Lesnov F.P. (2000d) The concentration of the rare-earth elements in plagioclases from rocks of different compositions and origins. *Petrology of igneous and metamorphic complexes*. Tomsk: Publishing House of Tomsk State University, 38–42 (in Rus.).

Lesnov F.P. (2001a) Regularity of distribution of rare-earth elements in orthopyroxenes. *Memoirs of the All-Russian Mineralogical Society. Part 130* (1), 3–20 (in Rus.).

Lesnov F.P. (2001b) Regularity of distribution of rare-earth elements in clinopyroxenes. *Memoirs of the All-Russian Mineralogical Society. Part 130* (4), 78–97 (in Rus.).

Lesnov F.P. (2001c) Geochemistry of rare-earth elements in plagioclase. *Russian Geology and Geophysics. 42* (6), 917–936 (in Rus.).

Lesnov F.P. (2005) Basic regularities of distribution of rare-earth elements in the main types of gabbroic rocks from mafic–ultramafic massifs. *The state and development of natural resources of Tuva and adjacent regions of Central Asia. Geoecology of the environment and society. Issue 8*. Kyzyl: Publishing House of the Tuvinian Institute of Integrated Development of Natural Resources, Siberian Branch of the Russian Academy of Sciences, 56–79 (in Rus.).

Lesnov F.P. (2007a) The contact relationship between ultramafic and mafic rocks in the Dovyren pluton and some debatable questions of its genesis (northern Baikal region, Russia). *Geochemistry and mineralization of radioactive, noble, and rare metals in endogenous and exogenous processes*. Ulan-Ude: Publishing House of the Geological Institute, Siberian Branch of the Russian Academy of Sciences, 140–142 (in Rus.).

Lesnov F.P. (2007b) Rare-earth elements in ultramafic and mafic rocks and their minerals. Book 1. The main rock types. Rock-forming minerals. Novosibirsk: Academic Publishing House Geo, 403 p. (in Rus.).

Lesnov F.P. (2007c) A new way of graphic interpretation of petrochemical data of ultramafic and mafic rocks by MORB-normalized trends. *Problems of geology, mineral resources and geoecology of western Transbaikalia. International Scientific-Practical conference*. Ulan-Ude: Publishing House of the Geological Institute, Siberian Branch of Russian Academy of Sciences, 94–96 (in Rus.).

Lesnov F.P. (2007d) The Nogontsav massif as an example of Hercynian protrusions of ultramafic restites in molasse deposits (southwestern Mongolia). *Ultrabasic–basic complexes of folded regions. Proceedings of the International Conference*. Irkutsk: Publishing House of Irkutsk State Technical University, 313–317 (in Rus.).

Lesnov F.P. (2007e) Spatiotemporal and contact relationships between ultramafic restites and gabbroids and questions of genesis of polygenic mafic–ultramafic massifs. *Ultrabasic–basic complexes of folded regions. Proceedings of International Conference*. Irkutsk: Publishing House of Irkutsk State Technical University, 317–321 (in Rus.).

Lesnov F.P. (2009a) Rare-earth elements in ultramafic and mafic rocks and their minerals. Book 2. Minor and accessory minerals. Novosibirsk: Academic Publishing House Geo. 190 p. (in Rus.).

Lesnov F.P. (2009b) Contrast fractionation of rare-earth and platinum group elements by partial melting and depletion of the upper mantle. *Geology and mineral resources of European Northeastern Russia. Volume 2*. Syktyvkar: Publishing House of the Institute of Geology, Komi Science Center, Uralian Branch of the Russian Academy of Sciences, 345–347 (in Rus.).

Lesnov F.P. (2009c) The structure and composition of mafic–ultramafic massifs as evidence for their polygenic formation. *Ultrabasic–basic complexes of folded regions and related deposits. Volume 2.* Yekaterinburg: Publishing House of the Institute of Geology and Geochemistry, Uralian Branch of the Russian Academy of Sciences, 20–23 (in Rus.).

Lesnov F.P. (2009d) The geochemical systematics of rocks of different-type mafic–ultramafic massifs of Tuva, Southern Urals, and western Mongolia on the basis of the distribution of Zr, Hf, Y, Nb, and other trace elements. *Petrology of igneous and metamorphic complexes.* Tomsk: Publishing House of Tomsk State University, 206–214 (in Rus.).

Lesnov F.P. (2009e) New data on petrochemistry of the Berezovka and other polygenic mafic–ultramafic massifs of the East Sakhalin ophiolite association. *Petrology of igneous and metamorphic complexes.* Tomsk: Publishing House of Tomsk State University, 215–220 (in Rus.).

Lesnov F.P. (2009f) Intrusions as a consequence of their polygenic multistage formation. *Large Igneous Provinces of Asia, Mantle Plumes and Metallogeny. International Symposium.* Novosibirsk: Publishing House of the Institute of Geology and Mineralogy, Siberian Branch of the Russian Academy of Sciences, 191–193.

Lesnov F.P. (2010a) The composition of coexisting minerals and distribution of the components between them in rocks of mafic–ultramafic massifs of the East Sakhalin ophiolite association. *Uralian Mineralogical School–2010.* Yekaterinburg: Publishing House of the Institute of Geology and Geochemistry, Uralian Branch of the Russian Academy of Sciences, 102–105 (in Rus.).

Lesnov F.P. (2010b) Rare Earth Elements in Ultramafic and Mafic Rocks and Their Minerals. Main types of rocks. Rock-forming minerals. Boca Raton–London–New York–Leiden: Publishing House CRC Press. Taylor & Francis Group, 580 p.

Lesnov F.P. (2011a) The structure and the geological model of formation of polychronous and polygenic mafic–ultramafic massifs. *Bulletin of Moscow Society of Naturalists. Volume 86, Issue 5,* 99–100 (in Rus.).

Lesnov F.P. (2011b) Isomorphism of rare-earth elements in rock-forming minerals of ultramafic and mafic rocks. *Mineralogical prospects–2011.* Syktyvkar: Publishing House of the Institute of Geology, Komi Science Center, Uralian Branch of the Russian Academy of Sciences. 89–90 (in Rus.).

Lesnov F.P. (2011c) Isomorphism of rare-earth elements in zircons and conditions of their crystallization. *Crystallochemistry, radiography, and spectroscopy of minerals. Proceedings of the 17th International Meeting.* St. Petersburg: Publishing House of St. Petersburg State University, 176–177 (in Rus.).

Lesnov F.P. (2012a) Geochemistry and isotopic age of zircons from rocks of the Berezovka polygenic mafic–ultramafic massif (Sakhalin Island, Russia). *Geochronometric isotope systems, methods of their study, and chronology of geologic processes.* Moscow: Publishing House of IGEM, Russian Academy of Sciences, 219–222 (in Rus.).

Lesnov F.P. (2012b) Rare Earth Elements in Ultramafic and Mafic Rocks and Their Minerals. Minor and accessory minerals. Boca Raton–London–New York–Leiden: Publishing House CRC Press. Taylor & Francis Group. 314 p.

Lesnov F.P. (2013) The geology, geochemistry, petrology and mineralogy of mafic–ultramafic complexes. Saarbrücken: LAP LAMBERT Academic Publishing House. 492 p. (in Rus.).

Lesnov F.P. and Agafonov L.V. (1976) On the composition of chromite ores of the South Shmidt and Berezovka plutons (Sakhalin Island). *Preorogenic metallogeny of eugeosynclines.* Sverdlovsk, 21–22 (in Rus.).

Lesnov F.P. and Anoshin G.N. (2011) Correlations between Re and PGE concentrations in rocks, ores, and minerals of mafic–ultramafic associations. *Doklady Earth Sciences, 437* (2), 228–234 (in Rus.).

Lesnov F.P. and Gora M.P. (1996) Rare-earth elements in coexisting pyroxenes from polytypic mafic–ultramafic rocks. *Problems of geology of Siberia. Volume 3.* Tomsk: Publishing House of Tomsk State University, 13–14 (in Rus.).

Lesnov F.P. and Gora M.P. (1997) The distribution coefficients of REE between coexisting clino- and orthopyroxenes from rocks of mafic–ultramafic massifs of Sakhalin. *Structure and evolution of the mineral world.* Syktyvkar: Publishing House of the Institute of Geology, Komi Science Center, Uralian Branch of the Russian Academy of Sciences, 155–157 (in Rus.).

Lesnov F.P. and Gora M.P. (1998a) Geochemistry of rare-earth elements in coexisting pyroxenes of mafic–ultramafic rocks of different types. *Geochemistry International* (9), 899–918 (in Rus.).

Lesnov F.P. and Gora M.P. (1998b) The distribution of rare-earth elements between coexisting pyroxenes of different types of mafic–ultramafic rocks. *Problems of petrology of igneous and metamorphic rocks.* Novosibirsk: Publishing House of UIGGM, Siberian Branch of the Russian Academy of Sciences, 19–20 (in Rus.).

Lesnov F.P. and Kolmogorov Yu.P. (2009) Geochemical studies of ultramafic and mafic rocks of the East Sakhalin ophiolite association by the XRF-SR method (the first data). *Petrology and mineralization. 14th readings in honor of A.N. Zavaritskii.* Yekaterinburg: Publishing House of the Institute of Geology and Geochemistry, Uralian Branch of the Russian Academy of Sciences, 192–198 (in Rus.).

Lesnov F.P. and Korolyuk V.N. (1977) The first data on the distribution of isomorphous iron impurities in plagioclases of mafic–ultramafic plutons in folded regions of the USSR. *Doklady of the USSR Academy of Sciences,* 234 (4), 922–924 (in Rus.).

Lesnov F.P. and Palesskii S.V. (2013) The distribution of Re and Mo in zircons from rocks of the Berezovka polygenic mafic–ultramafic massif (Sakhalin Island). *Rhenium. Scientific research, technological development, and industrial application.* Moscow: Publishing House of the Institute GINTSVETMET, 121–122 (in Rus.).

Lesnov F.P. and Stepashko A.A. (2010) New data on the chemical composition of the rocks of mafic–ultramafic massifs of the East Sakhalin ophiolite association. *Metallogeny of the ancient and modern oceans-2010. Models of mineralization and evaluation of deposits.* Miass: Publishing House of the Institute of Mineralogy, Uralian Branch of the Russian Academy of Sciences, 234–239 (in Rus.).

Lesnov F.P., Pinus G.V., and Velinskii V.V. (1973) Relationships between ultramafic rocks and associated gabbroids in folded areas. *Problems of petrology of ultrabasic rocks in folded areas.* Novosibirsk: Publishing House of the Institute of Geology and Geophysics, Siberian Branch of the USSR Academy of Sciences, 44–56 (in Rus.).

Lesnov F.P., Agafonov L.V., and Kuznetsova I.K. (1976) Alkaline amphibole from the crossite-rhodusite group in albitites of the South Shmidt ultramafic massif (northern Sakhalin). *Proceedings on genetic and experimental mineralogy. Volume 10.* Novosibirsk: Publishing House Nauka, 85–92 (in Rus.).

Lesnov F.P., Melyakhovetskii A.A., and Bayarkhuu Zh. (1977) The Shishkhid-Gol ultramafic massif (northern Mongolia). *Proceedings on genetic and experimental petrology.* Novosibirsk: Publishing House Nauka, 130–146 (in Rus.).

Lesnov F.P., Simonov V.A., and Pospelova L.N. (1980) On the conditions of crystallization of gabbroids from the Kuyul basic–ultrabasic massif (Koryak highland). *Petrology of ultrabasic and basic rocks of Siberia, the Far East, and Mongolia.* Novosibirsk: Publishing House of the Institute of Geology and Geophysics, Siberian Branch of the USSR Academy of Sciences, 174–189 (in Rus.).

Lesnov F.P., Vasil'ev Yu.R., and Narizhnev V.V. (1982) The geology, petrography, and geochemistry of the Gishun basic–ultrabasic pluton (North Pamirs). *Ultrabasic associations of folded areas. Issue 1. Geology, petrography, petrochemistry, geochemistry.* Novosibirsk: Publishing House of the Institute of Geology and Geophysics, Siberian Branch of the USSR Academy of Sciences. 95–122 (in Rus.).

Lesnov F.P., Gora M.P., Bobrov V.A., and Kovaleva V.A. (1998a) The distribution of rare-earth elements and questions of genesis of the Berezovka mafic–ultramafic massif (Sakhalin Island). *Russian Journal of Pacific Geology, 17* (4), 42–58 (in Rus.).

Lesnov F.P., Gora M.P., Kovyazin S.V., and Fomina L.N. (1998b) Fluid components in orthopyroxenes of two-pyroxene mafic–ultramafic rocks and their relationship with the REE composition of the mineral. *Problems of petrology and minerageny of ultramafic–mafic complexes of Siberia.* Tomsk: Publishing House of Tomsk State University, 62–78 (in Rus.).

Lesnov F.P., Mongush A.A., Oidup Ch.K., and Popov V.A. (2005a) Structural and genetic relationship between ultramafic and gabbroid rocks in the Kurtushiba ophiolite association (West Sayan). *Ultramafic–mafic complexes of Precambrian folded areas.* Ulan-Ude: Publishing House of the Buryatia Science Center, Siberian Branch of the Russian Academy of Sciences, 59–61 (in Rus.).

Lesnov F.P., Chernyshov A.I., and Istomin V.E. (2005b) Geochemical characteristics and typomorphism of olivine from heterogeneous ultramafic rocks. *Geochemistry International,* (4), 395–414 (in Rus.).

Lesnov F.P., Podlipskii M.Yu., Polyakov G.V., and Palesskii S.V. (2008) Geochemistry of accessory chrome-spinellides from rocks of the Ergakskii chrome-bearing hyperbasite massif (western Sayan) and conditions of its formation. *Doklady Earth Sciences, 422* (5), 660–664 (in Rus.).

Lesnov F.P., Koz'menko O.A., Nikolaeva I.V., and Palesskii S.V. (2009a) Residence of incompatible trace elements in a large spinel lherzolite xenolith from alkali basalt of Shavaryn Tsaram-1 paleovolcano (western Mongolia). *Russian Geology and Geophysics, 50* (12), 1371–1384 (in Rus.).

Lesnov F.P., Palesskii S.V., Nikolaeva I.V., Koz'menko O.A., Kuchkin A.M., and Korolyuk V.N. (2009b) Detailed mineralogical–geochemical study of a large spinel lherzolite xenolith in alkali basalt of Shavaryn Tsaram paleovolcano, Mongolia. *Geochemistry International, 47* (1), 21–44 (in Rus.).

Lesnov F.P., Stepashko A.A., Rechkin A.N., and Gal'versen V.G. (2009c) Structural position, structure, and composition of the mafic–ultramafic massifs of the East Sakhalin ophiolite association. *Tectonics and deep structure of East Asia. 6th readings in honor of Kosygin.* Khabarovsk: Publishing House of the Institute of Tectonics and Geophysics, Far Eastern Branch of the Russian Academy of Sciences, 202–205 (in Rus.).

Lesnov F.P., Gal'versen V.G., Tsimbalist V.G., and Titov A.T. (2010a) First data on the U–Pb isotopic dating and original platinum content of the Berezovskii polygenic mafic–ultramafic massif (Sakhalin Island). *Doklady Earth Sciences, 433* (6), 792–795 (in Rus.).

Lesnov F.P., Kiseleva V.Yu., and Dokukina G.A. (2010b) Sr and Rb isotopes in the rocks of the mafic–ultramafic complex of the East Sakhalin ophiolite association. *19th Symposium on isotope geochemistry in honor of Academician A.P. Vinogradov. Abstracts.* Moscow: Publishing House of GEOKHI, 214–217 (in Rus.).

Lesnov F.P., Korolyuk V.V., and Lyakh A.V. (2010c) The chemical composition of coexisting minerals and distribution of components between them in the rocks of the Berezovka mafic–ultramafic massif on Sakhalin Island (the first data). *Metallogeny of ancient and modern oceans–2010. Models of mineralization and evaluation of deposits.* Miass: Publishing House of the Institute of Mineralogy, Uralian Branch of the Russian Academy of Sciences, 242–247 (in Rus.).

Lesnov F.P., Khlestov V.V., and Selyatitskii A.Yu. (2011) Multiparametric discrimination of ultramafic rocks by rare-earth elements in clinopyroxenes. *Doklady Earth Sciences*, *438* (5), 665–670 (in Rus.).

Lesnov F.P., Koz'menko O.A., Nikolaeva I.V., and Palesskii S.V. (2012) The distribution of platinum group elements and rhenium in the rocks and chromitites of the Berezovka and South Shmidt mafic–ultramafic massifs (Sakhalin Island). *Metallogeny of the ancient and modern oceans–2012. Hydrothermal fields and ores. Proceedings of the 18th Conference of Young Scientists.* Miass: Publishing House of the Institute of Mineralogy, Uralian Branch of the Russian Academy of Sciences, 229–233 (in Rus.).

Liberi F., Piluso E., and Langone A. (2011) Permo-Triassic thermal events in the lower Variscan continental crust section of the Northern Calabrian Arc, Southern Italy: Insights from petrological data and in situ U–Pb zircon geochronology on gabbros. *Lithos*, *124* (3–4), 291–307.

Liu S., Hu R., Gao S., Feng C., Coulson I.M., Feng G., Qi Y., Yang Y., Yang C., and Tang L. (2012) U–Pb zircon age, geochemical and Sr–Nd isotopic data as constraints on the petrogenesis and emplacement time of the Precambrian mafic dyke swarms in the North China Craton (NCC). *Lithos*, *140–141*, 38–52.

Lobodenko I.Yu. (2010) Holocene tectonic disturbances (paleoseismic dislocations) in the areas of the Hokkaido–Sakhalin and Central Sakhalin fractures. *Abstract of dissertation.* PhD. Moscow: Publishing House of Moscow State University, 18 p. (in Rus.).

Lutskina N.V. (1976) Concentrically zoned massif of ultramafic and mafic rocks of Feklistov Island (Shantar Islands). *Basic and ultrabasic rocks of the Far East.* Vladivostok: Publishing House of the Far Eastern Science Center, USSR Academy of Sciences, 52–67 (in Rus.).

Lutts B.G. (1980) Geochemistry of oceanic and continental magmatism. Moscow: Publishing House Nedra. 247 p. (in Rus.).

Malitch K.N., Efimov A.A., and Ronkin Yu.L. (2009) Archean U–Pb isotope age of zircon from dunite of the Nizhny Tagil massif (the Uralian Platinum Belt). *Doklady Earth Sciences*, *427* (1), 101–105 (in Rus.).

McBirney A.R. (1983) Effects of assimilation. *Evolution of the igneous rocks: Fiftieth anniversary perspectives.* Moscow: Publishing House Mir, 301–331 (in Rus.).

McDonough W.F. and Sun S.-s. (1995) The composition of the Earth. *Chem. Geol.*, *120*, 223–253.

Mekhonoshin A.S. and Kolotilina T.B. (2009) Formational typification of basic–ultrabasic complexes of the Alkhadyr terrane (southern Siberia) in connection with their ore potential. *Ultrabasic–basic complexes of folded areas and related deposits. Volume 2.* Yekaterinburg: Publishing House of the Institute of Geology and Geochemistry, Uralian Branch of the Russian Academy of Sciences, 51–54 (in Rus.).

Mekhonoshin A.S., Plyusnin G.S., Bognibov V.I., Polyakov G.V., Posokhov V.F., Sandimirova G.P., Shipitsin Yu.G., and Frolova L.P. (1986) Isotopes of strontium and rare-earth elements in the Dugda peridotite–pyroxenite–gabbro massif. *Soviet Geology and Geophysics* (8), 3–9 (in Rus.).

Miao L., Fan W., Liu D., Zhang F., Shi Y., and Guo F. (2008) Geochronology and geochemistry of the Hegenshan ophiolitic complex: Implications for late-stage tectonic evolution of the Inner Mongolia-Daxinganling Orogenic Belt, China. *Journal of Asian Earth Sciences*, *32* (5–6), 348–370.

Mikhalev Yu.A. and Popeko V.A. (1986) Isotope $^{87}Sr/^{86}Sr$ composition of ophiolite complexes of the Koryak ophiolite province. *Ophiolites on the eastern margin of East Asia.* Khabarovsk: Publishing House of the Institute of Tectonics and Geophysics, Far Eastern Science Center, USSR Academy of Sciences, 45–47 (in Rus.).

Miyashiro A. (1966) Some aspects of peridotite and serpentinite in orogenic belts. *Japan Journal Geol. Geogr.*, *37* (1), 45–61.

Morgan J.W. (1986) Ultramafic xenoliths: Clues to Earth's late accretionary history. *J. Geophysical Research*, *91*, (B12), 12375–12387.

Mysen B.O. (1983) Rare earth element partitioning between ($H_2O + CO_2$) vapor and upper mantle minerals: experimental data bearing on the conditions of formation of alkali basalt and kimberlite. *Neues Jahrb. Mineral. Abh.*, *146* (1), 41–65.

Neruchev S.S., Prasolov E.M., and Stepanov V.A. (1997) Fluid regime of basic and ultrabasic magmatism. *3rdInternational conference "New ideas in Earth sciences"*. Moscow, 104-105 (in Rus.).

Nikolaev G.S. (2009) Conditions of crystallization of primary-magmatic chromite in the formation of the Burakovo–Aganozero layered massif, Zaonezhye. *Ultrabasic–basic complexes of folded areas and related deposits. Volume 2.* Yekaterinburg: Publishing House of the Institute of Geology and Geochemistry, Uralian Branch of the Russian Academy of Sciences. 2009. 83–86 (in Rus.).

Oh C.W., Seo J., Choi S.G., Rajesh V.J., and Lee J.H. (2012) U–Pb SHRIMP zircon geochronology, petrogenesis, and tectonic setting of the Neoproterozoic Baekdong ultramafic rocks in the Hongseong Collision Belt, South Korea. *Lithos*, *128–131*, 100–112.

Oktyabr'skii R.A. (1971) Petrology of Paleozoic mafic intrusions of southern Primorye. *Abstract of dissertation.* Vladivostok. 25 p. (in Rus.).

Ostapenko A.B. (2003) Variations of the composition of the mantle peridotites as a reflection of the melting and reaction magma interaction in the suprasubduction areas. *Modern problems of formational analysis and petrology and ore content of magmatic formations.* Novosibirsk: Publishing House of the Siberian Branch of the Russian Academy of Sciences, Geo Subsidiary, 246–247 (in Rus.).

Pagé P., Barnes S.-J., Bédard J.H., and Zientek M.L. (2012) In situ determination of Os, Ir, and Ru in chromites formed from komatiite, tholeiite and boninite magmas: Implications for chromite control of Os, Ir and Ru during partial melting and crystal fractionation. *Chem. Geol.*, *302–303*, 3–15.

Palesskii S.V., Nikolaeva I.V., Koz'menko O.A., and Anoshin G.N. (2009) Determination of platinum-group elements and rhenium in standard geological samples by isotope dilution with mass-spectrometric ending. *Journal of analytical chemistry*, *64* (3), 287–291 (in Rus.).

Palesskii S.V., Koz'menko O.A., Nikolaeva I.V., and Anoshin G.N. (2010) Determination of nanoamounts of platinum group elements and rhenium in geological samples with isotope dilution and mass spectrometry ending. *Abstracts of the 19th International Chernyayev Conference on chemistry, analysis, and technology of platinum metals. Part 1.* Novosibirsk, 163 (in Rus.).

Peltonen P. and Mänttäri I. (2001) An ion microprobe U-Th-Pb study of zircon xenocrysts from the Lahtojoki kimberlite pipe, Eastern Finland. *Bulletin of the Geological Society of Finland. 73.* Parts 1–2, 47–58.

Pertsev A.N., Bortnikov N.S., Aranovich L.Ya., Vlasov E.A., Beltenev V.E., Ivanov V.N., and Simakin S.G. (2009) Peridotite-melt interaction under transitional conditions between the spinel and plagioclase facies beneath the Mid-Atlantic Ridge: Insight from peridotites at 13°N. *Petrology*, *17* (2), 139–153 (in Rus.).

Piccardo G.B., Zanetti A., Spagnolo G., and Poggi E. (2005) Recent researches on melt-rock interaction in the Lanzo south peridotite. *Ofioliti*, *30* (2), 135–160.

Pietruszka A.J. and Garcia M.O. (1999) A rapid fluctuation in the mantle source and melting history of Kilauea Volcano inferred from the geochemistry of its historical summit lavas (1790–1982). *J. Petrology. 40* (8), 1321–1342.

Pilot J., Werner C.-D., Haubrich F., and Baumann N. (1998) Palaeozoic and Proterozoic zircons from the Mid-Atlantic Ridge. *Nature, 393* (6686), 676–679.

Pinus G.V. and Sterligova V.E. (1973) The new belt of Alpine-type ultramafites in the Northeastern USSR and some geologic regularities of formation of ultramafic belts. *Soviet Geology and Geophysics* (12), 109–112 (in Rus.).

Pinus G.V., Kuznetsov V.A., and Volokhov I.M. (1958) Ultramafites of the Altai–Sayan folded area. Moscow: Publishing House Nauka, 296 p. (in Rus.).

Pinus G.V., Velinskii V.V., Lesnov F.P., Bannikov O.L., and Agafonov L.V. (1973) Alpine-type ultramafites of the Anadyr'–Koryak folded system. Novosibirsk: Publishing House Nauka, 320 p. (in Rus.).

Pinus G.V., Velinskii V.V., Lesnov F.P., Agafonov L.V., and Bayarkhuu Zh. (1976) Ultramafic belts of Central Asia and some general issues of petrology of ultramafites. *Problems of petrology*. Moscow: Publishing House Nauka, 94–105 (in Rus.).

Pinus G.V., Lesnov F.P., Agafonov L.V., and Bayarkhuu Zh. (1979) Alpine-type ultramafites of Mongolia and their metallogeny. *Geology and magmatism of Mongolia*. Moscow: Publishing House Nauka, 145–155 (in Rus.).

Pinus G.V., Agafonov L.V., and Lesnov F.P. (1984) Alpine-type ultramafites of Mongolia. Moscow: Publishing House Nauka, 200 p. (in Rus.).

Powell R. and Holland T.J.B. (1988) An internally consistent thermodynamic dataset with uncertainties and correlations: 3. Application to geobarometry, worked examples and a computer program. *J. Metamorph. Geol.*, 6, 173–204.

Raznitsyn Yu.N. (1975) Comparative tectonics of ultramafic belts of the Shmidt Peninsula (Sakhalin Island), Papua New Guinea, and Sabah (Borneo). *Geotectonics*, (2), 68–84 (in Rus.).

Raznitsyn Yu.N. (1978) Serpentinite mélange and olistostromes of the southeastern part of the East Sakhalin Mountains. *Geotectonics*, (2), 96–108 (in Rus.).

Raznitsyn Yu.N. (1982) Ophiolite allochthons and adjacent deep troughs in the Western Pacific. *Proceedings of the Geological Institute of the USSR Academy of Sciences*. Issue 371. Moscow: Publishing House Nauka, 108 p. (in Rus.).

Rechkin A.N. (1984) The role of ophiolites in the structure of Sakhalin. *Correlation of endogenous processes in the Soviet Far East*. Vladivostok: Publishing House of the Institute of Tectonics and Geophysics, Far Eastern Branch of the USSR Academy of Sciences, 102–120 (in Rus.).

Rechkin A.N., Semenov D.F., and Sheiko V.T. (1975) Ophiolite associations of Sakhalin and their structural position. *Questions of magmatism and tectonics of the Far East*. Vladivostok: Publishing House of the Institute of Tectonics and Geophysics, Far Eastern Branch of the USSR Academy of Sciences, 88–100 (in Rus.).

Rikhter A.V. (1985) Tectonic evolution of Sakhalin Island in the Mesozoic. *Abstract of dissertation*. PhD. Moscow: Publishing House of the Geological Institute of the Russian Academy of Sciences (in Rus.).

Ringwood A.E. (1972) Composition of the crust and upper mantle. *The Earth's crust and upper mantle*. Ed. by Hart., P. Moscow: Publishing House Mir, 7–26 (in Rus.).

Ringwood A.E. (1981) Composition and petrology of the Earth's mantle. Moscow: Publishing House Nedra, 584 p. (in Rus.).

Ringwood A.E., MacGregor I.D., and Boyd F.R. (1968) Petrographic composition of the upper mantle. *Petrology of the upper mantle*. Moscow: Publishing House Mir, 272–277 (in Rus.).

Roedder E. (1987) Fluid inclusions. Volume 1. Moscow: Publishing House Mir, 557 p. (in Rus.).

Rollinson H.R. (1993) Using Geochemical Data. Evaluation. Presentation. Interpretation. Singapore: Longman Scientific & Technical. 352 p.

Romanovich I.F. (1973) Some features of the relationship between gabbro and ultramafic rocks in Sikhote-Alin. *Petrology and metallogeny of basic rocks.* Moscow: Publishing House Nauka, 190–194 (in Rus.).

Ronkin Yu.L., Efimov A.A., Lepikhina G.A., Rodionov N.V., and Maslov A.V. (2013) U–Pb dating of the baddeleyite–zircon system from Pt-bearing dunite of the Konder massif, Aldan shield: New data. *Doklady Earth Sciences, 450* (5), 579–585 (in Rus.).

Rozhdestvenskii V.S. (1975) Shifts of northeastern Sakhalin. *Geotectonics,* (2), 85–97 (in Rus.).

Rozhdestvenskii V.S. (1987) Tectonic evolution of Sakhalin Island. *Russian Journal of Pacific Geology,* (3), 42–51 (in Rus.).

Rozhdestvenskii V.S. (1988) Geologic structure and tectonic evolution of the Schmidt Peninsula. *Russian Journal of Pacific Geology,* 62–71 (in Rus.).

Rozhdestvenskii V.S. and Rechkin A.N. (1982) Evolution of ophiolite magmatism of Sakhalin. *Russian Journal of Pacific Geology,* (2), 40–45 (in Rus.).

Rudnev, S.N., Izokh, A.E., Borisenko, A.S., Shelepaev, R.A., Orihashi, Y., Lobanov, K.V., Vishnevsky, A.V. (2012) Early Paleozoic magmatism in the Bumbat-Hairhan area of the Lake Zone in western Mongolia (geological, petrochemical, and geochronological data). *Russian Geology and Geophysics 53* (5), 557–578.

Ryabchikov I.D. (2012) Oxygen potential of high-magnesium magmas. *Doklady Earth Sciences,* 447 (6), 664–668 (in Rus.).

Saltykova A.K., Nikitina L.P., and Matukov L.I. (2008) U–Pb age of zircons from xenoliths of mantle peridotites in Cenozoic alkali basalts from the Vitim Plateau (Transbaikalia). Notes of the Russian Mineralogical Society. *Part 137* (3), 1–22 (in Rus.).

Savel'ev D.E., Snachev V.I., Savel'eva E.N., and Bazhin E.A. (2008) Geology, petrochemistry, and chromite content of gabbro–ultramafic massifs of the Southern Urals. Ufa: Publishing House Dizain Poligra fServis, 320 p. (in Rus.).

Savelieva G.N., Suslov P.V., and Larionov A.N. (2007) Vendian tectono-magmatic events in mantle ophiolitic complexes of the polar Urals: U–Pb dating of zircon from chromitite. *Geotectonics,* 41 (2), 23–33 (in Rus.).

Savelieva G.N., Shishkin M.A., Larionov A.N., Suslov P.V., and Berezhnaya N.G. (2006) Late Vendian tectonic and igneous events in the mantle ophiolite complexes of the Polar Urals (data from U–Pb zircon dating of chromites). *Ophiolites: geology, petrology, metallogeny, and geodynamics.* Ekaterinburg: Publishing House of the Institute of Geology and Geochemistry, Uralian Branch of the Russian Academy of Sciences, 160–164 (in Rus.).

Schuth S., Gornyy V.I., Berndt J., Shevchenko S.S., Sergeev S.A., Karpuzov A.F., and Mansfeldt T. (2012) Early Proterozoic U-Pb zircon ages from basement gneiss at the Solovetsky Archipelago, White Sea, Russia. *International Journal of Geosciences, 3* (2), 289–296.

Semenov D.F. (1967) Magmatic formations of the West Sakhalin Mountains. *Igneous and metamorphic complexes of the Far East of USSR.* Khabarovsk: Publishing House of the Far Eastern Geological Institute, USSR Academy of Sciences, 134–137 (in Rus.).

Semenov D.F. (1982) Magmatic formations of the Pacific folded areas (by the example of Sakhalin). Moscow: Publishing House Nauka, 168 p. (in Rus.).

Sharkov E.V. and Bogatikov O.A. (1985) The problem of differentiation of basic magmas. *Magmatic rocks. The mafic rocks. Chapter 10.* Moscow: Publishing House Nauka, 454–464 (in Rus.).

Shcheka S.A., Ishiwatari A., and Vrzhosek A.A. (2001) Geology and petrology of Cembrian Khanka ophiolite in Primorye (Far East Russia) with notes on its manganese-rich chromian spinel. *Earth Science (Chikyu Kagaku),* 55, 265–274.

Shervais J.W. (2001) Birth, death, and resurrection: The life cycle of suprasubduction zone ophiolites. *Geochemistry, Geophysics, Geosystems*, 2 (1), doi: 10.1029/2000GC000080.

Shmelev V.R. (2009) Petrogenesis and formation environment of ultramafic ophiolite complexes (for example, the Polar Urals). *Petrogenesis and mineralization*. Yekaterinburg: Publishing House of the Uralian Branch of the Russian Academy of Sciences, 93–96.

Shteinberg D.S., Zoloev K.K., and Chashchukhin I.S. (1986) Ophiolites of the Urals as representatives of the world type. *Ophiolites on the eastern margin of Asia*. Khabarovsk: Publishing House of the Institute of Tectonics and Geophysics, Far Eastern Science Center of the USSR Academy of Sciences, 4–6 (in Rus.).

Sidorov E.G. (2009) Platinum content of basic–ultrabasic complexes of the Koryak–Kamchatka region. *Abstract of dissertation*. Petropavlovsk-Kamchatskii, 46 p. (in Rus.).

Sklyarov E.V. (2001) Interpretation of geochemical data. Moscow: Publishing House Interment Engineering. 288 p. (in Rus.).

Skolotnev S.G., Bel'tenev V.E., Lepekhina E.N., and Ipat'eva I.S. (2009) Young and old zircons from rocks of the oceanic lithosphere in the Central Atlantic, geotectonic implications. *Geology of the oceans and seas. Proceedings of the 18th International Conference on marine geology. Volume 5*. Moscow: Publishing House GEOS, 251–255 (in Rus.).

Skolotnev S.G., Bel'tenev V.E., Lepekhina E.N., and Ipat'eva I.S. (2010) Younger and older zircons from rocks of the oceanic lithosphere in the Central Atlantic and their geotectonic implications. *Geotectonics*, 44 (6), 24–59 (in Rus.).

Slodkevich V.V. (1975a) Mafic–ultramafic intrusive complexes of Sakhalin. *Abstract of dissertation*. Leningrad: Publishing House of VSEGEI, 25 p. (in Rus.).

Slodkevich V.V. (1975b) The Shel'ting peridotite–pyroxenite–norite layered plutons in eastern Sakhalin. *Doklady of the USSR Academy of Sciences*, 222 (1), 946–949 (in Rus.).

Slodkevich V.V. (1977) The Alpine-type gabbro-peridotite formation of Sakhalin. *Magmatism and formations of the seabed, island arcs, and continental margins. Part 2*. Vladivostok: Publishing House of SakhKNII, Far Eastern Branch of the USSR Academy of Sciences, 86–98 (in Rus.).

Slodkevich V.V. and Lesnov F.P. (1976) Geology and some questions of petrology of the Berezovka mafic–ultramafic pluton (Sakhalin Island). *Proceedings on genetic and experimental mineralogy. Volume 10*. Novosibirsk: Publishing House of the Institute of Geology and Geophysics, Siberian Branch of the USSR Academy of Sciences, 53–63 (in Rus.).

Sobolev A.V. and Batanova V.G. (1995) Mantle lherzolites of the Troodos ophiolite complex, Cyprus Island: clinopyroxene geochemistry. *Petrology*, 3 (5), 487–495 (in Rus.).

Starozhilov V.T. (1990) The structural-tectonic zonation of the Pioner–Shel'ting area of the East Sakhalin mountains, Sakhalin Island. *Russian Journal of Pacific Geology*, (3), 90–96 (in Rus.).

Stepashko A.A. and Lesnov F.P. (2011) Variations of the composition and nature of the peridotites of the East Sakhalin ophiolite association. *Tectonics, magmatism, and geodynamics of East Asia. 7th readings in honor of Kosygin. Proceedings of the All-Russian conference*. Khabarovsk: Publishing House of the Institute of Tectonics and Geophysics, Far Eastern Branch of the Russian Academy of Sciences, 637–640 (in Rus.).

Stepashko A.A. and Lesnov F.P. (2013) Ultramafites of the East Sakhalin ophiolite association: structure, heterogeneity, and geodynamics of formation. *Condition and development of natural resources of Tuva and adjacent regions of Central Asia. Geoecology of the environment and society. Issue 12*. Kyzyl: Publishing House of the Tuvinian Institute of Integrated Development of Natural Resources, Siberian Branch of the Russian Academy of Sciences, 95–114 (in Rus.).

Sychev P.M. (1966) Features of the structure and evolution of the Earth's crust of Sakhalin Island and the surrounding water areas. Moscow: Publishing House Nauka, 124 p. (in Rus.).

Tsukanov N.V. and Skolotnev S.G. (2010) Data of SHRIMP U–Pb studies of zircons from the gabbro–ophiolite association of the Kamchatka Peninsula (eastern Kamchatka). *Bulletin of the Kamchatka Regional Association "Scientific Training Center". Earth Sciences Series. Volume 15* (2), 78–85 (in Rus.).

Turkina, O.M., Sergeev, S.A., and Kapitonov, I.N. (2014) The U–Pb age and Lu–Hf isotope composition of detrital zircon from metasedimentary rocks of the Onot greenstone belt (*Sharyzhalgay uplift, southern Siberian craton*). *Russian Geology and Geophysics 55* (11), 1581–1597.

Ultrabasic rocks in Japan. (1967) *Journal of Geol. Soc. Japan., 73* (12), 543–553.

Uysal I., Tarkian M., Sadiklar M.B., Zaccarini F., Meisel T., Garuti G., and Heidrich S. (2009) Petrology of Al- and Cr-rich ophiolitic chromitites from the Muğla, SW Turkey: implications from composition of chromite, solid inclusions of platinum-group mineral, silicate, and base-metal mineral, and Os-isotope geochemistry. *Contributions to Mineralogy and Petrology, 158* (5), 659–674.

Vasilenko N.F. and Prytkov A.S. (2012) GPS-based modeling of the interaction between the lithospheric plates in Sakhalin. *Russian Journal of Pacific Geology, 31* (1), 42–48 (in Rus.).

Velinskii V.V. (1979) Alpine-type ultrabasic rocks from ocean–continent transition zones. Novosibirsk: Publishing House Nauka, 264 p. (in Rus.).

Velinskii V.V., Pavlov A.L., and Bishaeva L.G. (1999) Thermodynamic justification of interaction of gabbro with ultrabasic rocks in ophiolites. *Problems of ophiolite petrology, mineralogy, geochemistry, and geology.* Novosibirsk: Publishing House of the Siberian Branch of the Russian Academy of Sciences, 74–84 (in Rus.).

Vergunov G.P. (1964) New data on ultrabasic rocks of Sakhalin and the Kurile Islands. *Doklady of the USSR Academy of Sciences, 158* (3), 629–632 (in Rus.).

Vil'danova E.Yu., Zaitsev V.P., Kravchenko L.I., Landa E.A., *et al.* (2002) The Koryak–Kamchatka region—a new Russian platinum province. St. Petersburg: Publishing House VSEGEI Map Factory. 384 p. (in Rus.).

Volchenko Yu.A., Koroteev V.A., and Neustroeva I.I. (2011a) The distribution and mode of occurrence of the platinum metals in chromite-bearing sections of mafic–ultramafic complexes of the Urals. *Platinum of the Urals. Selected works of Yu.A. Volchenko. Volume 2.* Yekaterinburg: Institute of Geology and Geochemistry, Uralian Branch of the Russian Academy of Sciences, 108–113 (in Rus.).

Volchenko Yu.A., Koroteev V.A., Neustroeva I.I. (2011b) Platinum content of ultramafic rocks and chromite ores in Alpine-type massifs of the main ophiolite belt of the Urals. *Platinum of the Urals. Selected works of Yu.A. Volchenko. Volume 2.* Yekaterinburg: Institute of Geology and Geochemistry, Uralian Branch of the Russian Academy of Sciences, 271–300 (in Rus.).

Vrublevskii, V.V., Gertner, I.F., Gutiérrez-Alonso, G., Hofmann, M., Grinev, O.M., and Tishin, P.A. (2014) Isotope (U–Pb, Sm–Nd, Rb–Sr) geochronology of alkaline basic plutons of the Kuznetsk Alatau. *Russian Geology and Geophysics 55* (11), 1598–1614.

Vysotskii S.V., Govorov G.I., Kemkin I.V., and Sapin V.I. (1998) The boninite–ophiolite association of eastern Sakhalin Island: geology and petrogenetic features. *Russian Journal of Pacific Geology, 17* (6), 3–15 (in Rus.).

Wager L.R. and Brown G.M. (1970) Layered igneous rocks. Moscow: Publishing House Mir, 552 p. (in Rus.).

Wedepohl K.H. (1981) Tholeiitic basalts from spreading ocean ridges: The growth of the oceanic crust. *Naturwissenschaften, 68* (3), 110–119.

Wittig N., Webb M., Pearson D.G., Dale, C.W., Ottley, C.J., Hutchison, M., Jensen, S.M., and Luguet, A. (2010) Formation of the North Atlantic Craton: Timing and mechanisms constrained from Re–Os isotope and PGE data of peridotite xenoliths from S.W. Greenland. *Chem. Geol.*, 276 (3–4), 166–187.

Wopenka B., Jolliff B.I., Zinner E., and Kremser D.T. (1996) Trace element zoning and incipient metamictization in a lunar zircon; application of three microprobe techniques. *American Mineralogist*, 81 (7–8), 902–912.

Yaroshevskii A.A., Bolikhovskaya S.V., and Koptev-Dvornikov E.V. (2006) Geochemical structure of the Ioko-Dovyren layered dunite–troctolite–gabbro–norite massif, northern Baikal area. *Geochemistry International*, 44 (10), 1027–1039 (in Rus.).

Yoder, H.S. (Ed.) (1983) The evolution of igneous rocks; fiftieth anniversary perspectives. Moscow: Publishing House Mir, 528 p. (in Rus.).

Yurkova R.M. (2012) Magmatism, metasomatism, and ore formation in connection with the rise of an ophiolite diapir. *Modern problems of magmatism and metamorphism. Volume 2.* St. Petersburg: Publishing House of St. Petersburg State University, 357–360 (in Rus.).

Zaccarini F., Campos L., Aiglsperger T., Garuti G., Thalhammer O.A.R., Proenza J.A., and Lewis J. (2009) The chromitites and the associated platinum-group minerals (PGM) of the Santa Elena ophiolite (Costa Rica): First and preliminary results. *Mafic–ultramafic complexes of folded regions and related mineral deposits. Volume 1.* Yekaterinburg: Publishing House of the Institute of Geology and Geochemistry, Uralian Branch of the Russian Academy of Sciences, 34–37.

Zavaritskii A.N. (1961) Igneous rocks. Moscow: Publishing House of the USSR Academy of Sciences. 480 p.

Zharikov V.A. and Yaroshevskii A.A. (2003) Geochemistry and its prolems (to the 50th anniversary of the Department of Geochemistry). *Herald of Moscow State University. Series 4. Geology* (4), 3–7 (in Rus.).

Zharov A.E. (2004) Accretionary tectonics and geodynamics of southern Sakhalin. *Geotectonics*, 38 (4). 45–63 (in Rus.).

Zimin S.S. (1973) Parageneses of ophiolites and the upper mantle. Moscow: Publishing House Nauka. 252 p. (in Rus.).

Zinger T.F., Bortnikov N.S., Sharkov E.V., Borisovskii S.E., and Antonov A.V. (2010) Influence of plastic deformations in zircon on its chemical composition: Evidence from gabbroids of the spreading zone of the Mid-Atlantic Ridge, Markov trough, 6 N. *Doklady Earth Sciences*, 433 (6), 785–791 (in Rus.).

Zonenshain L.P. and Kuz'min M.I. (1978) The Khantayshirin ophiolite complex of western Mongolia and the problem of ophiolites. *Geotectonics* (1), 19–42 (in Rus.).

Subject index

Note: page locators in bold indicate figures/tables.

Author index